Extracellular Vesicles - Role in Diseases, Pathogenesis and Therapy

Edited by Manash K. Paul

Published in London, United Kingdom

Extracellular Vesicles - Role in Diseases, Pathogenesis and Therapy
http://dx.doi.org/10.5772/intechopen.94822
Edited by Manash K. Paul

Contributors
Di Wu, Yanling Cai, Prince Amoah Barnie, Justice Afrifa, Eric Ofori Gyamerah, Benjamin Amoani, Weidong Zhou, Teng Zhang, Yikai Lian, Wenjing Zhang, Sushanta Barik, Keshar Kunja Mohanty, Sanghamitra Pati, Sashi Bhusan Mohapatra, Srikanta Jena, Srikanth Prasad Tripathy, Manash K. Paul, Deepak S. Chauhan, Priyanka Mudaliar, Soumya Basu, Jyotirmoi Aich, Veronika Kralj-Iglič, Gabriella Pocsfalvi, Aleš Iglič, Soldano Ferrone, Theresa L. Whiteside, Giorgia Pastorin, Jacob Yeo Xian Ping, Yub Raj Neupane, Olga Ostrovsky, Diego Aviles, David Warshal, Lauren Krill, Ryohei Sasaki, Ai Nakaoka, Kana Kobayashi, Mennaallah Mennaallah, Crețoiu Maria Sanda Maria, Gaspar Bogdan Severus, Ionescu Ruxandra Florentina, Enache Robert Mihai, Dobrică Elena Codruța, Crețoiu Dragoș, Voinea Silviu Cristian, Abdel Alli, Manjari Singh, Sapna Jain, Eman Helmy Thabet, Baranya Murugan, Suresh Sagadevan, Zeynep İşlek, Mehmet Hikmet Ucisik, Fikrettin Şahin, Batuhan Turhan Bozkurt

First published in London, United Kingdom, 2022 by IntechOpen
IntechOpen is the global imprint of INTECHOPEN LIMITED, registered in England and Wales, registration number: 11086078, 5 Princes Gate Court, London, SW7 2QJ, United Kingdom
Printed in Croatia

British Library Cataloguing-in-Publication Data
A catalogue record for this book is available from the British Library

Additional hard and PDF copies can be obtained from orders@intechopen.com

Extracellular Vesicles - Role in Diseases, Pathogenesis and Therapy
Edited by Manash K. Paul
p. cm.

This title is part of the Physiology Book Series, Volume 13
Topic: Cell Physiology
Series Editor: Tomasz Brzozowski
Topic Editor: Angel Catala

Print ISBN 978-1-80355-054-1
Online ISBN 978-1-80355-055-8
eBook (PDF) ISBN 978-1-80355-056-5
ISSN 2631-8261

For EU product safety concerns:
IN TECH d.o.o., Prolaz Marije Krucifikse Kozulić 3, 51000 Rijeka, Croatia,
info@intechopen.com or visit our website at intechopen.com.

IntechOpen Book Series

Physiology

Volume 13

Aims and Scope of the Series

Modern physiology requires a comprehensive understanding of the integration of tissues and organs throughout the mammalian body, including the cooperation between structure and function at the cellular and molecular levels governed by gene and protein expression. While a daunting task, learning is facilitated by identifying common and effective signaling pathways mediated by a variety of factors employed by nature to preserve and sustain homeostatic life. As a leading example, the cellular interaction between intracellular concentration of Ca+2 increases, and changes in plasma membrane potential is integral for coordinating blood flow, governing the exocytosis of neurotransmitters, and modulating gene expression and cell effector secretory functions. Furthermore, in this manner, understanding the systemic interaction between the cardiovascular and nervous systems has become more important than ever as human populations' life prolongation, aging and mechanisms of cellular oxidative signaling are utilised for sustaining life. Altogether, physiological research enables our identification of distinct and precise points of transition from health to the development of multimorbidity throughout the inevitable aging disorders (e.g., diabetes, hypertension, chronic kidney disease, heart failure, peptic ulcer, inflammatory bowel disease, age-related macular degeneration, cancer). With consideration of all organ systems (e.g., brain, heart, lung, gut, skeletal and smooth muscle, liver, pancreas, kidney, eye) and the interactions thereof, this Physiology Series will address the goals of resolving (1) Aging physiology and chronic disease progression (2) Examination of key cellular pathways as they relate to calcium, oxidative stress, and electrical signaling, and (3) how changes in plasma membrane produced by lipid peroxidation products can affect aging physiology, covering new research in the area of cell, human, plant and animal physiology.

Meet the Series Editor

Prof. Dr. Thomas Brzozowski works as a professor of Human Physiology and is currently a Chairman at the Department of Physiology and is V-Dean of the Medical Faculty at Jagiellonian University Medical College, Cracow, Poland. His primary area of interest is physiology and pathophysiology of the gastrointestinal (GI) tract, with a major focus on the mechanism of GI mucosal defense, protection, and ulcer healing. He was a postdoctoral NIH fellow at the University of California and the Gastroenterology VA Medical Center, Irvine, Long Beach, CA, USA, and at the Gastroenterology Clinics Erlangen-Nuremberg and Munster in Germany. He has published 290 original articles in some of the most prestigious scientific journals and seven book chapters on the pathophysiology of the GI tract, gastroprotection, ulcer healing, drug therapy of peptic ulcers, hormonal regulation of the gut, and inflammatory bowel disease.

Meet the Volume Editor

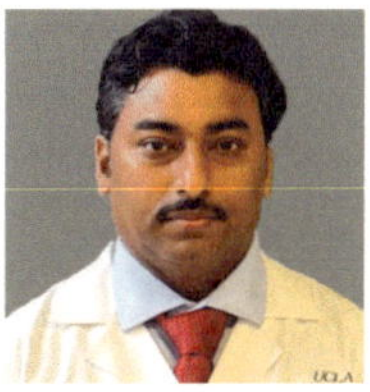

Manash K. Paul is a Principal Investigator and Scientist at the University of California Los Angeles. He has contributed significantly to the fields of stem cell biology, regenerative medicine, and lung cancer. His research focuses on various signaling processes involved in maintaining stem cell homeostasis during the injury-repair process, deciphering lung stem cell niche, pulmonary disease modeling, immuno-oncology, and drug discovery. He is currently investigating the role of extracellular vesicles in premalignant lung cell migration and detecting the metastatic phenotype of lung cancer via machine-learning-based analyses of exosomal signatures. Dr. Paul has published in more than fifty peer-reviewed international journals and is highly cited. He is the recipient of many awards, including the UCLA Vice Chancellor's award, a senior member of the Institute of Electrical and Electronics Engineers (IEEE), and an editorial board member for several international journals.

Contents

Preface

Extracellular vesicles (EVs) are attracting scientific interest, as an increasing number of clinical studies are evaluating exosomes for diagnostic and therapeutic uses. EVs are lipid bilayer-delimited nanoscale vesicles secreted by most cells and contain cargo, including proteins, nucleic acids, lipids, and metabolites from the parent cell and mediate a horizontal transport of the cargo to recipient cells. EVs facilitate cell-to-cell contact and communication under normal and pathological conditions and play a role in the pathogenesis of many diseases. Thus, EVs have inspired a new field of research in almost every aspect of biology, including developmental, host-pathogen interactions, tissue regeneration, and cancer. This comprehensive book presents current updates in EV biology and the relationship of EVs with disease diagnosis and treatment. It delves into the biogenesis of EVs, cargo loading, composition of EVs, their interactions with cell membranes, EV isolation, and future directions to overcome current hurdles associated with liquid biopsy. It further elaborates the scientific advances in characterizing and engineering EVs for biomarker discovery and disease diagnosis, prognosis, therapeutic application, and theranostics. The book also examines the role of EVs in the comprehension of inflammation, stress resistance, and vascular integrity. Chapters address the role of EVs in embryonic development, HIV-1, reproductive issues, and associated clinical translation. Additionally, the book examines the role of EVs produced from protozoan parasites in host immunomodulation, pathogenesis, and disease progression, and presents information on novel immunotherapeutic models.

Recent studies strongly emphasize the pathogenic and translational potential of EVs in cancer. This book describes the potential diagnostic implications and molecular characterization of EVs in various cancers and concepts for using exosomes as nanocarriers for therapeutic medicines. Cancer cells actively discharge EVs (tumor-derived EVs) into biological fluids, which mediate enhanced immunosuppression, angiogenesis, metastasis, and metabolic reprogramming. Liquid biopsy has enormous promise as a diagnostic and therapeutic monitoring tool and may soon replace invasively collected tissue samples-based diagnostics. The book highlights the significance of EVs in cancer treatment resistance, especially with radiotherapy and immunotherapy, and their potential role as prognostic and diagnostic biomarkers. It also discusses the status of EVs in clinical trials in multiple cancers like breast cancer and ovarian cancer. Finally, it also identifies the loopholes for clinical translation of EVs and points out potential future research directions for therapeutic translation and cancer therapy. This book is a useful resource for biologists, clinicians, and translational scientists.

Manash K. Paul
Department of Pulmonary and Critical Care Medicine,
University of California Los Angeles (UCLA),
Los Angeles, USA

Section 1

Introduction

Chapter 1

Introductory Chapter: Role of Extracellular Vesicles in Human Diseases and Therapy

Manash K. Paul

1. Introduction

Extracellular vesicles (EVs) are nanoscale vesicles secreted by cells that mediate horizontal cargo transport from donor to recipient cell, thereby establishing

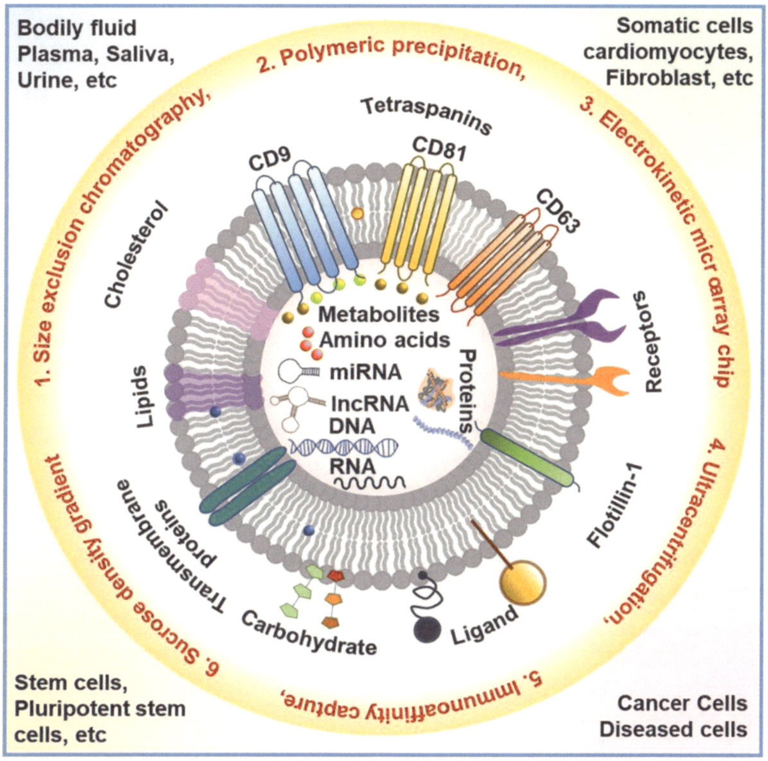

Figure 1.
Centre: showing exosome composition, including proteins, lipids, carbohydrates, nucleic acids, mRNA, miRNA, non-coding RNA, and DNA. Proteins in the exosome include heat shock proteins (HSP), cytoskeletal proteins (ESCRT components), membrane transporters, fusion proteins, growth factors and cytokines, Tetraspanins, Flotillin, ligands like TRAIL (TNF-related apoptosis-inducing ligand), FasL (Fas ligand) and receptors like TfR (transferrin receptor). The different methods commonly used for exosome isolation is shown in the yellow ring. Four corners: shows the different sources of exosomes, including bodily fluid, somatic cells, stem cells and diseased.

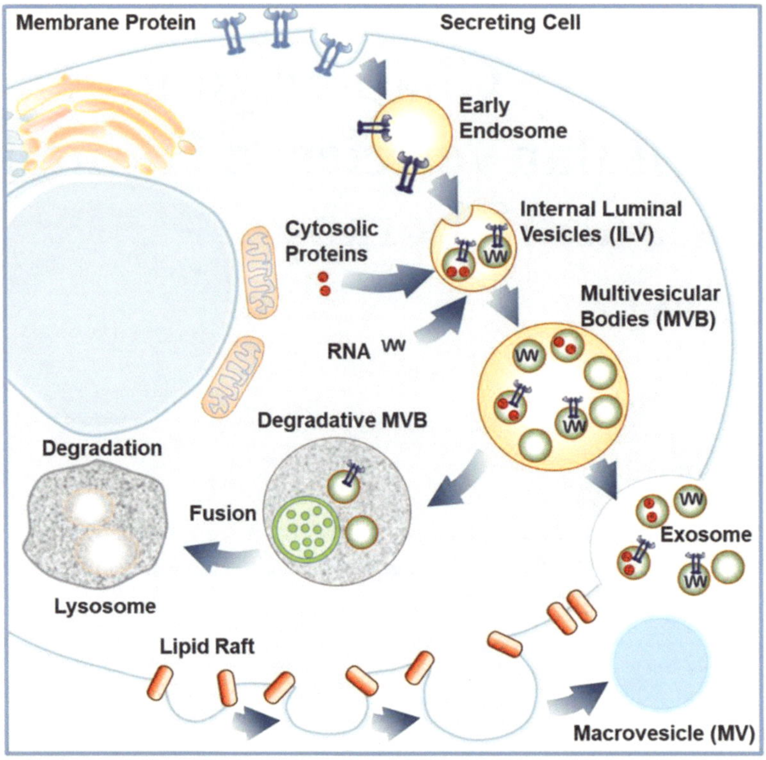

Figure 2.
Exosome biogenesis: exosome biogenesis initiates with the intraluminal vesicles (ILVs) formation within the multi-vesicular body (MVB), followed by the fusion of MVB with the plasma membrane. After fusion, these ILVs are secreted as exosomes, although some are chosen for destruction by the lysosome. Exosomes carry lipids, RNA, DNA, proteins, adhesion molecules, receptors, and other functional.

cell-cell communication and signaling [1]. EVs are lipid bilayer-delimited particles released spontaneously by almost all types of cells. The EVs contain cargo, including proteins, nucleic acids, lipids, metabolites, and even organelles, representing the parent cell's physiological state (**Figure 1**) [1, 2]. There are three primary subtypes of EVs identified by their size and biological processes; exosomes (~30–150 nm), microvesicles (~100–1000 nm), and apoptotic bodies (~1000–5000 nm). The exosome biogenesis process is quite intriguing where multivesicular bodies (MVBs) are specialized endosomal compartments containing multiple intraluminal vesicles (ILVs). ILVs are generated by the inward budding of endosomal membranes within MVBs and followed by MVBs fusion with the plasma membrane and extracellular exocytosis-based release as exosomes (**Figure 2**) [3]. Microvesicles are generated by the budding outward of the healthy cell's plasma membrane. In comparison, apoptotic bodies are plasma membrane blebs of cells that originate during apoptosis. Several EV subgroups have been hypothesized, including ectosomes, microparticles, oncosomes, and others, in addition to the three primary forms, but the dearth of established biomarkers and the lack of standardized isolation techniques have resulted in misconceptions in classifying EV subgroups [4]. This book presents a comprehensive overview of EVs and has three sections: the biology of extracellular vesicles; the second is the role of extracellular vesicles in human diseases, and third is extracellular vesicles and cancer.

2. Biology of extracellular vesicles

The mechanism of EV internalization in the target cell may be mediated by multiple mechanisms, EV-surface contact molecule interaction to boost juxtracrine downstream signaling, fuse with the membrane to deposit payloads into the cytosol, or are taken up by phagocytosis, macropinocytosis, or receptor-mediated endocytosis is unclear [1, 5]. EVs play a crucial role in facilitating the cell-cell environment under normal and pathological conditions and in the pathogenesis of many diseases [3]. This section reviews the relationship between EV composition and interactions with biological membranes before delivering EV cargo to the target cells and stimuli-induced EV release. This section presents EVs biogenesis, different cargo loading mechanisms of EVs, their release, and the role of G Proteins. This book also discusses the scientific advances that have made it feasible to address the existing bottlenecks associated with the isolation and characterization of EV subsets from body fluids. The roadmap for effective immunocapture and molecular characterization is presented along with the review on immunoaffinity-based techniques for separating specific EV subsets from plasma and biofluids.

Exosomes play an important role in cell-cell communication, signal transduction, immune response in normal and disease backgrounds. Their possible use as diagnostic and prognostic biomarkers and the leverage to use them as therapeutic carrier vehicles have sparked tremendous clinical interest. The main constituents of exosomes are proteins, nucleic acids, and metabolites, as summarized in **Figure 1**. Among the highly enriched proteins in exosomes, the tetraspanins (CD9, CD63, CD81, CD82) help exosome-cell fusion, while the Heat shock proteins (HSP70, HSP90) are involved in stress response and antigen-binding and presentation, and other proteins (Alix, TSG101) are involved in exosome release. Some of these proteins are involved in exosome biosynthesis (Alix, flotillin, and TSG101), whereas others are considered exosomal biomarkers (e.g., TSG101, HSP70, CD81, and CD63) (**Figure 1**).

An intriguing question is how the cellular cargo is selectively sorted in the exosomes? This section of the book addresses the EV biogenesis, cargo loading mechanisms, their release, and the role of G Proteins (**Figure 2**). Another critical aspect of EV biology is how they are taken up by recipient cells? Whether EVs naturally cross biological barriers or need genetic modifications? This section highlights several critical areas, including the interplay of EVs with biological membranes, EV target cell internalization mechanism, the relationship between EV composition and interactions with biological membranes, and stimuli-induced EV release. It is also essential to understand the biophysical aspect of cellular vesicles' morphology and formation mechanisms discussed in this section. **Figure 1** elaborates exosome's hallmarks and shows different isolation techniques, including size exclusion chromatography, polymeric precipitation, electrokinetic microarray chip, ultracentrifugation-based, immunoaffinity capture, and sucrose density gradient.

This book also specially emphasizes the bottlenecks associated with the isolation and characterization of EV subsets from plasma, thereby limiting a better understanding of their biological significance. A chapter reviews the immunoaffinity-based techniques for separating specific EV subsets from plasma and presents a roadmap for effective immunocapture and molecular characterization. This section also discusses other popular techniques of exosome production *in vitro* and suggests the challenges of *in vivo* physiological or pathological characterization of exosomes. The usual sources of exosomes are bodily fluids (plasma, saliva, urine, etc.), somatic cells (cardiomyocytes, fibroblast, pneumocyte, etc.), stem cells or pluripotent cells, and tumor or diseased cells (**Figure 1**). Red blood cell (RBC) contamination presents a significant challenge in EV isolation from urine for the non-invasive source of disease biomarkers. This section presents an Innovative method describing the removal of RBCs contamination from the urine fraction. This section elaborates the scientific advances that have made it feasible to characterize and engineer EVs, leading to their use as tools in biomarker discovery and disease diagnosis, prognosis, therapeutic application, and theranostics [6]. The potential of liquid biopsy is significant and can be essential for both diagnosis and therapy monitoring [7]. Blood and saliva EVs may assist achieve this without needing tissue samples.

3. Role of extracellular vesicles in human diseases

This section emphasizes emerging data confirming the role of EVs in the pathogenesis of diseases. The role of EVs in relation to inflammation, stress protection, and vascular integrity is also presented and may help better understand tissue resolution and vascular restoration. Another exciting area that has been discussed in this section is the role of EV-transmitted cargos in embryonic development and reproduction-related diseases and clinical translation. This fascinating chapter covers this critical subject and may aid fundamental understanding and clinical translation. Potential applications from circulating biomarkers for early illness detection to future therapeutic carriers for halting disease progression and regenerating damaged tissue/organs for potential regenerative medicine-based applications are also discussed. An emerging area is the studies related to the role of EVs in influencing HIV-1 pathogenesis, how HIV-1 factors target EVs, EVs as an antiretroviral therapy option, and their potential use as diagnostics prognostics, and theranostics in relation to H1V patient management [8]. Another fascinating area is the role of protozoan parasite-derived EVs in mediating host immunomodulation, pathogenesis, and parasite disease development, especially in the context of Leishmania, Toxoplasma, Plasmodium, and Trypanosoma [9]. This section deals with studies investigating new immunotherapeutic models based on protozoan parasite immunomodulation approaches and discusses many aspects of protozoan EV-based strategies to create innovative immunotherapeutic approaches.

4. Extracellular vesicles and cancer

This section presents recent studies on EVs' pathological and translational potential in malignancies. Cancer-derived EV payloads preserve their molecular features, and cancer cells actively discharge EVs into easily accessible body fluids [10]. The transport of cancer-associated biomolecules by EVs from cancer cells promotes cancer development and reflects changes in cancer status during treatment. EVs bearing tumor antigens are also studied as cancer vaccines to induce tumor-specific anti-tumor immunity. Tumor cell-derived EVs stimulate immunosuppression, angiogenesis, metastasis, metabolic reprogramming, and other processes in the tumor microenvironment. Tumor-derived exosomes (TEX's) ability to inhibit or boost the immune system is thrilling and intriguing. The occurrence or absence of immunological recipient cells in the TME may affect the outcome of TEX-driven interactions [11]. Transducers that create juxtracrine or paracrine signals may modify immunological recipient cell suppressive pathways, resulting in accelerated tumor development. Due to TEX-based antibody sequestration, immunotherapies may not work fully. TEX-induced immunostimulatory signals may alter the TME to promote immune activation rather than tumor development. TEXs are excellent diagnostic and prognostic biomarkers [12].

The diversity, cargo composition, and molecular mechanism of phenotypic transfer of TEX to recipient cells and vice versa is a critical question. Cancer-derived EV payloads preserve their molecular features, and cancer cells actively discharge EVs into easily accessible body fluids [13]. The transport of cancer-associated biochemicals by EVs from cancer cells promotes cancer development and reflects changes in cancer status during treatment. Moreover, this section discusses the role of EVs in resistance to treatment and diagnostics and being attractive indicators for assessing therapeutic response. EVs produced by disseminated tumor cells chemotactically attract circulating tumor cells (CTCs) and stimulate nearby stromal cells to produce extracellular matrix components like integrins, collagens, and laminin proteins to promote metastatic cell-extracellular matrix remodeling by modulating neighboring tumor cells and stromal cells, promoting tumor invasion and metastasis. Tumor-derived EVs carry molecular signatures specific to the tumor's genetic complexity and may be used as minimally invasive cancer immunotherapy biomarkers. Through secretory factors and miRNAs, tumor exosomes have been demonstrated to facilitate distant cell-cell contact, resulting in the creation of pro-tumorigenic microenvironments favorable to metastatic spread. EV-induced fibroblast activation, ECM synthesis, angiogenesis, and immunological regulation are essential for metastatic dissemination. This section presents many aspects of the EV-based mechanism involved in metastasis.

The role of EVs in resistance to cancer treatment and diagnostics and being attractive indicators for assessing therapeutic response. Radiation is now often coupled with immunotherapy [13, 14]. EVs may also reduce chemoresistance by carrying RNA forms, and therefore activity regulation of EVs may overcome immunotherapy resistance. Also discussed many aspects of EVs/exosomes and their potential in targeting chemoresistance, radio-resistance, and cancer management. Tumor-derived EVs serve as excellent diagnostic and prognostic biomarkers. The critical bioactivities of tumor-derived exosomes using examples of their cargo molecules are also presented. EVs are immune cell evaders and are currently being investigated as potential diagnostic biomarkers and drug delivery vehicles. Exosomal immune checkpoint regulators may serve as clinical predictors for treatment response or recurrence in a variety of different malignancies. It may be possible that exosome-based paracrine mediators will be necessary for tailoring immune-based therapies to different tumors.

This section also reviews the role of EVs and the potential to use them in the management of difficult to diagnose and treat cancers, like ovarian cancer and breast cancer. Oncologic malignancies such as ovarian cancer are difficult to diagnose, with dismal results, and critically need new treatments. This section describes EVs' role as a critical player in the spread of ovarian cancer, and EVs may help us learn more about ovarian cancer proliferation and metastasis while also revealing potential new therapeutics. Breast cancer is the most frequent cancer among women, and understanding the role of EVs in facilitating intercellular communication between cancer and stromal cells and its therapeutic possibilities for breast cancer therapy is critical. This book also discussed the information gaps for clinical translation of EVs and pointed out the current research projects on developing EVs as biomarkers or therapeutic delivery systems. The solutions to improve EVs' efficacy as cancer treatments are also presented. Moreover, the direct and indirect cell surface modification is discussed, emphasizing ongoing and finished clinical studies utilizing naturally generated EVs to treat breast cancer. This book also presents the loopholes for clinical translation of EVs and points out potential future research directions for therapeutic translation and cancer therapy. This anthology of chapters is presented with a broad audience in mind and will serve as a valuable must-have resource to basic biologists, translational scientists, and clinicians.

Author details

Manash K. Paul
Department of Pulmonary and Critical Care Medicine, David Geffen School of Medicine, University of California Los Angeles (UCLA), Los Angeles, California, USA

*Address all correspondence to: manashp@ucla.edu

References

[1] Kalluri R, LeBleu VS. The biology, function, and biomedical applications of exosomes. Science. 2020;**367**(6478): eaau6977

[2] Gurunathan S, Kang M-H, Jeyaraj M, Qasim M, Kim J-H. Review of the isolation, characterization, biological function, and multifarious therapeutic approaches of exosomes. Cells. 2019;**8**(4):307

[3] Gurung S, Perocheau D, Touramanidou L, Baruteau J. The exosome journey: From biogenesis to uptake and intracellular signalling. Cell Communication and Signaling. 2021; **19**(1):47

[4] Gurunathan S, Kang M-H, Kim J-H. A comprehensive review on factors influences biogenesis, functions, therapeutic and clinical implications of exosomes. International Journal of Nanomedicine. 2021;**16**:1281-1312

[5] McKelvey KJ, Powell KL, Ashton AW, Morris JM, SA MC. Exosomes: Mechanisms of uptake. Journal of Circulating Biomarkers. 2015;**4**:7

[6] Taghikhani A, Farzaneh F, Sharifzad F, Mardpour S, Ebrahimi M, Hassan ZM. Engineered tumor-derived extracellular vesicles: Potentials in cancer immunotherapy. Frontiers in Immunology. 2020;**11**:221. DOI: 10.3389/fimmu.2020.00221

[7] Zhou B, Xu K, Zheng X, Chen T, Wang J, Song Y, et al. Application of exosomes as liquid biopsy in clinical diagnosis. Signal transduction and targeted. Therapy. 2020;**5**:144

[8] Chen J, Li C, Li R, Chen H, Chen D, Li W. Exosomes in HIV Infection. Current Opinion in HIV and AIDS. 2021;**16**(5):262-270. DOI: 10.1097/COH.0000000000000694

[9] Wu Z, Wang L, Li J, Wang L, Wu Z, Sun X. Extracellular vesicle-mediated communication within host-parasite interactions. Frontiers in Immunology. 2019;**9**:3066

[10] Liu J, Ren L, Li S, Li W, Zheng X, Yang Y, et al. The biology, function, and applications of exosomes in cancer. Acta Pharmaceutica Sinica B. 2021;**11**(9): 2783-2797

[11] Knox MC, Ni J, Bece A, Bucci J, Chin Y, Graham PH, et al. A clinician's guide to cancer-derived exosomes: Immune interactions and therapeutic implications. Frontiers in Immunology. 2020;**11**:1612

[12] Nam GH, Choi Y, Kim GB, Kim S, Kim SA, Kim IS. Emerging prospects of exosomes for cancer treatment: From conventional therapy to immuno therapy. Advanced Materials. 2020; **32**:2002440

[13] Dai J, Su Y, Zhong S, Cong L, Liu B, Yang J, et al. Exosomes: Key players in cancer and potential therapeutic strategy. Signal transduction and targeted. Therapy. 2020;**5**(1):145

[14] Xu Z, Zeng S, Gong Z, Yan Y. Exosome-based immunotherapy: A promising approach for cancer treatment. Molecular Cancer. 2020; **19**:160

Section 2

Biology of Extracellular Vesicles

Chapter 2

Mechanisms of Extracellular Vesicle Biogenesis, Cargo Loading, and Release

Abdel A. Alli

Abstract

Extracellular vesicles (EVs) are carriers of various biomolecules including bioactive enzymes, lipids, proteins, nucleic acids, and metabolites. EVs are classified into three main types based on their size, biogenesis, and cargo. Exosomes originate from endosomal membranes and are the smallest type of EV. Microvesicles (MVs) or microparticles are larger in size, and like apoptotic bodies which represent the largest type of EVs, both of these vesicles originate from outward budding of the plasma membrane. As discussed in this chapter, cargo loading of EVs and their release into the extracellular space where they can be taken up by neighboring or distant cells plays an important role in physiology and pathophysiology. This chapter will outline specific mechanisms involved in the loading and enrichment of miRNAs, proteins, and lipids within EVs. As explained here, various external and biological stimuli play a role in EV release. Finally, recent studies have shown that the biogenesis, cargo loading, and release of EVs are governed by circadian rhythms. Although EVs were once thought to serve as garbage disposals of cells, the numerous roles they serve in physiology and pathophysiology are now being appreciated.

Keywords: extracellular vesicles, exosomes, microvesicles, microparticles

1. Introduction

Extracellular vesicles (EVs) represent a heterogenous population of vesicles that include exosomes and the plasma membrane shedding microvesicles (MVs) (also known as microparticles) and apoptotic bodies. Exosomes represent the smallest subtype of EVs and have spherical bodies with a lipid bilayer membrane. Exosome formation begins in endosomes as the budding of the endosomal membrane results in the formation of multivesicular bodies (MVBs) [1]. The fusion of MVBs with the inner leaflet of the plasma membrane results in the release of intraluminal vesicles (ILVs) as exosomes (**Figure 1**) [1]. The ILVs within the lumen of the endosomes have three fates [2]. First, the contents can be used for the biogenesis of specialized lysosome-related organelles such as melanosomes. Second, the ILVs may fuse with lysosomes. Third, the ILVs may fuse with the plasma membrane to release the content

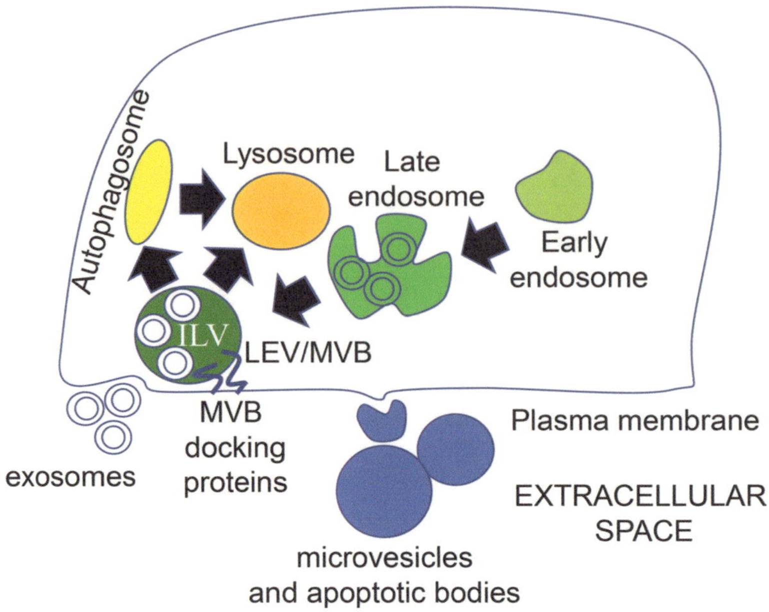

Figure 1.
Biogenesis of EV's. Microvesicles and apoptotic bodies are generated by outward budding of the cell's plasma membrane. Exosomes are formed from the endocytic pathway of the cell. In this process early endosomes undergo inward budding to form intraluminal vesicles (ILVs) inside the late endosomal vesicles (LEVs) or multivesicular bodies (MVBs). The fate of the MVBs include degradation by lysosomes, fusion with autophagosomes, or fusion with the plasma membrane of the cell resulting in the release of the ILVs to the extracellular space in the form of exosomes. As shown, MVBs and autophagosomes can be degraded by lysosomes.

into the extracellular space in which the vesicles will then be termed exosomes. There are a number of signaling and specialized proteins that contribute to the biogenesis of exosomes. The biogenesis of microvesicles and apoptotic bodies is different from that of exosomes as these vesicles are produced from the shedding or budding of the plasma membrane from the parent cell (**Figure 1**). The release of EVs from a cell can be triggered by a myriad of stimuli. Importantly, not all EVs are created equally and EVs found in biological fluids including plasma and urine represent a heterogenous mixture of subpopulations of EVs. The fate of EVs after they are released from their parent cell is to either be taken up by a neighboring or distant cell or to be excreted from the body.

2. ESCRT dependent and independent multivesicular body biogenesis

Endosomal sorting complex required for transport (ESCRT) proteins and ESCRT associated proteins including TSG101, ALG-2 interacting protein X (ALIX), SKD1, and Chmp4 are required for ESCRT dependent MVB biogenesis. ESCRT is a ubiquitin-dependent mechanism that contributes to the sorting of ubiquitinated proteins into exosomes. ESCRT consist of four complexes that are numbered according to their sequential action. For example, ESCRT-0 aka vacuolar protein sorting (VPS) associated protein 27/heat shock element 1 complex VPS27/HSE1 recruits ESCRT-1, which plays an essential role in MVB cargo sorting and bud formation. VPS36 associates with ESCRT-1 via a ubiquitin moiety while VPS25 is required for the assembly of the ESCRT-II complex [3]. ESCRT-II also plays a role in cargo sorting and regulates

ESCRT-III formation. VPS20 is involved in the assembly of the ESCRT-III complex and VPS24 completes the scission of the budding membrane [4].

In addition to sorting of cargo within multivesicular bodies, ESCRT-III participates in ESCRT recycling and binds ALIX to allow for cargo sorting [5]. ALIX also stimulates exosome secretion containing the tetraspanins CD63, CD9 and CD81 [5].

Alternatively, there are multiple ESCRT independent MVB biogenesis pathways that utilize various other proteins. This particular mechanism may be dependent on alternate mechanisms including the Rab associated proteins, ceramides, sphingomyelins, cholesterols, and tetraspanins. Members of the Rab family of small GTPases regulate various steps in the formation and transition of the endosomal transport network. Rab5 is thought to regulate the formation and fusion of early endosomes [6, 7]. Rab7 is thought to regulate the fusion of late endosomes/multivesicular endosomes with lysosomes resulting in degradation of the intraluminal vesicles [8, 9]. Rab27 is thought to regulate multivesicular endosome docking and fusion with the plasma membrane to allow for release of the intraluminal vesicles as exosomes [10]. One report provided evidence that Rab31 engages flotillin proteins from lipid rafts to orchestrate epidermal growth factor receptor entry into multivesicular endosomes to form intraluminal vesicles (ILV) and the release of exosomes [11].

3. Lipid rafts and exosome biogenesis

There are a number of lipid raft proteins that are associated with exosome biogenesis. Ceramide is enriched in the inner leaflet of the exosomal membrane [12]. One feature of ceramides is that they can initiate spontaneous membrane invagination which allows for ILV formation and the maintenance of vesicle shape and structure. One study showed exosome release was reduced after the inhibition of the enzyme that catalyzes the formation of ceramide from sphingomyelin, neutral sphingomyelinases [13]. Cholesterol is another lipid raft protein that is enriched in the exosomal membranes. The accumulation of cholesterol leads to the secretion of exosomes enriched in flotillins. In some cells the inhibition of sphingomyelinase results in the suppression of exosome production while exosome production is not affected by changes in sphingomyelinase activity in other cells. This suggest that there may be cell type specific mechanisms that regulate exosome production. For example, the ESCRT dependent mechanism may be favored over the lipid raft dependent mechanism in PC-3 cells since exosome production is not affected by inhibition of sphingomyelinase in these cells.

There are several other proteins that are involved in exosome biogenesis. Proteins associated with the ESCRT dependent pathway, glycosylphosphatidylinositol anchored proteins (GPIAPs), palmitoylated forms of transmembrane proteins, flotillins, annexins, and cholesterol binding proteins such as caveolins have been shown to be enriched in exosomes. Flotillin 1 and Flotillin 2 appear to play a role in EV composition as Phuyal et al. showed interfering RNA-mediated knockdown of flotillin 1 and flotillin 2 resulted in alteration of the EV composition in PC-3 cells [14]. The exosomal protein Annexin A2 may play an essential role in the organization of the exosomal membrane as it preferentially binds to phosphatidylinositiol 4,5 bisphosphate rich domains and cholesterols within the cytoplasmic leaflet of lipid raft membranes and influence raft dynamics of parent cells [15]. It is not surprising caveolins were found to play an essential role in EV generation and uptake since they are known to regulate multiple cellular processes including endocytosis, exocytosis, and maintaining the shape of the cell membrane. Accumulating data suggest caveolin 1 promotes the production and release of EVs while caveolin deletion results in a decrease in EV release [16].

4. Regulation of size distribution

The size of EVs may or may not depend on the amount and type of cargo enriched in the vesicles. Some studies have shown an increase in EV release without a change in EV size, while other studies have shown the contrary [17, 18]. Apoptotic bodies represent the largest type of EVs and contain organelles, in addition to nucleic acids, proteins, and lipids. Conversely, exosomes and microvesicles lack organelles, but contain specific molecules of interest.

Numerous studies have shown the same cell type can produce subpopulations of exosomes that are remarkable different in size. For example, polarized epithelial cells, such as those in the kidney tubule release exosomes across the apical plasma membrane and the basolateral plasma membrane (**Figure 2**). These two types of exosomes show distinct differences in their size and protein composition [19]. Interestingly, Matsui et al. showed two independent mechanisms for exosome release across the apical and basolateral plasma membranes [20]. This group showed that ceramide plays an essential role in basolateral exosome release whereas, ALIX is important for apical exosome release [20].

Although it is reasonable to expect the size of EVs to be directly proportional to the concentration of a particular molecule present in the EVs, this is not always the case. For example, Chacko et al. showed urinary EV concentration decreased after infusing Tempol in hypertensive 129Sv mice while the size of the EVs increased [21].

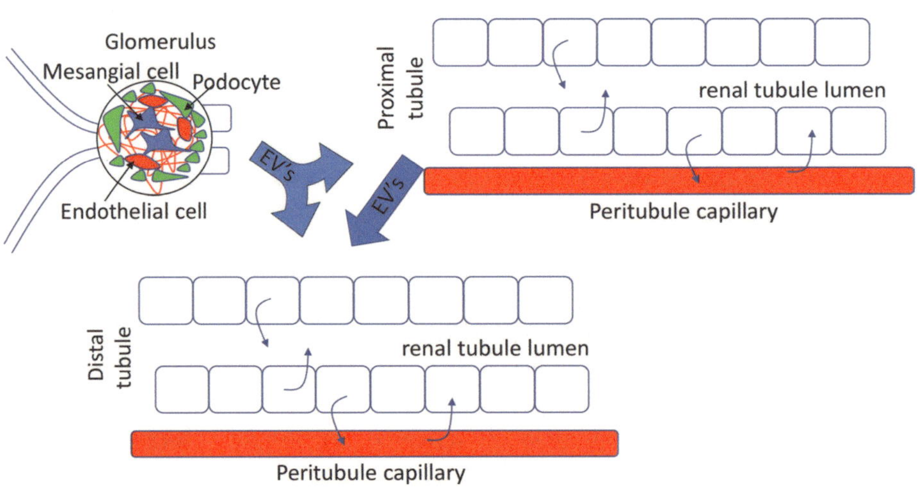

Figure 2.
Release of EV's from various kidney cell types. All cell types in the kidney including glomerular cells and cells within each segment of the nephron release EV's that can allow for intercellular communication, intracellular signaling, and the regulation of various mechanisms in health and disease. Polarized epithelial cells of the nephron release two distinct populations of EV's across the apical membrane that faces the lumen and the basolateral membrane that faces the peritubular capillaries. EV's that are released into the lumen containing the filtrate can be taken up by recipient cells downstream in the nephron or excreted from the body in the form of urine. EV's that are released across the basolateral membrane can be reabsorbed back into the blood.

The amount of the EV marker protein annexin A2, but not the amount of other EV marker proteins including TSG101 or flotillin-2 decreased after tempol treatment [21].

5. Cargo loading/sorting

The cargo that is loaded and carried within EVs include proteins, lipids, nucleic acids, metabolites, and miRNAs. The loading of specific molecules within different types of EVs is dependent on the parent cell and specific loading mechanisms. The enrichment of various types of cargo molecules is also dependent on a particular cell type being subject to pathological conditions, drugs, and various stimuli.

The loading of cargo within EVs may be a means of the cell getting rid of unneeded or excess material. For example, EVs from epithelial cells of the renal tubule are released across the luminal plasma membrane into the filtrate and are subsequently excreted from the body in the form of urine if the EVs are not taken up by recipient cells downstream in the nephron. In fact, a large amount of urinary EVs (uEVs) are those released from renal epithelial cells and bladder cells.

Multiple mechanisms have been proposed for the loading of miRNAs into EVs. One mechanism involves 3′ end uridylation of miRNAs [22]. A second mechanism involves the incorporation of a four nucleotide motif (GGAG) and its interaction with the ribonucleoprotein hnRNPA2B1 [22]. A third mechanism involves the RISC complex associated protein, AGO2 [22]. Another mechanism involves over-expression of nSMase2 [22]. The exosomal protein Annexin A2 was found to play a role in the packaging of miRNAs in EVs [23].

Proteins found to be enriched in EVs include those enriched in lipid rafts, metalloproteinases, tetraspanins, proteins associated with endosomes, heat shock proteins, glycosylphosphatidylinositol (GPI) anchored proteins, and those associated with the exosome biogenesis pathway. Proteomic studies have identified metalloproteinases including ADAMs (a disintegrin and metalloproteinase) with proteolytic activity enriched in EVs [24]. ADAMs are versatile proteins that are involved in cell adhesion, cellular communication, migration, and the regulation of growth factors and cytokines. EVs also contain members of the LAMP family proteins, LAMP-1 [25] and LAMP-2 [26]. These lysosome-associated proteins help maintain the integrity of the lysosomes by forming a glycocalix that protects the membranes from enzymatic degradation. Tetraspanins including CD9, CD63, and CD81 are enriched in exosomes [24]. Heat shock proteins (HSP) including HSP60, HSP70, and HSP90 have been found to be enriched in EVs [24]. Flotillin proteins were found to regulate exosome formation and cargo sorting. Other proteins such as syntenin has been found to contribute to the loading of exosomes with specific cargo [27]. GPI proteins including the complement regulator proteins CD55 and CD50 are enriched in exosomes [28].

Lipids that are enriched in EVs generally share the same features as the cells of origin. Lipids enriched in EVs include cholesterol, ceramides, cholesterol, and phosphatidylserine, which is also enriched in vesicles derived from MVBs. In one study, Glover found 13 ceramides that were significantly reduced in uEVs of hereditary α-tryptasemia (HαT) patients compared to healthy volunteers [29]. In another study, Nouri et al. showed sphingomyelin, phosphatidylethanolamine, and lysophosphatidylcholine were present in greater concentrations in EVs isolated from the conditioned media of human aortic endothelial cells compared to control EVs isolated from the complete growth media of these cells [18].

6. Incorporation of viral components into EVs

Accumulating experimental evidence suggest viral components are packaged in EVs during infection and this results in alterations within recipient cells. Virus proteins including nucleoproteins and glycoproteins packaged in EVs were found to induce apoptosis in recipient immune cells.

One study reported the presence of ebola virus VP40, nucleoprotein, and glycoprotein in EVs that leads to apoptosis in recipient cells [30]. Studies by the same group also showed that VP40 can become incorporated into exosomes and thereby negatively impact recipient T cells and monocytes [30–32]. Multiple groups showed Zika virus infected cells release EVs with infectious viral RNA and viral proteins [33, 34]. York et al. showed that Zika virus modifies EV density, cargo, and secretion [33]. Ning et al. showed RNA from SARS-CoV-2, the causative agent of Severe Acute Respiratory Syndrome, COVID-19 in plasma EVs just one day after infection which plateaued from day 6–28 in non-human primates and 20–60 days in young children [35]. In another study, Troyer et al. showed EVs released from cells expressing the CoV-2 spike protein contain multiple peptides originating from this protein [36]. This group reported EVs carrying these spike proteins can serve as a decoy for anti-spike neutralizing antibodies and therefore promote viral infection.

The pathogenic and deleterious effect of EVs containing viral components is not limited to acute disease but it can result in asymptomatic or recurrent infections. The delivery of various molecules packaged in EVs and the delivery of these EVs to healthy cells may allow the virus to remain latent.

7. EVs in promoting infectious disease

Protozoan parasites such as Leishmania, Plasmodium, and Toxoplasma contribute to significant morbidity and mortality in humans. Studies have shown EVs can modulate host immune cells [37]. *Leishmania* parasites are transmitted by sandflies and are responsible for infecting phagocytic cells inside the mammalian host. A study by Atayde et al. showed exosomes secreted within the midgut of *L. major* parasites enhance infection and contribute to the development of lesions after being injected in mice [38]. *Plasmodium* parasites are the etiological agents of malaria. Multiple studies have demonstrated an important role for EVs in *Plasmodium* infections. One study showed malaria-infected red blood cells use exosome-like vesicles to communicate which promotes differentiation to sexual forms [39]. Another study showed malaria antigens are enriched in microvesicles released from infected red blood cells which activate host monocytes and neutrophils [40]. *Toxoplasma* is acquired by ingestion of raw or uncooked meat and these parasites are common in virtually all species of warm-blooded vertebrates. Li et al. performed an analysis of differential exosomal miRNAs in dendritic cells induced by Toxoplasma gondii infection [41]. These studies also implicate a role for exosomes in toxoplasmosis and the ability to stimulate naïve recipient cells.

8. EV release in health and disease

A myriad of different mechanisms have been shown to contribute to EV release in various types of cells. These mechanisms include specialized roles of proteins including Rab proteins, SNARE proteins, small GTPases of the Rho/Rac/cdc42 family, and diacyl glycerol kinase α. These mechanisms also include posttranslational modifications of EV cargo proteins, inhibition of various kinases, and activation of cell surface receptors. Different Rab proteins play specialized roles in regulating exosome transport between different cellular compartments. For example, Rab5 and Rab7 are important in delivering cargo to the early endosomes, while Rab27 is involved in membrane docking to promote fusion, while Rab11 and Rab35 are involved in MVE secretion [42]. The process by which exosomes bud off the plasma membrane is accomplished with the activation of the ARF6 protein [43].

The specific inhibition of protein kinase C, but not protein kinase D in cultured human aortic endothelial cells resulted in both an increase in EV size and increase in EV release [18]. It is still not known whether the increase in EV release is a direct consequence of decreased Protein Kinase C (PKC) activity or a indirect consequence resulting from other proteins that are regulated by PKC activity. Also, it is likely that various mechanisms that regulate PKC activity such as diacylglycerol (DAG) also regulate EV size and release. This could provide a feedback mechanism by which EVs that are released by one cell type and taken up by another cell type can regulate the release of EVs in the recipient cells.

Pharmacological activation of specific G-protein coupled receptors (GPCRs) from trophoblast cells was shown to trigger the release of EVs [17]. Activation of CCKBR, TAS2R14, cholinergic muscarinic 1 and 3, and angiotensin II receptors, each increased EV release without affecting the overall size of the EVs [17]. Also, EV release by the calcium ionophore, A23187, was less robust when compared to receptor-mediated stimulation [17]. These finding warrant the investigation on whether activation of other GPCRs can mediate EV release.

9. EV release by various stimuli

EV release can be triggered by several types of stimuli. Many of these stimuli alter exosome release in an ADAM dependent manner. For example, EV release by p53 activation in tumor cells during radiation treatment has been reported [44]. Bacterial toxins including lipopolysaccharide have been shown to enhance the release of exosomes enriched in ADAM17 [24]. Calcium mobilization by treatment with calcium ionophores such as ionomycin has been reported to activate ADAM10 and trigger the secretion of EVs [24]. On the other hand, PMA has been reported to activate ADAM17 and induce its enrichment in exosomes [24]. Hypoxia has been reported to induce metalloproteinases and the release of EVs [24].

Akuthota et al. showed human eosinophils which are known to secrete chemokines, cytokines, and cationic proteins also secrete MVs, and the secretion of these vesicles increase in response to inflammatory stimuli such as tumor necrosis factor alpha (TNF-α) stimulation and eotaxin-1 (CCL11) [45]. In another study, Hunter et al. showed MVs are released from human brain microvascular endothelial cells in response to either thrombotic or inflammatory stimuli in a sex dependent manner [46]. Experimental evidence suggests nitric oxide (NO) negatively regulates EV release [47].

Numerous studies have demonstrated oxidative stress regulates the release of EVs from various cell types. One study showed oxidative stress triggers the release of microparticles by human alveolar cells and human bronchial epithelial cells [48]. Another study showed oxidative stress induces the release of membrane complement regulatory protein positive microparticles and this was blocked by the thiol antioxidant N-acetylcysteine amide [49]. Another study showed cigarette smoke extract induces exosome release by airway epithelial cells by depleting cell surface thiols but this is prevented by N-acetyl-L-cysteine and glutathione [50]. In another study, ATP mediated signaling through the purinergic P2X7 receptor was found to trigger macrophage activation of tissue factor activation procoagulant EVs and this was blocked by the ROS inhibitor N-acetyl cysteine or the flavoenzyme inhibitor diphenyleneiodonium but not nitric oxide synthase inhibitors [51]. In another study, elevated levels of carbon dioxide were reported to activate mitochondrial ROS in neutrophils resulting in an increase in microparticles release [52].

10. Circadian regulation of EVs

Circadian rhythms are seen in many physiological processes in several organ systems of various species and these rhythms are controlled by a master pacemaker located in the hypothalamus called the central suprachiasmatic nucleus and peripheral clocks in peripheral tissues. The circadian clocks within peripheral tissues not only regulate local physiological functions but they also contain essential core clock proteins such as Per1/2, Bmal1, Cry1/2, and Clock that work in concert to generate cell-autonomous oscillations and circadian rhythms.

Since the content of EVs represents a "snapshot" of a cell's internal environment at a given time, it is important to consider the regulation of EV biogenesis and cargo in a circadian dependent manner. Proteomic analysis of uEVs revealed these vesicles are enriched in at least 19 proteins that are associated with various renal diseases [53]. Proteins that were previously found to be packaged within uEVs include aquaporin-2, subunits of ENaC, the sodium chloride co-transporter (NCC) and the sodium potassium chloride co-transporter (NKCC2) [53–55]. Multiple studies have shown several of these proteins including ENaC and NCC are regulated by circadian clock proteins that are responsible for regulating circadian patterns and rhythm [56, 57].

At least in healthy male rats, the excretion of EVs into the urine appears to follow a circadian pattern [58]. This study reported that the highest excretion rate of EVs occurred during the active dark cycle (19,00–23,00). Additionally, the excretion rate of the EV associated protein TSG101 and the EV excretion rate showed a similar stoichiometry and circadian pattern. This finding suggest TSG101 can be used as a means of normalization of uEVs during time of day urine collections. The regulation of EV biogenesis and release in a circadian dependent manner for pathophysiological conditions is still currently under investigation. These studies are more complex since there are usually several variables associated with disease mechanisms that contribute to EV cargo loading and release.

11. Conclusions

Although EVs were once thought to serve as a garbage disposal system for cells, EVs are now regarded as important vehicles to shuttle biomolecules and allow for intercellular communication, cellular differentiation, intracellular signaling, and various other biological functions. The development of state-of the art equipment including advanced mass spectrometers for conducting lipidomic, proteomic, and metabolomic studies coupled to bioinformatics for performing pathway analysis has led to a better understanding for the role of EVs in physiology and pathophysiology. The identification of packaged molecules within EVs is not only important for biomarker discovery and the identification of novel drug targets, but it also provide clues to answering other questions related to molecular mechanism. For example, as discussed earlier, various protozoan parasites are able to hijack the host cellular machinery to promote their own survival and propagation. The use of engineered EVs to deliver small molecule drugs is gaining attention as it represents an efficient mechanism to treat various diseases including cancers and metabolic diseases.

Although several fundamental questions surrounding the biogenesis, loading, and release of EVs have been answered within the past 10 years there are several questions that remain. First, the specific mechanisms that regulate the balance between the release of exosomes and the fates of ILVs remain largely unknown. Second, EVs may have a dichotomous role in either inhibiting or promoting viral infection, but there are unanswered questions for each distinct process. Third, mechanisms for the cell type specific recognition and uptake of EVs is not completely understood.

Acknowledgements

The author thanks the University of Florida College of Medicine for providing funds to conduct EV research related to various disease mechanisms.

Conflict of interest

The author declare no conflict of interest.

Appendices and nomenclature

Non-standard abbreviations:

ADAM	a disintegrin and metalloproteinase
DAG	diacylglycerol
ESCRT	endosomal sorting complex required for transport
EVs	Extracellular vesicles
GPCRs	G-protein coupled receptors
GPIAP	glycosylphosphatidylinositol anchored proteins
HSPs	Heat shock proteins
ILVs	intraluminal vesicles
MVs	microvesicles
MVBs	multivesicular bodies
NO	nitric oxide
NCC	sodium chloride co-transporter
NKCC2	sodium potassium chloride co-transporter
PKC	protein kinase C
TNF-α	tumor necrosis factor alpha
VPS	vacuolar protein sorting

Author details

Abdel A. Alli
Department of Physiology and Functional Genomics, University of Florida College of Medicine, Gainesville, FL, USA

*Address all correspondence to: aalli@ufl.edu

References

[1] Gurunathan, S., et al., *Biogenesis, Membrane Trafficking, Functions, and Next Generation Nanotherapeutics Medicine of Extracellular Vesicles.* Int J Nanomedicine, 2021. **16**: p. 3357-3383.

[2] Edgar, J.R., *Q&A: What are exosomes, exactly?* BMC Biol, 2016. **14**: p. 46.

[3] Hierro, A., et al., *Structure of the ESCRT-II endosomal trafficking complex.* Nature, 2004. **431**(7005): p. 221-225.

[4] Hurley, J.H. and P.I. Hanson, *Membrane budding and scission by the ESCRT machinery: it's all in the neck.* Nat Rev Mol Cell Biol, 2010. **11**(8): p. 556-566.

[5] Larios, J., et al., *ALIX- and ESCRT-III-dependent sorting of tetraspanins to exosomes.* J Cell Biol, 2020. **219**(3).

[6] Murray, D.H., et al., *An endosomal tether undergoes an entropic collapse to bring vesicles together.* Nature, 2016. **537**(7618): p. 107-111.

[7] Simonsen, A., et al., *EEA1 links PI(3) K function to Rab5 regulation of endosome fusion.* Nature, 1998. **394**(6692): p. 494-498.

[8] Stenmark, H., *Rab GTPases as coordinators of vesicle traffic.* Nat Rev Mol Cell Biol, 2009. **10**(8): p. 513-525.

[9] Vanlandingham, P.A. and B.P. Ceresa, *Rab7 regulates late endocytic trafficking downstream of multivesicular body biogenesis and cargo sequestration.* J Biol Chem, 2009. **284**(18): p. 12110-12124.

[10] Ostrowski, M., et al., *Rab27a and Rab27b control different steps of the exosome secretion pathway.* Nat Cell Biol, 2010. **12**(1): p. 19-30; sup pp 1-13.

[11] Wei, D., et al., *RAB31 marks and controls an ESCRT-independent exosome pathway.* Cell Res, 2021. **31**(2): p. 157-177.

[12] Skryabin, G.O., et al., *Lipid Rafts in Exosome Biogenesis.* Biochemistry (Mosc), 2020. **85**(2): p. 177-191.

[13] Trajkovic, K., et al., *Ceramide triggers budding of exosome vesicles into multivesicular endosomes.* Science, 2008. **319**(5867): p. 1244-1247.

[14] Phuyal, S., et al., *Regulation of exosome release by glycosphingolipids and flotillins.* FEBS J, 2014. **281**(9): p. 2214-2227.

[15] Babiychuk, E.B. and A. Draeger, *Annexins in cell membrane dynamics. Ca(2+)-regulated association of lipid microdomains.* J Cell Biol, 2000. **150**(5): p. 1113-1124.

[16] Ni, K., et al., *The Evolving Role of Caveolin-1: A Critical Regulator of Extracellular Vesicles.* Med Sci (Basel), 2020. **8**(4).

[17] Conrad, K.P., et al., *Activation of multiple receptors stimulates extracellular vesicle release from trophoblast cells.* Physiol Rep, 2020. **8**(20): p. e14592.

[18] Nouri, M.Z., et al., *Increased endothelial sodium channel activity by extracellular vesicles in human aortic endothelial cells: Putative role of MLP1 and bioactive lipids.* Am J Physiol Cell Physiol, 2021.

[19] Dang, V.D., et al., *Lipidomic and proteomic analysis of exosomes from mouse cortical collecting duct cells.* FASEB J, 2017. **31**(12): p. 5399-5408.

[20] Matsui, T., et al., *ALIX and ceramide differentially control polarized small extracellular vesicle release from epithelial cells.* EMBO Rep, 2021. **22**(5): p. e51475.

[21] Chacko, K.M., et al., *The Effect of Tempol on Systolic Blood Pressure in Salt-loaded Hypertensive 129Sv Mice.* FASEB Journal, 2019. **33**(S1).

[22] Abels, E.R. and X.O. Breakefield, *Introduction to Extracellular Vesicles: Biogenesis, RNA Cargo Selection, Content, Release, and Uptake.* Cell Mol Neurobiol, 2016. **36**(3): p. 301-312.

[23] Hagiwara, K., et al., *Commitment of Annexin A2 in recruitment of microRNAs into extracellular vesicles.* FEBS Lett, 2015. **589**(24 Pt B): p. 4071-8.

[24] Shimoda, M. and R. Khokha, *Metalloproteinases in extracellular vesicles.* Biochim Biophys Acta Mol Cell Res, 2017. **1864**(11 Pt A): p. 1989-2000.

[25] Luketic, L., et al., *Antigen presentation by exosomes released from peptide-pulsed dendritic cells is not suppressed by the presence of active CTL.* J Immunol, 2007. **179**(8): p. 5024-5032.

[26] Leone, D.A., et al., *Surface LAMP-2 Is an Endocytic Receptor That Diverts Antigen Internalized by Human Dendritic Cells into Highly Immunogenic Exosomes.* J Immunol, 2017. **199**(2): p. 531-546.

[27] Baietti, M.F., et al., *Syndecan-syntenin-ALIX regulates the biogenesis of exosomes.* Nat Cell Biol, 2012. **14**(7): p. 677-685.

[28] Lopez-Cobo, S., C. Campos-Silva, and M. Vales-Gomez, *Glycosyl-Phosphatidyl-Inositol (GPI)-Anchors and Metalloproteases: Their Roles in the Regulation of Exosome Composition and NKG2D-Mediated Immune Recognition.* Front Cell Dev Biol, 2016. **4**: p. 97.

[29] Glover, S.C., et al., *Lipidomic analysis of urinary exosomes from hereditary alpha-tryptasemia patients and healthy volunteers.* FASEB Bioadv, 2019. **1**(10): p. 624-638.

[30] Pleet, M.L., et al., *Ebola Virus VP40 Modulates Cell Cycle and Biogenesis of Extracellular Vesicles.* J Infect Dis, 2018. **218**(suppl_5): p. S365-S387.

[31] Pleet, M.L., et al., *The Role of Exosomal VP40 in Ebola Virus Disease.* DNA Cell Biol, 2017. **36**(4): p. 243-248.

[32] Pleet, M.L., et al., *Extracellular Vesicles and Ebola Virus: A New Mechanism of Immune Evasion.* Viruses, 2019. **11**(5).

[33] York, S.B., et al., *Zika Virus Hijacks Extracellular Vesicle Tetraspanin Pathways for Cell-to-Cell Transmission.* mSphere, 2021: p. e0019221.

[34] Zhou, W., et al., *Exosomes mediate Zika virus transmission through SMPD3 neutral Sphingomyelinase in cortical neurons.* Emerg Microbes Infect, 2019. **8**(1): p. 307-326.

[35] Ning, B., et al., *Liposome-mediated detection of SARS-CoV-2 RNA-positive extracellular vesicles in plasma.* Nat Nanotechnol, 2021.

[36] Troyer, Z., et al., *Extracellular vesicles carry SARS-CoV-2 spike protein and serve as decoys for neutralizing antibodies.* J Extracell Vesicles, 2021. **10**(8): p. e12112.

[37] Mantel, P.Y. and M. Marti, *The role of extracellular vesicles in Plasmodium and other protozoan parasites.* Cell Microbiol, 2014. **16**(3): p. 344-354.

[38] Atayde, V.D., et al., *Exosome Secretion by the Parasitic Protozoan Leishmania within the Sand Fly Midgut.* Cell Rep, 2015. **13**(5): p. 957-967.

[39] Regev-Rudzki, N., et al., *Cell-cell communication between malaria-infected red blood cells via exosome-like vesicles.* Cell, 2013. **153**(5): p. 1120-1133.

[40] Mantel, P.Y., et al., *Malaria-infected erythrocyte-derived microvesicles mediate cellular communication within the parasite population and with the host immune system.* Cell Host Microbe, 2013. **13**(5): p. 521-534.

[41] Li, D.L., et al., *Analysis of the Differential Exosomal miRNAs of DC2.4 Dendritic Cells Induced by Toxoplasma gondii Infection.* Int J Mol Sci, 2019. **20**(21).

[42] van Niel, G., G. D'Angelo, and G. Raposo, *Shedding light on the cell biology of extracellular vesicles.* Nat Rev Mol Cell Biol, 2018. **19**(4): p. 213-228.

[43] Schlienger, S., S. Campbell, and A. Claing, *ARF1 regulates the Rho/MLC pathway to control EGF-dependent breast cancer cell invasion.* Mol Biol Cell, 2014. **25**(1): p. 17-29.

[44] Jelonek, K., P. Widlak, and M. Pietrowska, *The Influence of Ionizing Radiation on Exosome Composition, Secretion and Intercellular Communication.* Protein Pept Lett, 2016. **23**(7): p. 656-663.

[45] Akuthota, P., et al., *Extracellular Microvesicle Production by Human Eosinophils Activated by "Inflammatory" Stimuli.* Front Cell Dev Biol, 2016. **4**: p. 117.

[46] Hunter, L.W., M. Jayachandran, and V.M. Miller, *Sex differences in the expression of cell adhesion molecules on microvesicles derived from cultured human brain microvascular endothelial cells treated with inflammatory and thrombotic stimuli.* Biol Sex Differ, 2019. **10**(1): p. 26.

[47] Nolan, S., et al., *Nitric oxide regulates neutrophil migration through microparticle formation.* Am J Pathol, 2008. **172**(1): p. 265-273.

[48] Novelli, F., et al., *Procoagulant, tissue factor-bearing microparticles in bronchoalveolar lavage of interstitial lung disease patients: an observational study.* PLoS One, 2014. **9**(4): p. e95013.

[49] Carver, K.A. and D. Yang, *N-Acetylcysteine Amide Protects Against Oxidative Stress-Induced Microparticle Release From Human Retinal Pigment Epithelial Cells.* Invest Ophthalmol Vis Sci, 2016. **57**(2): p. 360-371.

[50] Benedikter, B.J., et al., *Cigarette smoke extract induced exosome release is mediated by depletion of exofacial thiols and can be inhibited by thiol-antioxidants.* Free Radic Biol Med, 2017. **108**: p. 334-344.

[51] Furlan-Freguia, C., et al., *P2X7 receptor signaling contributes to tissue factor-dependent thrombosis in mice.* J Clin Invest, 2011. **121**(7): p. 2932-2944.

[52] Thom, S.R., et al., *Increased carbon dioxide levels stimulate neutrophils to produce microparticles and activate the nucleotide-binding domain-like receptor 3 inflammasome.* Free Radic Biol Med, 2017. **106**: p. 406-416.

[53] Hoorn, E.J., et al., *Prospects for urinary proteomics: exosomes as a source of urinary biomarkers.* Nephrology (Carlton), 2005. **10**(3): p. 283-290.

[54] Hu, C.C., et al., *Pre-eclampsia is associated with altered expression of the renal sodium transporters NKCC2, NCC and ENaC in urinary extracellular vesicles.* PLoS One, 2018. **13**(9): p. e0204514.

[55] Qi, Y., et al., *Activation of the Endogenous Renin-Angiotensin-Aldosterone*

System or Aldosterone Administration Increases Urinary Exosomal Sodium Channel Excretion. J Am Soc Nephrol, 2016. **27**(2): p. 646-656.

[56] Alli, A., et al., *Direct and indirect inhibition of the circadian clock protein Per1: effects on ENaC and blood pressure.* Am J Physiol Renal Physiol, 2019. **316**(5): p. F807-F813.

[57] Richards, J., et al., *A role for the circadian clock protein Per1 in the regulation of the NaCl co-transporter (NCC) and the with-no-lysine kinase (WNK) cascade in mouse distal convoluted tubule cells.* J Biol Chem, 2014. **289**(17): p. 11791-11806.

[58] Koritzinsky, E.H., et al., *Circadian variation in the release of small extracellular vesicles can be normalized by vesicle number or TSG101.* Am J Physiol Renal Physiol, 2019. **317**(5): p. F1098-F1110.

Chapter 3

Extracellular Vesicles and Their Interplay with Biological Membranes

Jacob Yeo Xian Ping, Yub Raj Neupane and Giorgia Pastorin

Abstract

Most cells secrete vesicles into the extracellular environment to interact with other cells. These extracellular vesicles (EVs), have undergone a paradigm shift upon the discovery that they also transport important material including proteins, lipids and nucleic acids. As natural cargo carriers, EVs are not recognised by the immune system as foreign substances, and consequently evade removal by immune cells. These intrinsic biological properties of EVs have led to further research on utilising EVs as potential diagnostic biomarkers and drug delivery systems (DDSs). However, the internalisation of EVs by target cells is still not fully understood. Moreover, it is unclear whether EVs can cross certain biological membranes like the blood-brain barrier (BBB) naturally, or require genetic modifications to do so. Hence, this review aims to evaluate the relationship between the composition of EVs and their association with different biological membranes they encounter before successfully releasing their cargo into target cells. This review identifies specific biomarkers detected in various EVs and important biological barriers present in the gastrointestinal, placental, immunological, neurological, lymphatic, pulmonary, renal and intracellular environments, and provides a recommendation on how to engineer EVs as potential drug carriers based on key proteins and lipids involved in crossing these barriers.

Keywords: biological barriers, diagnostic biomarkers, drug delivery, engineering, extracellular vesicles, bioengineering

1. Introduction

Extracellular vesicles (EVs), phospholipid bilayer-enclosed vesicles consisting of proteins, lipids and nucleic acids, were once thought of as merely how cells may discard their waste materials and debris. However, recent discoveries have proven them to be indispensable to cells even in normal physiological functions and as diagnostic biomarkers for various diseases [1]. EVs are secreted by various cells and can be isolated from diverse biological sources like saliva, breast milk and blood serum [2].

Over the years, EVs have been researched as promising diagnostic biomarkers for pathological conditions. This is because their concentration and composition correlate with disease progression, a unique characteristic that sets them apart from other types of paracrine secretions [3, 4]. EVs have also been explored as possible carriers for drug delivery. Recent studies have shown promising results regarding the utilisation of EVs as drug delivery systems (DDSs) to treat various conditions, such as cardiovascular diseases [2, 5], osteoporosis [2, 6] and brain tumours [2, 7]. In light of this, EVs are seen as a more desirable strategy for drug delivery compared to other conventional nanoparticles like liposomes, micelles and polymeric nanoparticles [8, 9]. Conventional DDSs have been extensively used for their ability to protect drugs from inactivation in the external environment. However, plasma proteins risk adsorbing onto the surfaces of these non-EV nanoparticles upon injection into the body, making them an easy target of immune cells and decreasing their uptake by their target cells [10]. Although these nanoparticles may undergo modification to avoid immune cell removal, they still lack biocompatibility due to their non-biological origins. EVs, on the other hand, can evade phagocytosis by immune cells naturally, in addition to being highly selective for designated target sites, due to their biological origins and cell-specific surface properties inherited from the parent cells that secrete them.

Although EVs are promising in their diagnostic and therapeutic applications, it is still unclear whether they can cross membranes like the blood-brain barrier (BBB) naturally or when genetically modified, or only when the membranes become more permeable in certain conditions like injury [11, 12]. Furthermore, the uptake of EVs by target cells is still not fully understood at a microscopic level, be it *via* endocytosis, membrane fusion or other mechanisms [3]. The ability to pass through biological membranes is an important factor to consider when engineering EVs to deliver drugs to specific cells. As there remains a lack of understanding on how EVs can cross significant biological membranes before reaching their target sites, this review aims to identify potential key proteins and lipids that play a dominant role in the functions of EVs, and evaluate the relationship of these key components on EVs with different biological membranes, so that a recommendation can be given on how to best engineer EVs as potential drug carriers.

2. EVs—classification and key components

Classified by their biogenesis, size, morphology and function, there are three main EV categories—exosomes, microvesicles and apoptotic bodies (**Figure 1**) [16–18].

Although exosomes, microvesicles and apoptotic bodies are distinct from one another, there is a partial overlap among their respective size range and composition. Although many different methods have been previously deployed to isolate EVs from their sample sources (a notable example being ultracentrifugation in isolating and purifying exosomes and microvesicles [19–22]), these methods are unable to provide an accurate attribution of unique characteristics to each EV category. This is due to the complex nature of EVs, such that different size ranges can be derived from the same EV source depending on the isolation technique used [23]. As such, this review will mainly elaborate on EVs in general, unless otherwise stated.

Apart from biogenesis, size and morphology, each EV category possesses its own unique set of key proteins, lipids and nucleic acids (**Table 1**). Being able to differentiate EV categories based on their key components is vital in understanding their specific roles in both normal and pathological conditions. In general, all EVs possess cell adhesion proteins [13, 14, 17, 18, 24–28], heat-shock proteins [13, 14, 18, 25, 28–30], biogenesis-associated proteins [13, 14, 17, 18, 24, 25, 28], fusion proteins [13, 14, 18, 25], cell-type specific proteins [13, 14, 18, 27, 28], cytoskeletal proteins [13, 18], signalling molecules [13, 14, 28, 31], enzymes [13, 25, 28], messenger ribonucleic acid (mRNA), micro ribonucleic acid (miRNA), non-coding ribonucleic

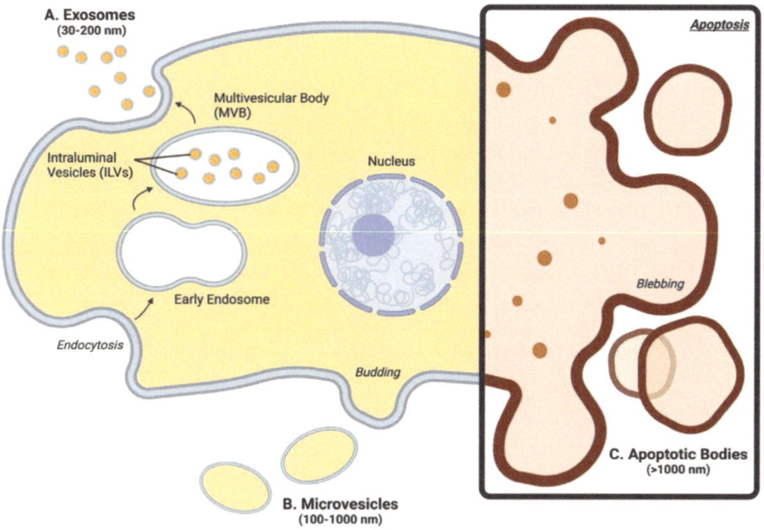

Figure 1.
Biogenesis, size, morphology and function of exosomes, microvesicles and apoptotic bodies. (A) ***Exosomes*** *(spheroid shape, 30–200 nm) are mainly involved in regulating intercellular communication. Their formation begins when the plasma membrane undergoes endocytosis to generate an early endosome. Intraluminal vesicles (ILVs) within the endosome are then formed from the inward budding of the endosomal membrane, resulting in a multivesicular body (MVB). This process is facilitated by either endosomal sorting complex required for transport (ESCRT)-dependent or -independent mechanisms [1, 13]. The MVB finally fuses with the plasma membrane to release the ILVs as exosomes. (B)* ***Microvesicles*** *(irregular shape, 100–1000 nm), like exosomes, also regulate intercellular communication. They are formed via budding from the plasma membrane directly without going through endocytic processes. (C)* ***Apoptotic bodies*** *(variable shape, usually >1000 nm but can be as small as 50 nm [14, 15]) are formed only during cell apoptosis, during which the post-apoptotic cell bulges outwards to form vesicles for easier removal by macrophages (created with BioRender.com).*

acid (RNA), phosphatidylethanolamine, sphingolipids and higher levels of phosphatidylserine (PS) than the cell plasma membrane [24, 25, 28, 35].

The distinct protein, lipid and nucleic acid profiles of each category might be correlated with its formation processes and functions. Both exosomes and microvesicles consist of key protein components which are responsible for cell-to-cell communication [18], such as glycoproteins [18, 28], membrane signalling receptors, growth factors and cytokines [18, 25], while apoptotic bodies do not. This is most likely because exosomes and microvesicles are meant to reach target cells, while apoptotic bodies are merely the means for discarding dead cells. Microvesicles and apoptotic bodies consist of other cytoplasmic proteins which seem to be less prominent in exosomes [13]. This might be due to the similar "budding/bulging" nature of the biogenesis of microvesicles and apoptotic bodies from the cytoplasmic membrane directly, a characteristic that differs from the endocytic-driven biogenesis of exosomes. Unlike exosomes and microvesicles, apoptotic bodies are composed of chromosomal deoxyribonucleic acid (DNA) fragments, chromatin remnants, cytosol portions, degraded proteins and cell organelles from dead cells [25, 35], indicative of their role in removing dead cells.

EVs also possess additional key features according to the specific cell line they originate from (**Table 2**). In general, cancer cells consist of higher levels of sphingolipids, glycerophospholipids, sterol lipids, ceramide, phosphatidic acid and matrix metalloproteinases like a disintegrin and metalloproteinase domain-containing protein 10 (ADAM10), while non-cancer cells consist of higher levels of prenol lipids, glycerolipids and fatty acids [24, 83].

Components	Exosomes	Microvesicles	Apoptotic bodies	Reference(s)
Proteins				
Tetraspanins	CD9, CD63, CD81, CD37, CD82, CD53, TSPAN 6, TSPAN 8, TSPAN 29, TSPAN 30	CD40 ligands, CD82	CD40 ligands, CD82	[13, 14, 24, 25]
Cell adhesion proteins	Integrins (integrin-alpha, integrin-beta), selectins (P-selectin), lactadherin, ICAM	integrins, selectins (P-selectin), fibronectin, PECAM-1	integrins, fibronectin, PECAM-1	[13, 14, 17, 18, 24–28]
Heat shock proteins	Hsc70, Hsp20, Hsp27, Hsp60, Hsp70, Hsp90	Hsp70, Hsp90	Hsp70, Hsp90	[13, 14, 18, 25, 28–30]
Biogenesis-associated proteins	ESCRT proteins (Alix, Tsg101), VPS4, clathrin, ubiquitin, syntenin, VPS32, PLD	ESCRT proteins (Alix, Tsg101), VPS4, ERK, PLD	VPS4, ERK, PLD	[13, 14, 17, 18, 24, 25, 28]
Fusion proteins	Flotillin 1 and 2, annexins, GTPases, Rab GTPases, dynamin, syntaxin	Flotilin-2, Rab GTPases, annexins	Rab GTPases, annexins (Annexin V)	[13, 14, 18, 25]
Cell-type specific proteins	MHC class I, MHC class II, APP, PMEL, TCR, CXCR4, HSPG, CD86, PrP, WNT	MHC class I, MHC class II, LFA1, CD14	MHC class I, LFA1, CD14	[13, 14, 17, 18, 27, 28]
	Actin, tubulin, cofilin	Actin, tubulin	Actin, tubulin	[13, 18]
Cytoskeletal proteins	Protein kinases, beta-catenin, 14-3-3, G proteins	ARF6, Rab11, ROCK	ARF6, Rab11, ROCK	[13, 14, 28, 31]
Signalling molecules	PLA2, peroxidases, pyruvate kinase, enolase, GADPH, ATPases	GADPH	GADPH	[13, 25, 28]
Other enzymes	Glycoproteins • e.g. beta-galactosidase, O-linked glycans, N-linked glycans	Glycoproteins • e.g. Glycoprotein Ib		[18, 28]
Additional proteins	Growth-factors and cytokines • e.g. TNF-α, TGF-β, TNF-related apoptosis-inducing ligand	Growth factors and cytokines		[18, 25]
	Membrane signalling receptors • e.g. FasL, TNF receptor, TfR	Membrane signalling receptors		
	Phosphoproteins	High phosphoproteins		[28, 32]
	Ribosomal proteins	GTP-binding protein ARF6		[28, 33]
	Lysosomal proteins • e.g. Lamp2b	Chemokines		[25]

Components	Exosomes	Microvesicles	Apoptotic bodies	Reference(s)
Lipids	High phosphatidylserine	High phosphatidylserine		[14, 25]
	Phosphatidylethanolamine	Phosphatidylethanolamine		[24, 25, 28, 32]
	Sphingolipids • e.g. sphingomyelin, gangliosides	Sphingolipids		[18]
	High cholesterol			[14, 28, 34]
	High diacylglycerol			[14]
	Ceramides			[13, 24, 28]
	Phosphatidylcholine			[28]
	Phosphatidylinositol			[18]
	LBPA			[13]
Nucleic acids	mRNA		mRNA	[25, 35]
	miRNA		miRNA	
	Non-coding RNA • e.g. small nucleolar RNAs, microRNAs, piwi-interacting RNAs, other long non-coding RNAs		Non-coding RNA • e.g. small nucleolar RNAs, microRNAs, piwi-interacting RNAs, other long non-coding RNAs	
	DNA with histones		Chromosomal DNA fragments with histones, chromatin remnants, cytosol portions, degraded proteins, cell organelles	

Abbreviations: ADP: adenosine diphosphate, APP: amyloid-beta precursor protein, ARF: ADP ribosylation factor, CXCR: CXC chemokine receptor, DNA: deoxyribonucleic acid, ERK: extracellular signal-regulated kinase, ESCRT: endosomal sorting complex required for transport, FasL: Fas ligand, GTP: guanosine triphosphate, HSPG: heparan sulphate proteoglycan, ICAM: intercellular adhesion molecule, Lamp: lysosome-associated membrane protein, LFA: lymphocyte function-associated antigen, MHC: major histocompatibility complex, PECAM: platelet endothelial cell adhesion molecule, piwi: P-element induced wimpy testis, PLA2: phospholipase A2, PLD: phospholipase D, PMEL: premelanosome protein, PrP: prion protein, Rab: Ras-associated binding, TCR: T-cell receptor, RNA: ribonucleic acid, ROCK: Rho-associated protein kinase, TSPAN: tetraspanin, Tsg: tumour suppressor gene, VPS: vacuolar protein sorting-associated protein, WNT: wingless/integrated, GADPH: glyceraldehyde 3-phosphate dehydrogenase, TDP: transactive response DNA-binding protein, TfR: transferrin receptor, TGF: transforming growth factor, TNF: tumour necrosis factor.

Table 1.
Classification of key components of EVs by their main categories.

EV source	EV source subtype	Component(s)	Reference(s)
Bacteria	Gram-positive	ABC transporters, mobility-related proteins (FliC, PilQ), multidrug efflux pumps, porins (Omps, OprF, PorA, PorB)	[24, 36]
	Gram-negative	Beta-lactamase, coagulation factor, penicillin-binding protein	[24, 37–39]
	Myxobacteria	Chaperonin GroEL1, GroEL2, hydrolase, peptidase	[24, 40, 41]
Blood cells	Platelets	CD31, CD41, CD42a, CD62, C-type lectin, CXCR4, GPIIb/IIIa, PF4, SDF-1α	[24, 42–44]
	Erythrocytes	Glycophorin A, stomatin	[24, 34]
	Reticulocytes	Galectin-5	[42, 45]
Bone cells	Osteoblasts	Cadherin-11	[42, 46]
Cancer cell lines	Breast cancer cells (MM231, MM231LN)	Rab-5b, actin, integrin beta 1, cavolin-1	[47]
	Breast cancer cells (MCF7)	Actin, Rab-5b	[47]
	Breast cancer cells (MCF10A)	Integrin beta 1	[47]
	Cervical cancer cells (HeLa)	EGF	[42, 48, 49]
	Colon cancer cells (LIM1863—EpCAM apical exosomes)	CD44, CD46, CD59, CLDN7, EpCAM, HMGB2, HMGB3, Muc-13, sucrase isomaltase	[50]
	Colon cancer cells (LIM1863—A33 basolateral exosomes)	ADP-ribosylation factor, AP1G1, AP1M1, AP1M2, AP3B1, CLSTN1, CLTA, CLTB, COPB2, EEA1, GPA33, HLA-A, HLA-B, HLA-C, HLA-E, HLA-A29.1, Rab-13, REEP6	[50]
	Colorectal cancer cells (CRC line SW403, CRC28462)	Carcinoembryonic antigen, class I HLA	[51]
	Hepatoblastoma cancer cells (HepG2, K562)	TfR1, TfR2	[42, 52]
	Hepatocellular cancer cells (HKCI-C3, HKCI-8, MHCC97L, MIHA)	ADAM10, ARHGEF18, BROX, CAV1, CAV2, CD44, CDC42, CLDN3, EDIL3, EIF4A3, GNA11, GNA13, GNAQ, GNAS, GRB2, MET, RHOG, RRAS, SNTA1, TNFRSF21, TNFAIP2	[53]
	Myeloma cancer cells (RPMI-8226, CAG)	Fibronectin	[54]
	Nasopharyngeal cancer cells (C15)	Galectin-9, LMP1	[55]
	Nasopharyngeal cancer cells (C17)	Galectin-9	[55]
	Ovarian cancer cells (IGROV1, OVCAR-3)	Beta-actin, EpCAM, hnRNPA1, hnRNPK	[56]
	Prostate cancer cells (PC3)	Rab-5b, integrin beta 1, cavolin-1	[47]
	Prostate cancer cells (PC-3 M-luc)	Rab-5b, actin, Integrin beta 1	[47]
	Prostate cancer cells (22Rv1)	Rab-5b, actin	[47]
	Prostate cancer cells (PNT2)	Actin, integrin beta 1	[47]

EV source	EV source subtype	Component(s)	Reference(s)
Endothelial and epithelial cells		C-type lectin, galectin-3, Muc-1	[42]
Immune cells	B-cells	A2,3-linked sialic acid, CD169	[42, 57]
	T-cells	CXCR4, SDF-1α	[42, 48, 58, 59]
	Dendritic cells	FLOT1, galectins, Lamp-1, MFG-E8, MHC class I and II, TNFR1, TNFR2	[24, 42, 60–62]
	Macrophages	C-type lectin, LFA-1	[42, 63, 64]
	Natural killer cells	Granzyme B, perforin	[65]
Mesenchymal stem cells		Alternative splicing and Golgi apparatus component mRNA encoding transcription factors, CD54, CD73, CD86, CD90, CD105, CD166, MHC class I and II, sialic acids	[24, 28, 42, 66–75]
Milk cells	Bovine milk cells	β-casein, β-lactoglobulin mRNA, CD59, MFG-E8, miR-30a, miR-92a, miR-223, Rab-1b, Rab-11a	[24, 76–80]
	Human breast milk cells	miR-17, miR-181a	[81]
Nervous cells	Astrocytes	MCP-1, MMP3, MMP9, TIMP-1	[65]
	Microglia	CD13, CD107a, CD107b	[65]
Placental cells		MHC class I chain-related proteins A and B, placental alkaline phosphatase, placental leucine aminopeptidase, pregnancy specific glycoprotein 3, RAET1 proteins/ULBP1–5, TGFβ1, TRAIL, trophoblast glycoprotein 5 T4	[82]

Abbreviations: ABC: adenosine triphosphate-binding cassette, ADAM: A disintegrin and metalloproteinase domain-containing protein, ADP: adenosine diphosphate, AP: adaptor related protein complex, ARFGEF: Rho/Rac guanine nucleotide exchange factor, BROX: BRO1 domain and CAAX motif containing, CA: carbohydrate antigen, CAV: caveolin, CLDN: claudin, CLSTN: calsyntenin, CLT: clathrin light chain, COP: coatomer protein complex, CXCR: CXC chemokine receptor, EDIL: EGF like repeats and discoidin domains, EEA: early endosome antigen, EGFR: epidermal growth factor receptor, EGF: Epidermal growth factor, EpCAM: epithelial cell adhesion molecule, FLOT: flotillin, GN: guanine nucleotide-binding protein, GP: glycoprotein, GRB: growth factor receptor-bound protein, HLA: human leukocyte antigen, HMG: high-mobility group, HNRNP: heterogeneous nuclear ribonucleoprotein, Lamp: lysosome-associated membrane protein, LDLR: low-density lipoprotein receptor, LDL: low-density lipoprotein, LFA: lymphocyte function-associated antigen, LMP1: Epstein-Barr virus latent membrane protein 1, MAPK: mitogen-activated protein kinase, MCP: membrane cofactor protein, MET: mesenchymal-epithelial transition facror, MFG-E: milk fat globule-EGF factor, MHC: major histocompatibility complex, miRNA: microribonucleic acid, MMP: matrix metalloproteinase, Muc: mucin, PF: platelet factor, RAET: retinoic acid early transcript, Rab: Ras-associated binding, REEP: receptor expression-enhancing protein, RHOG: Ras homolog family member G, RRAS: RAS-related protein R-Ras, SDF: stromal cell-derived factor, SNT: syntrophin, TfR: transferrin receptor, TNF: tumour necrosis factor, TNFR: tumour necrosis factor receptor, TNFRSF: TNF receptor superfamily, TNFAIP: TNF alpha-induced protein, TRAIL: tumour necrosis factor-related apoptosis-inducing ligand, TYRP: tyrosinase-related protein, ULBP: UL16 binding protein.

Table 2.
Classification of additional key components of EVs by their specific cell lines.

3. The EV journey—overcoming biological barriers

To reach their target sites, EVs need to overcome various biological barriers (**Figure 2**). Complementing these barriers are blood vessels (capillaries in particular). EVs can enter and extravasate from these vessels *via* diffusion, due to their lipidic nature which enables them to pass through the highly-lipidic capillary endothelium and their small size that enables them to pass or squeeze through fenestrations in the capillary wall [84].

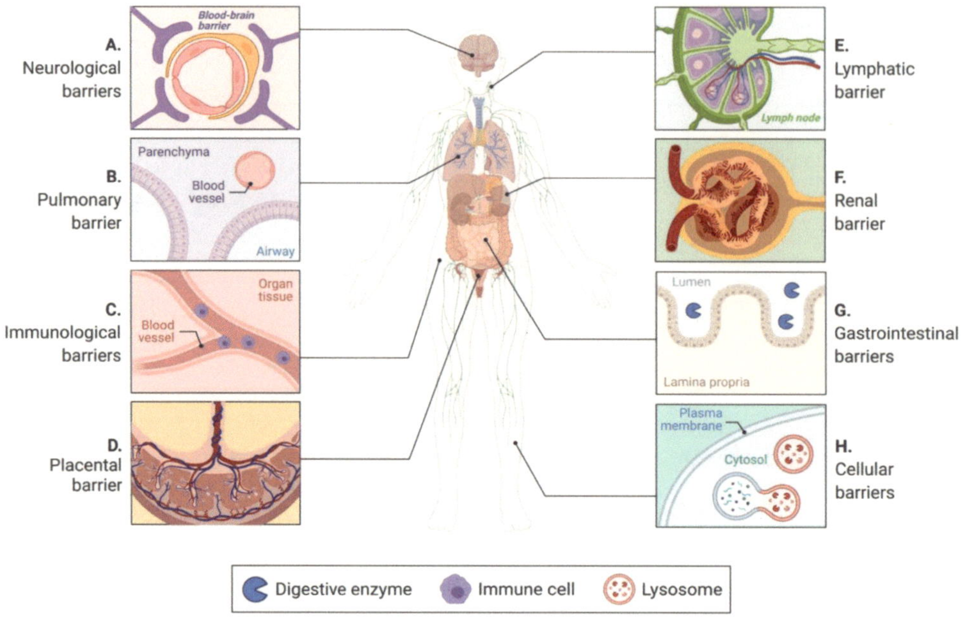

Figure 2.
Biological barriers encountered by extracellular vesicles (EVs). (A) ***Neurological barriers*** *include the blood-brain barrier (BBB), blood-labyrinth barrier (BLaB) and blood-retinal barrier (BRB). (B)* ***The pulmonary barrier****, or blood-air barrier (BAB), guards against the invasion of pathogens in the lungs via its immune cell-rich lung mucosa, lung epithelial cells and ciliary action. (C)* ***Immunological barriers*** *eliminate pathogens and perceived foreign substances from the body via the mononuclear phagocyte system and the adaptive immune system. (D)* ***The placental barrier*** *consists of an inner blood-vessel-rich layer with the syncytiotrophoblast facing the bloodstream and an outer layer of trophoblasts. (E)* ***The lymphatic barrier****, or blood-lymph barrier (BLyB), is regulated by various mechanisms including extravasation, overcoming the interstitium, diffusion, and passage through the mucosal barrier. The collagen reticular network (RN) also hinders soluble substances from passing through. (F)* ***The renal barrier****, or glomerular filtration barrier, composes of the fenestrated endothelium, glomerular basement membrane and glomerular epithelium, and this hinders the passage of large molecules across the barrier. (G)* ***Gastrointestinal barriers*** *are associated with digestive enzyme degradation, harsh stomach acidic conditions and the small intestinal barrier. (H)* ***Cellular barriers*** *of the target cell include the plasma membrane, endosomal membrane and lysosomal membrane. EVs internalised by cells via endocytosis are packaged into endosomes which may risk fusing with lysosomes to undergo degradation (created with BioRender.com).*

3.1 Gastrointestinal barriers

EVs administered orally need to overcome digestive enzymatic degradation, harsh stomach acidic conditions and the small intestinal barrier before entering the bloodstream for systemic absorption. As milk and plant-derived EVs are delivered into the body naturally *via* oral consumption, they might provide key insights into how EVs can be used and/or engineered for oral administration. *In vivo* evidence in rodents showed that unmodified bovine milk-derived EVs naturally containing immune-active proteins were able to cross the intestinal barrier *via* endocytosis to treat inflammatory bowel disease (IBD) [85], and were distributed significantly in the bloodstream 24 h post-oral consumption [86]. EVs can pass through the intestinal barrier *via* intestinal epithelial cell (IEC) mediated transendocytosis, a process that requires surface glycoproteins on both EVs and target cells, based on *in vitro* findings of skimmed bovine milk-derived EVs being internalised by human colon carcinoma Caco-2 cells and rodent small intestinal IEC-6 cells [87, 88]. Paracellular translocation is another possible mechanism by which EVs may cross the intestinal epithelium, through tight junctions between adjacent epithelial cells [76, 87, 89]. Although *in vivo* evidence is lacking, it is possible that EVs might cross the intestinal epithelium paracellularly to a greater extent in pathological conditions like IBD as the tight junctions would be disrupted [90], making the intestinal epithelium more penetrable.

Milk-derived EVs have been shown to withstand acidic and enzymatic conditions [87, 91]. However, their ability to do so might be dependent on the milk source, as EVs from processed milk would have undergone more damage than those from unprocessed milk and hence possess less integrity [87, 92–97]. Although bovine milk-derived EV surface proteins CD9 and CD81 were found to be partially degraded by acidification at pH 4.6 in one study [98], these findings did not demonstrate whether these EVs can survive stomach acidic conditions, which are usually characterised by a much lower pH. Moreover, the study was focused on evaluating the effectiveness of acidification in ultracentrifugation to isolate EVs. Thus, these conditions would have differed vastly from true gastrointestinal conditions. Although the underlying mechanism is unclear, the ability of both processed and unprocessed milk-derived EVs to withstand harsh conditions might be correlated with milk calcium content [87]. This could be due to the adhering of milk calcium to the surface of EVs, which might strengthen their membrane integrity against acidic and enzymatic degradation. Another hypothesis is that calcium might influence milk-derived EV biogenesis pathways in alveoli cells to increase the expression of certain proteins or transporters in secreted EVs that enable them to withstand gastrointestinal conditions.

Fruit and vegetable-derived EVs have been shown to withstand gastrointestinal conditions and eventually be internalised by rodent intestinal tissue *in vivo*, though their passage across the intestinal barrier into the bloodstream cannot be concluded in some studies [77, 99–102]. Grape EVs derived *via* cold-pressing have been discovered to enter rodent IECs *via* macropinocytosis [100], while a previous analysis of grapefruit EVs derived *via* homogenization revealed their internalisation by intestinal macrophages *via* macropinocytosis and clathrin-mediated endocytosis [101]. Watermelon EVs were also observed to be taken up by human IECs in an *in vitro* experiment *via* clathrin-mediated endocytosis, causing the cells to multiply rapidly and their basal secretome to change [103]. Ginger EVs were found to accumulate in rodent liver tissue 12 h post-oral consumption, implying that the EVs were able to withstand gastrointestinal conditions and cross the intestinal barrier into the bloodstream while remaining intact [104]. Though unconfirmed, the uptake of plant-derived EVs *via* clathrin-mediated endocytosis and macropinocytosis probably indicates that these EVs possess receptor tyrosine kinases, G protein-coupled receptors (GPCRs) and transferrin receptors [105], while passage across the intestinal barriers into the bloodstream might imply that these plant-derived EVs undergo transendocytosis like milk-derived EVs, a mechanism which requires EVs to possess surface glycoproteins [87, 88]. The ability of milk and plant-derived EVs to withstand and overcome gastrointestinal conditions and barriers makes them highly suitable as DDSs *via* the oral route as a non-invasive alternative to intravenous DDSs.

3.2 Placental barrier

The placenta supports foetal growth and development while secreting female hormones [106–111]. The placental barrier (PB) is suggested to be selectively penetrable, given that drugs administered to pregnant women can either cause adverse side effects in both the mother and the fetus or not penetrate the PB at all. It consists of an inner blood-vessel-rich layer with the syncytiotrophoblast facing the bloodstream and an outer layer of trophoblasts [106, 112–114]. Occurring in large amounts during pregnancy [115, 116], placental exosomes exert their functions during foetal growth and development, being involved in processes like angiogenesis regulation and cell migration [106, 116–126]. This implies that they can overcome the PB, though the underlying mechanism is unclear. Placental exosomes have also been tested as diagnostic biomarkers for foetal development [106, 115] and gestational diabetes [106, 127].

Although placental EVs may be used to pass through the PB, the use of non-placental EVs to deliver drugs across the PB is a potential area for exploration. The placenta can respond to signals from immune cells and exert an inflammatory response during infection. An *in vitro* study revealed that THP-1 monocyte-derived exosomes were internalised by human placental trophoblast cells *via* clathrin-mediated endocytosis, exerting a pro-inflammatory effect that caused the cells to release cytokines [128]. Provided that this mechanism can be proven *in vivo*, packaging drugs in EVs derived from immune cells might be one way to deliver drugs across the PB. Another possible method to deliver drug-containing EVs across the PB might be *via* administering EVs that target IECs instead of placental cells, as IECs can communicate with the placenta [103]. IECs that internalise watermelon EVs can secrete watermelon EV contents *via* the formation of intestinal exosomes, which are shown to be taken up by placental cells *via* clathrin-mediated endocytosis [103]. This concept, however, is deduced from a few *in vitro* studies and has yet to be proven in a single *in vivo* experiment. Nevertheless, being able to deploy non-placenta-derived EVs to treat placental pathological conditions like chorioamnionitis may offer some flexibility in EV engineering, as researchers would not need to adhere strictly to using placental EVs.

3.3 Immunological barriers

The body is heavily guarded by immune cells responsible for eliminating pathogens and perceived foreign substances. As such, nanoparticles injected into the bloodstream risk being removed by phagocytes of the mononuclear phagocyte system (including those in the liver and spleen), or the adaptive immune system *via* antibody production [8]. Thus, in conventional non-EV drug therapy, immune cells can potentially hinder the therapeutic effects of nanoparticles by decreasing their systemic circulation half-life [10]. EVs, on the other hand, can evade removal by immune cells naturally. CD47 is a prominent component found on EVs that binds to signal regulatory protein alpha (SIRPα) on dendritic cells and macrophages, which inhibits phagocytosis *via* a "don't eat me" signal [129–131]. Other EV components found on both cancer and non-cancer cell-derived EVs like CD24, CD31 and PD-L1 have been associated with exerting a similar "don't eat me" signal, with PD-L1 also inhibiting T-cell activation [130, 132–134].

A recent study on an *in vivo* rodent tumour model seems to suggest that it may be possible for EVs to be phagocytosed by Kupffer cells in the liver and eliminated *via* biliary excretion, given that the fluorescent markers tagged to the U937 human myeloid leukaemia EVs used in the study were found to accumulate in the liver and eventually detected in the faeces [10]. However, these fluorescent markers were also predominantly detected in CT26 mouse colon adenocarcinoma cells targeted by the EVs, probably because the EVs might have already undergone disintegration in the cells, and the florescent marker component might have been excreted *via* exocytosis before being transported to the liver *via* systemic circulation. In other words, the accumulation of dyes associated with EVs in the liver is not synonymous with a definite uptake of EVs by Kupffer cells. Nevertheless, EVs may still be removed by immune cells, as shown in another study where melanoma, myoblast, fibroblast, aortic endothelial and macrophage-like cell exosomes from rodents were eliminated by rodent liver macrophages *in vivo*, most likely due to the presence of PS on EVs which is recognised by macrophages [135]. As to whether these EVs possess high amounts of CD47, CD24, CD31 or PD-L1, the study did not include such findings.

3.4 Neurological barriers

The blood-brain barrier (BBB) is characterized by an innermost layer of endothelial cells (which prevents blood and extracellular fluid from mixing), pericytes surrounding the endothelial cells and astrocyte end-feet acting as a sheath in the outermost layer. Though the movement of substances across the BBB is tightly regulated [136], different EVs have been observed to cross the BBB. One study demonstrated the ability of exosomes to carry miR-193b-3p across the BBB to exert an anti-inflammatory effect on rodent brain cells with subarachnoid haemorrhage [137], although the mechanism of crossing was unclear. Other studies involving the BBB in zebrafish showcased the ability of various human breast cancer cell EVs and brain endothelial cell EVs to cross the BBB *via* clathrin-mediated endocytosis [7, 138] and macropinocytosis [138], a notable surface protein that enabled clathrin-mediated endocytosis being CD63 [7]. Another study conducted on rodent BBB showed that human and rodent EVs derived from both cancer and non-cancer cells were able to cross the BBB *via* adsorptive-mediated transcytosis, which correlated with the presence of CD46 on the surface of EVs [11].

Modifications have also been made to EVs to enhance their ability to cross the BBB. In one experiment, after overexpressing the rabies virus glycoprotein (RVG) peptide on their surface, dendritic exosomes became significantly localized in rodents' brain cells [139]. Mouse L929 fibroblastic cell exosomes loaded with methotrexate and functionalized with LDL peptide in another experiment showed enhanced BBB exosome extravasation in rodents [140]. When miR-210-loaded mesenchymal stromal cell-derived exosomes were coupled with c(RGDyK) peptide in another experiment, they displayed enhanced targeting of rodent ischaemic brain cells, indicating greater angiogenesis and improving animal survival significantly [141]. Another experiment showed that RGE-Exo EVs demonstrated greater accumulation and duration of accumulation in murine glioma tumour cells than free exosomes [142].

Apart from surface components, the size of EVs might also be a crucial factor in determining whether EVs can cross the BBB, as deduced from another study where intranasal administration of exosomes to rodent microglial cells *via* the extra-neuronal pathway showed rapid translocation of exosomes to target cells, in contrast to larger microparticles of at least 500 nm in diameter which did not reach these target cells [143]. However, surface components of EVs might be a more vital factor than the size of EVs in enabling passage across the BBB, as proven by how larger brain endothelial EVs can penetrate the BBB better than smaller EVs of the same cell source, due to the higher levels of CD63 in the larger EVs [7, 106].

The blood-labyrinth barrier (BLaB) and blood-retinal barrier (BRB) are two other neurological barriers pertaining to the ear and eye, respectively. The BLaB consists of five layers, namely, the blood-endolymph barrier, blood-perilymph barrier, cerebrospinal-fluid-perilymph barrier, middle-ear-labyrinth barrier and endolymph-perilymph barrier [106, 144]. The BRB consists of the retinal vascular endothelium and the retinal pigment epithelium (RPE) [106]. These two barriers share similarities with each other and the BBB, though the number of EV studies on these two barriers is smaller than that involving passage across the BBB [106]. Nevertheless, the utilization of EVs as potential drug carriers targeting the ear and eye with negligible side effects is worth further research, especially when current drug treatments have resulted in adverse side effects [106]. EVs from RPE cells are involved in the progression of age-related macular degeneration *via* regulating the production of pigment granule and lipid balance in RPE cells [106, 145]. They also promote vascular leakage *via* miR-105 which interferes with the tight cellular junctions of barriers [106, 146]. It is hoped that these seemingly destructive EV mechanisms can be manipulated to enable drug delivery across the BLaB and BRB, by modifying these EVs in a way that does not harm the barriers yet still permits their passage across them.

3.5 Lymphatic barrier

The process of crossing the blood-lymph barrier (BLyB) is regulated by various mechanisms including extravasation, overcoming of the interstitium, diffusion and passage through the mucosal barrier [106, 147]. In addition, the collagen reticular network (RN) hinders soluble substances from passing through [106, 148–153]. However, EVs possess certain characteristics that enable them to cross the BLyB. For instance, human ovarian cancer cell exosomes were found to be able to travel from the periphery to the lymph node in just a matter of minutes in rodents due to their small size [106, 154], and their lipidic rather than soluble nature seemed to enable them to cross the RN [106, 155].

Although EVs already possess intrinsic advantages that enable them to cross the BLyB, methods like microfluidic surface engineering have been conducted on EVs to modify them further as potential drug carriers for lymphoma treatment or other diseases related to the lymphatic system [106, 156, 157]. A recent study explored the modification of exosomes derived from bovine serum. α-D-mannose was added to the exosomes containing immune stimulators to enable them to interact with the mannose receptors on dendritic cells for uptake, and the exosomes were PEGylated. This has been found to enhance the internalisation of the exosomes by murine dendritic cells and to increase their localisation in the lymph nodes, paving the way for efficient delivery of immune stimulators *via* EVs *in vivo* as a potential form of drug therapy [158].

3.6 Pulmonary barrier

Located in the lungs, the blood-air barrier (BAB) possesses characteristics that enable it to guard against pathogenic invasion. For instance, the lung mucosa is a rich source of immune cells [106, 159], and lung epithelial cells can sense a wide range of bacteria and viruses *via* a broad array of membrane-bound, endosomal and cytosolic pattern-recognition receptors (PRR) [106, 160]. In response to the presence of pathogens, the BAB regulates paracellular flow, cell-to-cell communication, synthesis of mucus and the composition of periciliary fluid [160], which complements the removal of foreign substances *via* ciliary action [106, 161]. While the passage of EVs across the BAB is still largely unexplored [106], exosomes derived from bronchoalveolar lavage fluid (BALF) have been discovered to possess a similar protein profile to mesenchymal derived dendritic cells, given that they carry CD54, CD63, CD86 and in particular, MHC classes I and II [162, 163], implying their involvement in triggering an immune response against pathogenic invasion in the BAB. In light of this, BALF exosomes might potentially be used as diagnostic biomarkers for pathogenic detection, in addition to engineering them as personalised medicine for effective drug delivery across the BAB.

3.7 Renal barrier

The nephron's ability to efficiently filter out waste materials from the blood into the urine is attributed to the high pressure in the glomerulus due to high blood flow, as well as the presence of the glomerular filtration barrier consisting of three layers—fenestrated endothelium, glomerular basement membrane and glomerular epithelium [66, 164–166]. Despite the tiny pores (2.5–2.8 nm) of the glomerular basement membrane [166, 167] which are smaller than the smallest EVs (30 nm [16–18]), and the presence of filter proteins lining the slits in the glomerular epithelium [165, 166, 168], the urine is surprisingly a rich source of EVs from both renal and non-renal sources. While the majority of EVs found in urine originate from the kidney, urinary bladder, testis, prostate, epididymis and seminal vesicle [169–172], studies have also identified EVs from outside the urinary tract, such as those carrying biomarkers of acute myocardial infarction [173]. EVs injected into the bloodstream of rodents in one study were found to accumulate in the kidneys, and eventually the urine, without undergoing degradation, as indicated by their ability to be internalized by HEK293 cells after being retrieved from the urine and introduced to the cells [166]. The presence of EVs in urine might imply that EVs can squeeze through the tiny pores of the glomerular filtration barrier due to their fluid membranes, or undergo mechanisms like transcytosis to reach the glomerular filtrate. It is also logical to deduce that EVs can cross the glomerular filtration barrier better in pathological conditions like diabetic nephropathy when the glomerular filtration barrier becomes more porous due to injury [166].

3.8 Cellular barriers

EVs that eventually reach the target cell has to overcome the plasma membrane, escape the endosome and evade lysosomal degradation to release their cargo into the cytosol (**Figure 3**).

The plasma membrane is an intricate structure consisting of various domains formed *via* different mechanisms. Some of these mechanisms include the formation of plasma membrane protein fences to reduce lateral diffusion in the plasma membrane, the arrangement of plasma membrane proteins into a scaffold that interacts with certain plasma membrane lipids, and protein-lipid interaction to form lipid rafts [174]. These domains represent the first barrier that a freshly-secreted EV needs to cross and determine the way the target cell internalizes EVs. Environmental factors also influence the interaction of EVs with the plasma membrane. For instance, the uptake of EVs *via* fusion with the cell membrane is observed to occur at a higher rate under acidic conditions [175], while endocytosis is shown to be hindered by neutral pH or high cholesterol levels [14, 176].

EVs can be internalised by target cells *via* endocytosis, be it caveolae-dependent endocytosis, flotillin-dependent endocytosis, ARF6-dependent endocytosis or other forms of endocytosis [105]. Clathrin-mediated endocytosis (CME), or "receptor-mediated endocytosis", plays an especially prominent role in the uptake of small EVs [105],

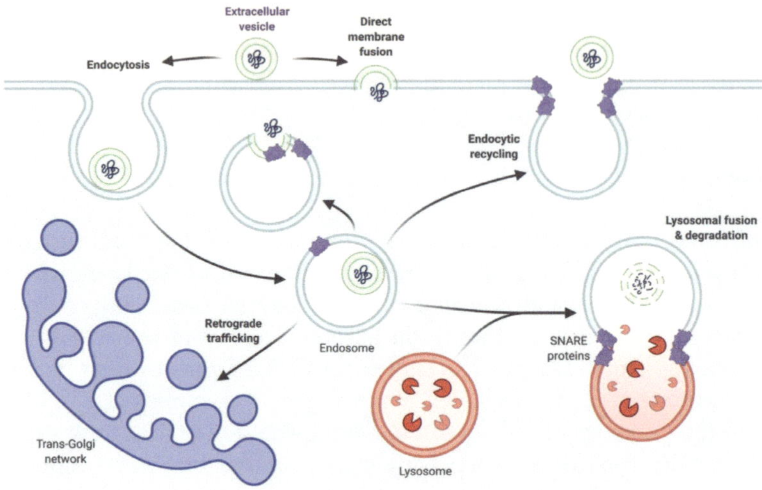

Figure 3.
Internalisation, lysosomal degradation and lysosomal escape mechanisms of extracellular vesicles (EVs) in target cells. Upon reaching the target cell, EVs may be internalised by the cell via endocytosis or direct fusion with the plasma membrane. EVs that are internalised via endocytosis are packaged into endosomes, which may fuse with lysosomes to degrade the EVs. To escape lysosomal degradation, endocytosed EVs may undergo retrograde trafficking to the trans-Golgi network, endocytic recycling to be secreted out of the cell, or another mechanism altogether. Endocytosed EVs that do not undergo lysosomal degradation fuse with the endosomal membrane via the mediation of soluble N-ethylmaleimide sensitive factor attachment protein receptor (SNARE) proteins to release their cargo into the cytosol. EVs that fuse directly with the plasma membrane to release their cargo into the cytosol evade endosomal and lysosomal activity completely (created with BioRender.com).

as supported by recent studies on the uptake of human epidermoid carcinoma EVs [177] and rat pheochromocytoma EVs [178] by human cervical carcinoma (HeLa) cells and rat bone marrow-derived mesenchymal stromal cells respectively. During CME, a temporary membrane scaffold forms as a result of membrane binding of Bin/amphiphysin/Rvs (BAR) domain-containing proteins which recruit clathrin. Clathrin then binds to the cytoplasmic tails of membrane proteins with the help of adaptor proteins, resulting in a clathrin-coated pit that internalises the EV [105]. EVs internalised *via* endocytosis are packaged into endosomes. These EVs then proceed to release their cargo into the cytoplasm by fusion of their membranes with the endosomal membrane, a process mediated by soluble *N*-ethylmaleimide sensitive factor attachment protein receptor (SNARE) proteins which join the cytosolic sides of the EV and endosomal membranes together [3].

In general, the CME of EVs is initiated through the mediation of lectins, tetraspanins, cell adhesion proteins and other receptor-ligand interactions [18, 179]. For instance, exosomes from macrophages possess C-type lectin, which interacts with the C-type lectin receptor found on dendritic and brain endothelial cells [179, 180]. Galectin-5 on red blood cell (RBC) EVs enables them to be internalised by macrophages [45, 179]. Integrins on tumour EVs have been associated with the uptake of these EVs by lung fibroblasts and liver macrophages [42, 179]. Exosomes and target cells can exploit the interaction between intercellular adhesion molecule 1 (ICAM-1) and lymphocyte function-associated antigen 1 (LFA-1) in exosomal uptake [179–181]. Heparan sulphate proteoglycans on target cells bind to EV fibronectin to facilitate uptake of EVs [54, 179, 182]. The high levels of outward-facing PS on the surface of exosomes also enable the recognition and uptake of these exosomes by antigen-presenting cells *via* T-cell immunoglobulin and mucin domain (TIM) receptors located on the antigen-presenting cells' surface [63, 179, 183].

EVs internalised *via* endocytosis might risk being degraded by lysosomes in the cytosol [3]. The fusion of the membrane of endosomes containing the endocytosed EV with the lysosomal membrane is mediated by SNARE proteins and involves the active transport of vesicles along the cytoskeleton [18]. EV size may play a role in determining the fate of EVs upon uptake *via* endocytosis, as EVs larger than 100 nm may require macropinocytosis for their uptake, which tends towards lysosomal degradation more than other internalisation mechanisms accessible to smaller EVs [177, 184, 185]. Endocytosed EVs might escape lysosomes *via* pathways similar to those of viruses, like the CD81 positive lysosome-associated membrane protein 1 (Lamp-1) negative route in dendritic cells which resembles that of HIV-1 uptake [186]. A study showed that HEK293 exosomes internalised by human fibroblastic, hepatic and renal cells were transported to the endoplasmic reticulum where they released their cargo, a pathway that might be a potential escape route from lysosomal degradation [187]. EVs may also evade lysosomal degradation *via* endocytic recycling out of the cell [188] or retrograde trafficking from the endosomal pathway to the trans-Golgi network [189].

EVs have also been reported to deliver their cargo into target cells *via* direct membrane fusion with the cell membrane, with EV surface proteins syncytin-1 and syncytin-2 seemingly playing a significant role in this [190–192]. Originally found on the plasma membranes of placental trophoblast cells [190, 191], gamete cells [190, 193] and various cancerous and non-cancerous cells known to fuse directly with other cells [190, 194–197], these proteins have also been detected on EVs secreted from these cells [190, 192]. In light of this, incorporating these surface proteins into EVs to increase their uptake *via* direct membrane fusion might be a possible way to evade endocytosis and lysosomal degradation completely.

4. Recommendation—EVs as drug carriers

In providing a recommendation to engineer EVs as DDSs, EV engineering methods to overcome specific barriers can be deployed. Natural evasion of immune cells is a highly favourable quality and should mark all engineered EVs regardless of the barrier(s) they target. CD47, CD24, CD31 and PD-L1 are prominent surface proteins that achieve this quality and should be incorporated into engineered EVs in high amounts if not originally present [129–134]. Fusing EVs with liposomes to create hybrid DDSs can also increase their drug loading capacity without risking cargo aggregation [198, 199].

The choice of the source of EVs depends on its availability and the ability of its EVs to overcome respective biological barriers associated with the disease. Milk and plant-derived EVs, which are highly available in nature and able to overcome gastrointestinal barriers [76, 77, 85–105], can be engineered for drug delivery *via* the oral route to treat IBD. The patient's own EVs might also be used as a form of personalised medicine. EVs from the patient may be chosen based on whether their cell of origin matches the target cell for better selectivity, but there are exceptions. For instance, immune and intestinal cell-derived EVs can be internalized by placental cells [103, 128]. Breast milk might also be a possible EV drug carrier source to treat both gastrointestinal and neurological conditions, as milk-derived EVs can cross gastrointestinal barriers [76, 85–90] and the BBB [87, 96] respectively. Human Type O RBC EVs loaded with antisense oligonucleotides were also found to target human leukaemia and breast cancer cells *in vitro* and *in vivo*. This is advantageous as RBCs are widely accessible from blood banks and lack DNA, which ensures that no oncogenic material is transferred from EVs to target cells [200]. This offers diverse compatible EV sources to choose from for a single ailment, enhancing the flexibility of the engineering process.

A summary of the recommendation is shown in **Figure 4**.

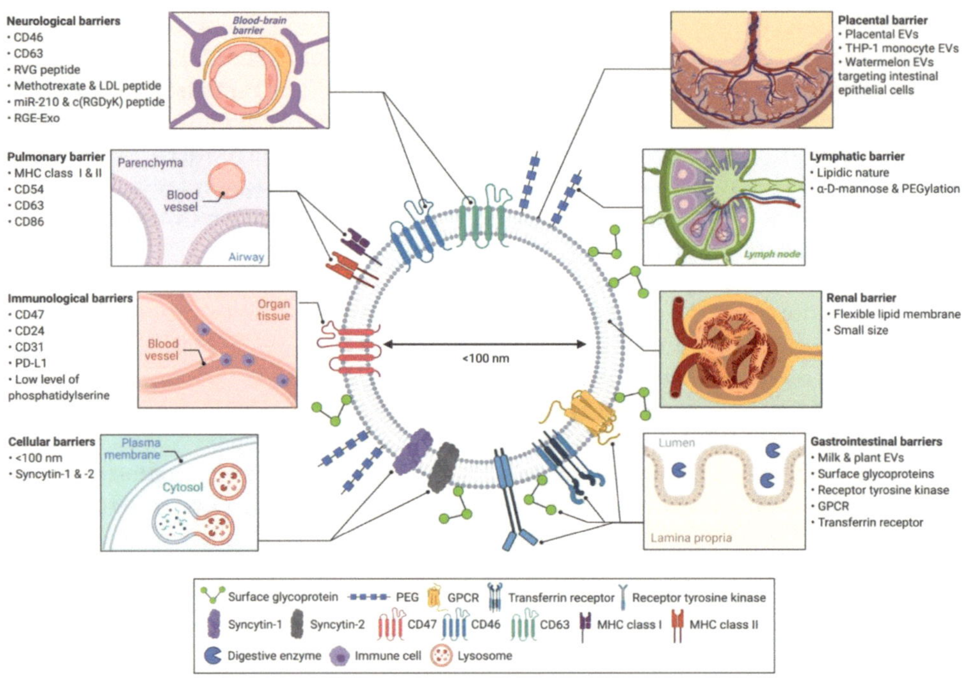

Figure 4.
Recommendation on how to engineer extracellular vesicles (EVs) to overcome various biological barriers for drug delivery applications. (1) ***Gastrointestinal barriers*** *can be overcome by deploying milk and plant-derived EVs to withstand the harsh acidic and enzymatic conditions, while also ensuring that EVs possess surface glycoproteins to enable trans-endocytosis across the small intestinal barrier. (2)* ***The placental barrier*** *can be overcome directly via the use of placental and THP-1 monocyte EVs, or indirectly via engineering watermelon EVs to target intestinal epithelial cells (IECs) which can communicate with the placenta. (3)* ***Immunological barriers*** *can be overcome by having a high proportion of CD47, CD24, CD31 and PD-L1 to produce the "don't eat me" signal, and probably a low level of phosphatidylserine (PS) on the surface of EVs to minimize the chances of being engulfed by macrophages. (4)* ***Neurological barriers*** *can be overcome minimally by incorporating high amounts of CD46 and CD63 into EVs as the quantity of these tetraspanins are positively correlated with the ability of EVs to cross the blood-brain barrier (BBB). (5)* ***The lymphatic barrier*** *can be overcome by EVs naturally as their lipidic nature enables them to cross the reticular network of the blood-lymph barrier (BLyB). Adding α-D-mannose and PEGylating EVs may also enhance their passage across the barrier. (6)* ***The pulmonary barrier*** *may be targeted by EVs derived from bronchoalveolar lavage fluid (BALF), which possess MHC classes I and II, CD54, CD63 and CD86. (7)* ***The renal barrier*** *can be overcome by EVs naturally, probably due to their small size and fluid lipid membranes which might allow them to squeeze through the tiny pores of the glomerular filtration barrier. (8)* ***Cellular barriers*** *can be overcome by EVs naturally via retrograde trafficking, endocytic recycling, direct fusion with the plasma membrane or other mechanisms. Engineering EVs with a size of <100 nm might help to reduce the chances of EVs being internalised via macropinocytosis which tends to lead to lysosomal degradation more than other mechanisms accessible to smaller EVs. Incorporating syncytin-1 and -2 into EVs might also enable them to fuse with the plasma membrane directly, allowing them to evade the endosomal and lysosomal membranes completely (created with BioRender.com).*

5. Conclusion—the future of EVs

Through critically analysing the relationship between the key components of EVs and the biological barriers EVs overcome, this review is the first to put together a recommendation in such a manner (**Figure 4**) to engineer EVs as suitable DDSs based on various studies. The implementation of EV drug carriers would revolutionise the global worldview of therapeutic treatments, as EVs unlock a whole new realm of endless possibilities in achieving the ideal therapeutic for patients, one of maximum efficacy and biocompatibility with negligible side effects. Even now, efforts have been made to transform the notion of personalised medicine into a reality, and having EVs as fully-approved personalised DDSs is worth pursuing. As past findings are limited due to the complex nature of EVs and various biological membranes, it is hoped that the mechanisms of EVs and their interactions with various biological membranes can continue to be more fully delved into, and that EV engineering can be carried out *via* efficacious yet sustainable methods, bearing in mind the availability and accessibility of natural EV sources, and hence the cost-effectiveness of the engineering processes.

Author details

Jacob Yeo Xian Ping, Yub Raj Neupane and Giorgia Pastorin*†
Department of Pharmacy, National University of Singapore, Singapore

*Address all correspondence to: phapg@nus.edu.sg

† Equal contribution.

References

[1] Abels ER, Breakefield XO. Introduction to extracellular vesicles: Biogenesis, RNA cargo selection, content, release, and uptake. Cellular and Molecular Neurobiology. 2016;**36**(3): 301-312

[2] Meng W, He C, Hao Y, Wang L, Li L, Zhu G. Prospects and challenges of extracellular vesicle-based drug delivery system: Considering cell source. Drug Delivery. 2020;**27**(1):585-598

[3] Mathieu M, Martin-Jaular L, Lavieu G, Théry C. Specificities of secretion and uptake of exosomes and other extracellular vesicles for cell-to-cell communication. Nature Cell Biology. 2019;**21**(1):9-17

[4] Yáñez-Mó M, Siljander PR-M, Andreu Z, Zavec AB, Borràs FE, Buzas EI, et al. Biological properties of extracellular vesicles and their physiological functions. Journal of Extracellular Vesicles. 2015;**4**(1):27066

[5] Beltrami C, Besnier M, Shantikumar S, Shearn AIU, Rajakaruna C, Laftah A, et al. Human pericardial fluid contains exosomes enriched with cardiovascular-expressed MicroRNAs and promotes therapeutic angiogenesis. Molecular Therapy. 2017;**25**(3):679-693

[6] Lu J, Yang J, Zheng Y, Chen X, Fang S. Extracellular vesicles from endothelial progenitor cells prevent steroid-induced osteoporosis by suppressing the ferroptotic pathway in mouse osteoblasts based on bioinformatics evidence. Scientific Reports. 2019;**9**(1):16130

[7] Yang T, Martin P, Fogarty B, Brown A, Schurman K, Phipps R, et al. Exosome delivered anticancer drugs across the blood-brain barrier for brain cancer therapy in *Danio rerio*. Pharmaceutical Research. 2015;**32**(6):2003-2014

[8] Busatto S, Pham A, Suh A, Shapiro S, Wolfram J. Organotropic drug delivery: Synthetic nanoparticles and extracellular vesicles. Biomedical Microdevices. 2019;**21**(2):46

[9] Elsharkasy OM, Nordin JZ, Hagey DW, de Jong OG, Schiffelers RM, Andaloussi SE, et al. Extracellular vesicles as drug delivery systems: Why and how? Advanced Drug Delivery Reviews. 2020;**159**:332-343

[10] Goh WJ, Zou S, Ong WY, Torta F, Alexandra AF, Schiffelers RM, et al. Bioinspired cell-derived nanovesicles versus exosomes as drug delivery systems: A cost-effective alternative. Scientific Reports. 2017;**7**(1):14322

[11] Banks WA, Sharma P, Bullock KM, Hansen KM, Ludwig N, Whiteside TL. Transport of extracellular vesicles across the blood-brain barrier: Brain pharmacokinetics and effects of inflammation. International Journal of Molecular Sciences. 2020;**21**(12):4407

[12] Saint-Pol J, Gosselet F, Duban-Deweer S, Pottiez G, Karamanos Y. Targeting and crossing the blood-brain barrier with extracellular vesicles. Cell. 2020;**9**(4):851

[13] van Niel G, D'Angelo G, Raposo G. Shedding light on the cell biology of extracellular vesicles. Nature Reviews. Molecular Cell Biology. 2018;**19**(4): 213-228

[14] Antimisiaris SG, Mourtas S, Marazioti A. Exosomes and exosome-inspired vesicles for targeted drug delivery. Pharmaceutics. 2018;**10**(4):218

[15] Ha D, Yang N, Nadithe V. Exosomes as therapeutic drug carriers and delivery vehicles across biological membranes: Current perspectives and future challenges. Acta Pharmaceutica Sinica B. 2016;**6**(4):287-296

[16] Colombo M, Raposo G, Théry C. Biogenesis, secretion, and intercellular interactions of exosomes and other extracellular vesicles. Annual Review of Cell and Developmental Biology. 2014; **30**(1):255-289

[17] El Andaloussi S, Mäger I, Breakefield XO, MJA W. Extracellular vesicles: Biology and emerging therapeutic opportunities. Nature Reviews. Drug Discovery. 2013;**12**(5): 347-357

[18] Gurung S, Perocheau D, Touramanidou L, Baruteau J. The exosome journey: From biogenesis to uptake and intracellular signalling. Cell Communication and Signaling: CCS. 2021;**19**(1):47

[19] Doyle LM, Wang MZ. Overview of extracellular vesicles, their origin, composition, purpose, and methods for exosome isolation and analysis. Cell. 2019;**8**(7):727

[20] Cvjetkovic A, Lötvall J, Lässer C. The influence of rotor type and centrifugation time on the yield and purity of extracellular vesicles. Journal of Extracellular Vesicles. 2014;**3**(1): 23111

[21] Duong P, Chung A, Bouchareychas L, Raffai RL. Cushioned-density gradient ultracentrifugation (C-DGUC) improves the isolation efficiency of extracellular vesicles. PLoS One. 2019;**14**(4):e0215324

[22] Takov K, Yellon DM, Davidson SM. Comparison of small extracellular vesicles isolated from plasma by ultracentrifugation or size-exclusion chromatography: Yield, purity and functional potential. Journal of Extracellular Vesicles. 2019;**8**(1):1560809

[23] Kanada M, Bachmann MH, Hardy JW, Frimannson DO, Bronsart L, Wang A, et al. Differential fates of biomolecules delivered to target cells via extracellular vesicles. Proceedings of the National Academy of Sciences of the United States of America. 2015;**112**(12): E1433-E1442

[24] Dang XTT, Kavishka JM, Zhang DX, Pirisinu M, Le MTN. Extracellular vesicles as an efficient and versatile system for drug delivery. Cell. 2020;**9**(10):2191

[25] Sluijter JPG, Verhage V, Deddens JC, van den Akker F, Doevendans PA. Microvesicles and exosomes for intracardiac communication. Cardiovascular Research. 2014;**102**(2): 302-311

[26] Heijnen HF, Schiel AE, Fijnheer R, Geuze HJ, Sixma JJ. Activated platelets release two types of membrane vesicles: Microvesicles by surface shedding and exosomes derived from exocytosis of multivesicular bodies and alpha-granules. Blood. 1999;**94**(11):3791-3799

[27] van Dommelen SM, Vader P, Lakhal S, Kooijmans SAA, van Solinge WW, Wood MJA, et al. Microvesicles and exosomes: Opportunities for cell-derived membrane vesicles in drug delivery. Journal of Controlled Release. 2012;**161**(2):635-644

[28] Zaborowski MP, Balaj L, Breakefield XO, Lai CP. Extracellular vesicles: Composition, biological relevance, and methods of study. Bioscience. 2015;**65**(8):783-797

[29] Géminard C, De Gassart A, Blanc L, Vidal M. Degradation of AP2 during

reticulocyte maturation enhances binding of hsc70 and Alix to a common site on TFR for sorting into exosomes: AP2 degradation induces TfR sorting in reticulocyte exosomes. Traffic. 2004;**5**(3):181-193

[30] Théry C, Boussac M, Véron P, Ricciardi-Castagnoli P, Raposo G, Garin J, et al. Proteomic analysis of dendritic cell-derived exosomes: A secreted subcellular compartment distinct from apoptotic vesicles. Journal of Immunology. 2001;**166**(12): 7309-7318

[31] van der Pol E, Böing AN, Harrison P, Sturk A, Nieuwland R. Classification, functions, and clinical relevance of extracellular vesicles. Pharmacological Reviews. 2012;**64**(3):676-705

[32] Vickers KC, Palmisano BT, Shoucri BM, Shamburek RD, Remaley AT. MicroRNAs are transported in plasma and delivered to recipient cells by high-density lipoproteins. Nature Cell Biology. 2011;**13**(4):423-433

[33] Di Vizio D, Morello M, Dudley AC, Schow PW, Adam RM, Morley S, et al. Large oncosomes in human prostate cancer tissues and in the circulation of mice with metastatic disease. The American Journal of Pathology. 2012; **181**(5):1573-1584

[34] Llorente A, Skotland T, Sylvänne T, Kauhanen D, Róg T, Orłowski A, et al. Molecular lipidomics of exosomes released by PC-3 prostate cancer cells. Biochimica et Biophysica Acta. 2013;**1831**(7):1302-1309

[35] György B, Szabó TG, Pásztói M, Pál Z, Misják P, Aradi B, et al. Membrane vesicles, current state-of-the-art: Emerging role of extracellular vesicles. Cellular and Molecular Life Sciences. 2011;**68**(16):2667-2688

[36] Lee E-Y, Choi D-Y, Kim D-K, Kim J-W, Park JO, Kim S, et al. Gram-positive bacteria produce membrane vesicles: Proteomics-based characterization of *Staphylococcus aureus*-derived membrane vesicles. Proteomics. 2009;**9**(24): 5425-5436

[37] Bottero D, Gaillard ME, Errea A, Moreno G, Zurita E, Pianciola L, et al. Outer membrane vesicles derived from *Bordetella parapertussis* as an acellular vaccine against *Bordetella parapertussis* and *Bordetella pertussis* infection. Vaccine. 2013;**31**(45):5262-5268

[38] Jang SC, Kim SR, Yoon YJ, Park K-S, Kim JH, Lee J, et al. In vivo kinetic biodistribution of nano-sized outer membrane vesicles derived from bacteria. Small. 2015;**11**(4):456-461

[39] Lee E-Y, Choi D-S, Kim K-P, Gho YS. Proteomics in gram-negative bacterial outer membrane vesicles. Mass Spectrometry Reviews. 2008;**27**(6): 535-555

[40] Berleman JE, Allen S, Danielewicz MA, Remis JP, Gorur A, Cunha J, et al. The lethal cargo of *Myxococcus xanthus* outer membrane vesicles. Frontiers in Microbiology. 2014;**5**:474

[41] Goes A, Lapuhs P, Kuhn T, Schulz E, Richter R, Panter F, et al. Myxobacteria-derived outer membrane vesicles: Potential applicability against intracellular infections. Cell. 2020;**9**(1):194

[42] Hoshino A, Costa-Silva B, Shen T-L, Rodrigues G, Hashimoto A, Tesic Mark M, et al. Tumour exosome integrins determine organotropic metastasis. Nature. 2015;**527**(7578):329-335

[43] Tao S-C, Guo S-C, Zhang C-Q. Platelet-derived extracellular vesicles: An emerging therapeutic approach.

International Journal of Biological Sciences. 2017;**13**(7):828-834

[44] Zaldivia MTK, McFadyen JD, Lim B, Wang X, Peter K. Platelet-derived microvesicles in cardiovascular diseases. Frontiers in Cardiovascular Medicine. 2017;**4**:74

[45] Barrès C, Blanc L, Bette-Bobillo P, André S, Mamoun R, Gabius H-J, et al. Galectin-5 is bound onto the surface of rat reticulocyte exosomes and modulates vesicle uptake by macrophages. Blood. 2010;**115**(3):696-705

[46] Bilen MA, Pan T, Lee Y-C, Lin S-C, Yu G, Pan J, et al. Proteomics profiling of exosomes from primary mouse osteoblasts under proliferation versus mineralization conditions and characterization of their uptake into prostate cancer cells. Journal of Proteome Research. 2017;**16**(8):2709-2728

[47] Yoshioka Y, Konishi Y, Kosaka N, Katsuda T, Kato T, Ochiya T. Comparative marker analysis of extracellular vesicles in different human cancer types. Journal of Extracellular Vesicles. 2013;**2**(1):20424

[48] Nakase I, Kobayashi NB, Takatani-Nakase T, Yoshida T. Active macro-pinocytosis induction by stimulation of epidermal growth factor receptor and oncogenic Ras expression potentiates cellular uptake efficacy of exosomes. Scientific Reports. 2015; **5**(10300). https://www.nature.com/articles/srep10300

[49] Kooijmans SAA, Aleza CG, Roffler SR, van Solinge WW, Vader P, Schiffelers RM. Display of GPI-anchored anti-EGFR nanobodies on extracellular vesicles promotes tumour cell targeting. Journal of Extracellular Vesicles. 2016;**5**(1):31053

[50] Tauro BJ, Greening DW, Mathias RA, Mathivanan S, Ji H, Simpson RJ. Two distinct populations of exosomes are released from LIM1863 colon carcinoma cell-derived organoids. Molecular & Cellular Proteomics. 2013;**12**(3):587-598

[51] Huber V, Fais S, Iero M, Lugini L, Canese P, Squarcina P, et al. Human colorectal cancer cells induce T-cell death through release of proapoptotic microvesicles: Role in immune escape. Gastroenterology. 2005;**128**(7):1796-1804

[52] Calzolari A, Raggi C, Deaglio S, Sposi NM, Stafsnes M, Fecchi K, et al. TfR2 localizes in lipid raft domains and is released in exosomes to activate signal transduction along the MAPK pathway. Journal of Cell Science. 2006;**119**(Pt 21): 4486-4498

[53] He M, Qin H, Poon TC, Sze SC, Ding X, Co NN. Hepatocellular carcinoma-derived exosomes promote motility of immortalized hepatocyte through transfer of oncogenic proteins and RNAs. Carcinogenesis. 2015;**36**: 1008-1018

[54] Purushothaman A, Bandari SK, Liu J, Mobley JA, Brown EE, Sanderson RD. Fibronectin on the surface of myeloma cell-derived exosomes mediates exosome-cell interactions. The Journal of Biological Chemistry. 2016;**291**(4):1652-1663

[55] Keryer-Bibens C, Pioche-Durieu C, Villemant C, Souquère S, Nishi N, Hirashima M, et al. Exosomes released by EBV-infected nasopharyngeal carcinoma cells convey the viral latent membrane protein 1 and the immunomodulatory protein galectin 9. BMC Cancer. 2006;**6**(1):283

[56] Liang B, Peng P, Chen S, Li L, Zhang M, Cao D, et al. Characterization and proteomic analysis of ovarian cancer-derived exosomes. Journal of Proteomics. 2013;**80**:171-182

[57] Saunderson SC, Dunn AC, Crocker PR, McLellan AD. CD169 mediates the capture of exosomes in spleen and lymph node. Blood. 2014;**123**(2):208-216

[58] Blanchard N, Lankar D, Faure F, Regnault A, Dumont C, Raposo G, et al. TCR activation of human T cells induces the production of exosomes bearing the TCR/CD3/zeta complex. Journal of Immunology. 2002;**168**(7):3235-3241

[59] Pienimaeki-Roemer A, Kuhlmann K, Böttcher A, Konovalova T, Black A, Orsó E, et al. Lipidomic and proteomic characterization of platelet extracellular vesicle subfractions from senescent platelets: PLT extracellular vesicle fractions. Transfusion. 2015;**55**(3):507-521

[60] Miksa M, Wu R, Dong W, Das P, Yang D, Wang P. Dendritic cell-derived exosomes containing milk fat globule epidermal growth factor-factor VIII attenuate proinflammatory responses in sepsis. Shock. 2006;**25**(6):586-593

[61] Morelli AE, Larregina AT, Shufesky WJ, Sullivan MLG, Stolz DB, Papworth GD, et al. Endocytosis, intracellular sorting, and processing of exosomes by dendritic cells. Blood. 2004;**104**(10):3257-3266

[62] Obregon C, Rothen-Rutishauser B, Gerber P, Gehr P, Nicod LP. Active uptake of dendritic cell-derived exovesicles by epithelial cells induces the release of inflammatory mediators through a TNF-alpha-mediated pathway. The American Journal of Pathology. 2009;**175**(2):696-705

[63] Baglio SR, van Eijndhoven MAJ, Koppers-Lalic D, Berenguer J, Lougheed SM, Gibbs S, et al. Sensing of latent EBV infection through exosomal transfer of 5'pppRNA. Proceedings of the National Academy of Sciences of the United States of America. 2016;**113**(5): E587-E596

[64] Chakraborty S, Núñez D, Hu S-Y, Domingo MP, Pardo J, Karmenyan A, et al. FRET based quantification and screening technology platform for the interactions of leukocyte function-associated antigen-1 (LFA-1) with intercellular adhesion molecule-1 (ICAM-1). PLoS One. 2014;**9**(7):e102572

[65] Mueller SK, Nocera AL, Bleier BS. Exosome function in aerodigestive mucosa. Nanomedicine. 2018;**14**(2): 269-277

[66] Kwon SH. Extracellular vesicles in renal physiology and clinical applications for renal disease. The Korean Journal of Internal Medicine. 2019;**34**(3):470-479

[67] Eirin A, Riester SM, Zhu X-Y, Tang H, Evans JM, O'Brien D, et al. MicroRNA and mRNA cargo of extracellular vesicles from porcine adipose tissue-derived mesenchymal stem cells. Gene. 2014;**551**(1):55-64

[68] Haga H, Yan IK, Borrelli DA, Matsuda A, Parasramka M, Shukla N, et al. Extracellular vesicles from bone marrow-derived mesenchymal stem cells protect against murine hepatic ischemia/ reperfusion injury. Liver Transplantation. 2017;**23**(6):791-803

[69] Katakowski M, Buller B, Zheng X, Lu Y, Rogers T, Osobamiro O, et al. Exosomes from marrow stromal cells expressing miR-146b inhibit glioma growth. Cancer Letters. 2013;**335**(1): 201-204

[70] Lou G, Song X, Yang F, Wu S, Wang J, Chen Z, et al. Exosomes derived from miR-122-modified adipose tissue-derived MSCs increase chemosensitivity of hepatocellular carcinoma. Journal of Hematology & Oncology. 2015;**8**(1):122

[71] Ramos TL, Sánchez-Abarca LI, Muntión S, Preciado S, Puig N, López-Ruano G, et al. MSC surface markers (CD44, CD73, and CD90) can identify human MSC-derived extracellular vesicles by conventional flow cytometry. Cell Communication and Signaling: CCS. 2016;**14**(1):2

[72] Munoz JL, Bliss SA, Greco SJ, Ramkissoon SH, Ligon KL, Rameshwar P. Delivery of functional anti-miR-9 by mesenchymal stem cell-derived exosomes to glioblastoma multiforme cells conferred chemosensitivity. Molecular Therapy: Nucleic Acids. 2013;**2**(e126):e126

[73] Pascucci L, Coccè V, Bonomi A, Ami D, Ceccarelli P, Ciusani E, et al. Paclitaxel is incorporated by mesenchymal stromal cells and released in exosomes that inhibit in vitro tumor growth: A new approach for drug delivery. Journal of Controlled Release. 2014;**192**:262-270

[74] Shimbo K, Miyaki S, Ishitobi H, Kato Y, Kubo T, Shimose S, et al. Exosome-formed synthetic microRNA-143 is transferred to osteosarcoma cells and inhibits their migration. Biochemical and Biophysical Research Communications. 2014;**445**(2):381-387

[75] Shimoda A, Tahara Y, Sawada S-I, Sasaki Y, Akiyoshi K. Glycan profiling analysis using evanescent-field fluorescence-assisted lectin array: Importance of sugar recognition for cellular uptake of exosomes from mesenchymal stem cells. Biochemical and Biophysical Research Communications. 2017;**491**(3):701-707

[76] Berk M, Williams LJ, Jacka FN, O'Neil A, Pasco JA, Moylan S, et al. So depression is an inflammatory disease, but where does the inflammation come from? BMC Medicine. 2013;**11**(1):200

[77] Mu J, Zhuang X, Wang Q, Jiang H, Deng Z-B, Wang B, et al. Interspecies communication between plant and mouse gut host cells through edible plant derived exosome-like nanoparticles. Molecular Nutrition & Food Research. 2014;**58**(7):1561-1573

[78] Arntz OJ, Pieters BCH, Oliveira MC, Broeren MGA, Bennink MB, de Vries M, et al. Oral administration of bovine milk derived extracellular vesicles attenuates arthritis in two mouse models. Molecular Nutrition & Food Research. 2015;**59**(9): 1701-1712

[79] Manca S, Giraud D, Zempleni J. The bioavailability and distribution of bovine milk exosomes is distinct from that of their cargos in mice. The FASEB Journal. 2017;**31**(S1). DOI: 10.1096/fasebj.31.1_supplement.148.2. https://www.researchgate.net/publication/352244149_The_Bioavailability_and_Distrubution_of_Bovine_Milk_Exosomes_is_Distinct_from_that_of_their_Cargos_in_Mice

[80] Reinhardt TA, Lippolis JD, Nonnecke BJ, Sacco RE. Bovine milk exosome proteome. Journal of Proteomics. 2012;**75**(5):1486-1492

[81] Kosaka N, Izumi H, Sekine K, Ochiya T. microRNA as a new immune-regulatory agent in breast milk. Silence. 2010;**1**(1):7

[82] Mincheva-Nilsson L, Baranov V. The role of placental exosomes in reproduction: Placental exosomes in reproduction. American Journal of Reproductive Immunology. 2010;**63**(6) 520-533

[83] Brzozowski JS, Jankowski H, Bond DR, McCague SB, Munro BR, Predebon MJ, et al. Lipidomic profiling of extracellular vesicles derived from

prostate and prostate cancer cell lines. Lipids in Health and Disease. 2018;**17**(1):211

[84] Barral J-P, Croibier A. Circulatory physiology. In: Visceral Vascular Manipulations. Elsevier; 2011. pp. 27-45

[85] Tong L, Hao H, Zhang Z, Lv Y, Liang X, Liu Q, et al. Milk-derived extracellular vesicles alleviate ulcerative colitis by regulating the gut immunity and reshaping the gut microbiota. Theranostics. 2021;**11**(17):8570-8586

[86] Munagala R, Aqil F, Jeyabalan J, Gupta RC. Bovine milk-derived exosomes for drug delivery. Cancer Letters. 2016;**371**(1):48-61

[87] Sanwlani R, Fonseka P, Chitti SV, Mathivanan S. Milk-derived extracellular vesicles in inter-organism, cross-species communication and drug delivery. Proteomes. 2020;**8**(2):11

[88] Wolf T, Baier SR, Zempleni J. The intestinal transport of bovine milk exosomes is mediated by endocytosis in human colon carcinoma Caco-2 cells and rat small intestinal IEC-6 cells. The Journal of Nutrition. 2015;**145**(10): 2201-2206

[89] Tulkens J, Vergauwen G, Van Deun J, Geeurickx E, Dhondt B, Lippens L, et al. Increased levels of systemic LPS-positive bacterial extracellular vesicles in patients with intestinal barrier dysfunction. Gut. 2020;**69**(1):191-193

[90] Lee B, Moon KM, Kim CY. Tight junction in the intestinal epithelium: Its association with diseases and regulation by phytochemicals. Journal of Immunology Research. 2018;**2018**:1-11

[91] Izumi H, Kosaka N, Shimizu T, Sekine K, Ochiya T, Takase M. Bovine milk contains microRNA and messenger RNA that are stable under degradative conditions. Journal of Dairy Science. 2012;**95**(9):4831-4841

[92] Baddela VS, Nayan V, Rani P, Onteru SK, Singh D. Physicochemical biomolecular insights into buffalo milk-derived nanovesicles. Applied Biochemistry and Biotechnology. 2016;**178**(3):544-557

[93] Baier SR, Nguyen C, Xie F, Wood JR, Zempleni J. MicroRNAs are absorbed in biologically meaningful amounts from nutritionally relevant doses of cow milk and affect gene expression in peripheral blood mononuclear cells, HEK-293 kidney cell cultures, and mouse livers. The Journal of Nutrition. 2014;**144**(10): 1495-1500

[94] Howard KM, Jati Kusuma R, Baier SR, Friemel T, Markham L, Vanamala J, et al. Loss of miRNAs during processing and storage of cow's (*Bos taurus*) milk. Journal of Agricultural and Food Chemistry. 2015;**63**(2):588-592

[95] Kirchner B, Pfaffl MW, Dumpler J, von Mutius E, Ege MJ. microRNA in native and processed cow's milk and its implication for the farm milk effect on asthma. The Journal of Allergy and Clinical Immunology. 2016;**137**(6): 1893-1895, e13

[96] Manca S, Upadhyaya B, Mutai E, Desaulniers AT, Cederberg RA, White BR, et al. Milk exosomes are bioavailable and distinct microRNA cargos have unique tissue distribution patterns. Scientific Reports. 2018;**8**(1). DOI: 10.1038/s41598-018-29780-1. https://www.nature.com/articles/s41598-018-29780-1

[97] Zhao Z, Yu S, Xu M, Li P. Effects of microwave on extracellular vesicles and microRNA in milk. Journal of Dairy Science. 2018;**101**(4):2932-2940

[98] Rahman MM, Shimizu K, Yamauchi M, Takase H, Ugawa S, Okada A, et al. Acidification effects on isolation of extracellular vesicles from bovine milk. PLoS One. 2019;**14**(9):e0222613

[99] Deng Z, Rong Y, Teng Y, Mu J, Zhuang X, Tseng M, et al. Broccoli-derived nanoparticle inhibits mouse colitis by activating dendritic cell AMP-activated protein kinase. Molecular Therapy. 2017;**25**(7):1641-1654

[100] Ju S, Mu J, Dokland T, Zhuang X, Wang Q, Jiang H, et al. Grape exosome-like nanoparticles induce intestinal stem cells and protect mice from DSS-induced colitis. Molecular Therapy. 2013;**21**(7):1345-1357

[101] Wang B, Zhuang X, Deng Z-B, Jiang H, Mu J, Wang Q, et al. Targeted drug delivery to intestinal macrophages by bioactive nanovesicles released from grapefruit. Molecular Therapy. 2014;**22**(3):522-534

[102] Zhang M, Viennois E, Prasad M, Zhang Y, Wang L, Zhang Z, et al. Edible ginger-derived nanoparticles: A novel therapeutic approach for the prevention and treatment of inflammatory bowel disease and colitis-associated cancer. Biomaterials. 2016;**101**:321-340

[103] Timms KM. Investigating the Effect of Plant-Derived Extracellular Vesicles on Human Placental Function. New York: The University of Manchester; 2018

[104] Zhuang X, Deng Z-B, Mu J, Zhang L, Yan J, Miller D, et al. Ginger-derived nanoparticles protect against alcohol-induced liver damage. Journal of Extracellular Vesicles. 2015;**4**(1): 28713

[105] Doherty GJ, McMahon HT. Mechanisms of endocytosis. Annual Review of Biochemistry. 2009;**78**(1): 857-902

[106] Elliott RO, He M. Unlocking the power of exosomes for crossing biological barriers in drug delivery. Pharmaceutics. 2021;**13**(1):122

[107] Wood CE, Keller-Wood M. The critical importance of the fetal hypothalamus-pituitary-adrenal axis. F1000Research. 2016;**5**:115

[108] Condon JC, Jeyasuria P, Faust JM, Mendelson CR. Surfactant protein secreted by the maturing mouse fetal lung acts as a hormone that signals the initiation of parturition. Proceedings of the National Academy of Sciences of the United States of America. 2004;**101**(14): 4978-4983

[109] Gao L, Rabbitt EH, Condon JC, Renthal NE, Johnston JM, Mitsche MA, et al. Steroid receptor coactivators 1 and 2 mediate fetal-to-maternal signaling that initiates parturition. The Journal of Clinical Investigation. 2015;**125**(7): 2808-2824

[110] Genbacev O, Zhou Y, Ludlow JW, Fisher SJ. Regulation of human placental development by oxygen tension. Science. 1997;**277**(5332):1669-1672

[111] Mikaelsson MA, Constância M, Dent CL, Wilkinson LS, Humby T. Placental programming of anxiety in adulthood revealed by Igf2-null models. Nature Communications. 2013;**4**:2311

[112] Houser BL. Decidual macrophages and their roles at the maternal-fetal interface. The Yale Journal of Biology and Medicine. 2012;**85**(1):105-118

[113] Pemathilaka RL, Reynolds DE, Hashemi NN. Drug transport across the human placenta: Review of placenta-on-a-chip and previous approaches. Interface Focus. 2019;**9**(5):20190031

[114] Robbins JR, Zeldovich VB, Poukchanski A, Boothroyd JC, Bakardjiev AI. Tissue barriers of the human placenta to infection with *Toxoplasma gondii*. Infection and Immunity. 2012;**80**(1):418-428

[115] Miranda J, Paules C, Nair S, Lai A, Palma C, Scholz-Romero K, et al. Placental exosomes profile in maternal and fetal circulation in intrauterine growth restriction—Liquid biopsies to monitoring fetal growth. Placenta. 2018;**64**:34-43

[116] Sarker S, Scholz-Romero K, Perez A, Illanes SE, Mitchell MD, Rice GE, et al. Placenta-derived exosomes continuously increase in maternal circulation over the first trimester of pregnancy. Journal of Translational Medicine. 2014;**12**(1):204

[117] Burnett LA, Nowak RA. Exosomes mediate embryo and maternal interactions at implantation and during pregnancy. Frontiers in Bioscience (Scholar Edition). 2016;**8**(1):79-96

[118] Chang G, Mouillet J-F, Mishima T, Chu T, Sadovsky E, Coyne CB, et al. Expression and trafficking of placental microRNAs at the feto-maternal interface. The FASEB Journal. 2017;**31**(7):2760-2770

[119] Makiya R, Stigbrand T. Placental alkaline phosphatase has a binding site for the human immunoglobulin-G Fc portion. European Journal of Biochemistry. 1992;**205**(1):341-345

[120] Makiya R, Stigbrand T. Placental alkaline phosphatase is related to human IgG internalization in HEp2 cells. Biochemical and Biophysical Research Communications. 1992;**182**(2):624-630

[121] Mitchell MD, Peiris HN, Kobayashi M, Koh YQ, Duncombe G, Illanes SE, et al. Placental exosomes in normal and complicated pregnancy. American Journal of Obstetrics and Gynecology. 2015;**213**(4):S173-S181

[122] Salomon C, Ryan J, Sobrevia L, Kobayashi M, Ashman K, Mitchell M, et al. Exosomal signaling during hypoxia mediates microvascular endothelial cell migration and vasculogenesis. PLoS One. 2013;**8**(7):e68451

[123] Salomon C, Torres MJ, Kobayashi M, Scholz-Romero K, Sobrevia L, Dobierzewska A, et al. A gestational profile of placental exosomes in maternal plasma and their effects on endothelial cell migration. PLoS One. 2014;**9**(6): e98667

[124] She QB, Mukherjee JJ, Chung T, Kiss Z. Placental alkaline phosphatase, insulin, and adenine nucleotides or adenosine synergistically promote long-term survival of serum-starved mouse embryo and human fetus fibroblasts. Cellular Signalling. 2000;**12**(9-10):659-665

[125] She QB, Mukherjee JJ, Huang JS, Crilly KS, Kiss Z. Growth factor-like effects of placental alkaline phosphatase in human fetus and mouse embryo fibroblasts. FEBS Letters. 2000; **469**(2-3):163-167

[126] Stefaner I, Stefanescu A, Hunziker W, Fuchs R. Expression of placental alkaline phosphatase does not correlate with IgG binding, internalization and transcytosis. The Biochemical Journal. 1997;**327**(2):585-592

[127] Liu J, Wang S-Z, Wang Q-L, Du J-G, Wang B-B. Gestational diabetes mellitus is associated with changes in the concentration and bioactivity of placental exosomes in the maternal circulation across gestation. European Review for Medical and Pharmacological Sciences. 2018;**22**(7):2036-2043

[128] Holder B, Jones T, Sancho Shimizu V, Rice TF, Donaldson B, Bouqueau M, et al. Macrophage exosomes induce placental inflammatory cytokines: A novel mode of maternal-placental messaging. Traffic. 2016;**17**(2):168-178

[129] Matozaki T, Murata Y, Okazawa H, Ohnishi H. Functions and molecular mechanisms of the CD47-SIRPalpha signalling pathway. Trends in Cell Biology. 2009;**19**(2):72-80

[130] Parada N, Romero-Trujillo A, Georges N, Alcayaga-Miranda F. Camouflage strategies for therapeutic exosomes evasion from phagocytosis. Journal of Advanced Research. 2021;**31**:61-74

[131] Weiskopf K. Cancer immunotherapy targeting the CD47/SIRPα axis. European Journal of Cancer. 2017;**76**:100-109

[132] Barkal AA, Brewer RE, Markovic M, Kowarsky M, Barkal SA, Zaro BW, et al. CD24 signalling through macrophage Siglec-10 is a target for cancer immunotherapy. Nature. 2019;**572**(7769):392-396

[133] Brown S, Heinisch I, Ross E, Shaw K, Buckley CD, Savill J. Apoptosis disables CD31-mediated cell detachment from phagocytes promoting binding and engulfment. Nature. 2002;**418**(6894): 200-203

[134] Gordon SR, Maute RL, Dulken BW, Hutter G, George BM, McCracken MN, et al. PD-1 expression by tumour-associated macrophages inhibits phagocytosis and tumour immunity. Nature. 2017;**545**(7655):495-499

[135] Charoenviriyakul C, Takahashi Y, Morishita M, Matsumoto A, Nishikawa M, Takakura Y. Cell type-specific and common characteristics of exosomes derived from mouse cell lines: Yield, physicochemical properties, and pharmacokinetics. European Journal of Pharmaceutical Sciences. 2017;**96**: 316-322

[136] Ballabh P, Braun A, Nedergaard M. The blood-brain barrier: An overview: Structure, regulation, and clinical implications. Neurobiology of Disease. 2004;**16**(1):1-13

[137] Lai N, Wu D, Liang T, Pan P, Yuan G, Li X, et al. Systemic exosomal miR-193b-3p delivery attenuates neuroinflammation in early brain injury after subarachnoid hemorrhage in mice. Journal of Neuroinflammation. 2020;**17**(1):74

[138] Morad G, Carman CV, Hagedorn EJ, Perlin JR, Zon LI, Mustafaoglu N, et al. Tumor-derived extracellular vesicles breach the intact blood-brain barrier via transcytosis. ACS Nano. 2019;**13**(12): 13853-13865

[139] Alvarez-Erviti L, Seow Y, Yin H, Betts C, Lakhal S, Wood MJA. Delivery of siRNA to the mouse brain by systemic injection of targeted exosomes. Nature Biotechnology. 2011;**29**(4):341-345

[140] Ye Z, Zhang T, He W, Jin H, Liu C, Yang Z, et al. Methotrexate-loaded extracellular vesicles functionalized with therapeutic and targeted peptides for the treatment of glioblastoma multiforme. ACS Applied Materials & Interfaces. 2018;**10**(15):12341-12350

[141] Zhang H, Wu J, Wu J, Fan Q, Zhou J, Wu J, et al. Exosome-mediated targeted delivery of miR-210 for angiogenic therapy after cerebral ischemia in mice. Journal of Nanobiotechnology. 2019;**17**(1):29

[142] Jia G, Han Y, An Y, Ding Y, He C, Wang X, et al. NRP-1 targeted and cargo-loaded exosomes facilitate

simultaneous imaging and therapy of glioma in vitro and in vivo. Biomaterials. 2018;**178**:302-316

[143] Zhuang X, Xiang X, Grizzle W, Sun D, Zhang S, Axtell RC, et al. Treatment of brain inflammatory diseases by delivering exosome encapsulated anti-inflammatory drugs from the nasal region to the brain. Molecular Therapy. 2011;**19**(10): 1769-1779

[144] Sun W, Wang W. Advances in research on labyrinth membranous barriers. Journal of Otology. 2015;**10**(3):99-104

[145] Klingeborn M, Dismuke WM, Bowes Rickman C, Stamer WD. Roles of exosomes in the normal and diseased eye. Progress in Retinal and Eye Research. 2017;**59**:158-177

[146] Zhou W, Fong MY, Min Y, Somlo G, Liu L, Palomares MR, et al. Cancer-secreted miR-105 destroys vascular endothelial barriers to promote metastasis. Cancer Cell. 2014;**25**(4): 501-515

[147] Thomas SN, Schudel A. Overcoming transport barriers for interstitial-, lymphatic-, and lymph node-targeted drug delivery. Current Opinion in Chemical Engineering. 2015;7:65-74

[148] Anderson AO, Anderson ND. Lymphocyte emigration from high endothelial venules in rat lymph nodes. Immunology. 1976;**31**(5):731-748

[149] Anderson AO, Anderson ND. Studies on the structure and permeability of the microvasculature in normal rat lymph nodes. The American Journal of Pathology. 1975;**80**(3):387-418

[150] Anderson AO, Shaw S. T cell adhesion to endothelium: The FRC conduit system and other anatomic and molecular features which facilitate the adhesion cascade in lymph node. Seminars in Immunology. 1993;**5**(4):271-282

[151] Gretz JE, Anderson AO, Shaw S. Cords, channels, corridors and conduits: Critical architectural elements facilitating cell interactions in the lymph node cortex. Immunological Reviews. 1997;**156**(1):11-24

[152] Gretz JE, Kaldjian EP, Anderson AO, Shaw S. Sophisticated strategies for information encounter in the lymph node: The reticular network as a conduit of soluble information and a highway for cell traffic. Journal of Immunology. 1996;**157**(2):495-499

[153] Sainte-Marie G, Peng FS. Diffusion of a lymph-carried antigen in the fiber network of the lymph node of the rat. Cell and Tissue Research. 1986;**245**(3): 481-486

[154] Srinivasan S, Vannberg FO, Dixon JB. Lymphatic transport of exosomes as a rapid route of information dissemination to the lymph node. Scientific Reports. 2016;**6**(1):24436

[155] Gretz JE, Norbury CC, Anderson AO, Proudfoot AEI, Shaw S. Lymph-borne chemokines and other low molecular weight molecules reach high endothelial venules via specialized conduits while a functional barrier limits access to the lymphocyte microenvironments in lymph node cortex. The Journal of Experimental Medicine. 2000;**192**(10):1425-1440

[156] Zhao Z, McGill J, Gamero-Kubota P, He M. Microfluidic on-demand engineering of exosomes towards cancer immunotherapy. Lab on a Chip. 2019;**19**(10):1877-1886

[157] Zhu Q, Heon M, Zhao Z, He M. Microfluidic engineering of exosomes:

Editing cellular messages for precision therapeutics. Lab on a Chip. 2018;**18**(12):1690-1703

[158] Choi ES, Song J, Kang YY, Mok H. Mannose-modified serum exosomes for the elevated uptake to Murine dendritic cells and lymphatic accumulation. Macromolecular Bioscience. 2019; **19**(7):e1900042

[159] Anderson MJ, Parks PJ, Peterson ML. A mucosal model to study microbial biofilm development and anti-biofilm therapeutics. Journal of Microbiological Methods. 2013;**92**(2): 201-208

[160] Leiva-Juárez MM, Kolls JK, Evans SE. Lung epithelial cells: Therapeutically inducible effectors of antimicrobial defense. Mucosal Immunology. 2018;**11**(1):21-34

[161] Ganesan S, Comstock AT, Sajjan US. Barrier function of airway tract epithelium. Tissue Barriers. 2013;**1**(4):e24997

[162] Admyre C, Grunewald J, Thyberg J, Gripenbäck S, Tornling G, Eklund A, et al. Exosomes with major histocompatibility complex class II and co-stimulatory molecules are present in human BAL fluid. The European Respiratory Journal. 2003;**22**(4): 578-583

[163] Admyre C, Telemo E, Almqvist N, Lötvall J, Lahesmaa R, Scheynius A, et al. Exosomes—Nanovesicles with possible roles in allergic inflammation. Allergy. 2008;**63**(4):404-408

[164] Kang YS, Li Y, Dai C, Kiss LP, Wu C, Liu Y. Inhibition of integrin-linked kinase blocks podocyte epithelial-mesenchymal transition and ameliorates proteinuria. Kidney International. 2010;**78**(4):363-373

[165] Patrakka J, Tryggvason K. Molecular make-up of the glomerular filtration barrier. Biochemical and Biophysical Research Communications. 2010;**396**(1):164-169

[166] Simeone P, Bologna G, Lanuti P, Pierdomenico L, Guagnano MT, Pieragostino D, et al. Extracellular vesicles as signaling mediators and disease biomarkers across biological barriers. International Journal of Molecular Sciences. 2020;**21**(7):2514

[167] Ota K, Ota Z, Shikata K, Makino H. The ultrastructural disruption of the glomerular basement membrane in diabetic nephropathy revealed by "tissue negative staining method". Journal of Diabetes and its Complications. 1995;**9**(4):285-287

[168] Wu X, Gao Y, Xu L, Dang W, Yan H, Zou D, et al. Exosomes from high glucose-treated glomerular endothelial cells trigger the epithelial-mesenchymal transition and dysfunction of podocytes. Scientific Reports. 2017;**7**(1):9371

[169] Kalra H, Simpson RJ, Ji H, Aikawa E, Altevogt P, Askenase P, et al. Vesiclepedia: A compendium for extracellular vesicles with continuous community annotation. PLoS Biology. 2012;**10**(12):e1001450

[170] Pathan M, Fonseka P, Chitti SV, Kang T, Sanwlani R, Van Deun J, et al. Vesiclepedia 2019: A compendium of RNA, proteins, lipids and metabolites in extracellular vesicles. Nucleic Acids Research. 2019;**47**(D1):D516-D519

[171] Svenningsen P, Sabaratnam R, Jensen BL. Urinary extracellular vesicles: Origin, role as intercellular messengers and biomarkers; efficient sorting and potential treatment options. Acta Physiologica (Oxford, England). 2020;**228**(1):e13346

[172] Uhlén M, Fagerberg L, Hallström BM, Lindskog C, Oksvold P, Mardinoglu A, et al. Proteomics. Tissue-based map of the human proteome. Science. 2015;**347**(6220):1260419

[173] Cheng Y, Wang X, Yang J, Duan X, Yao Y, Shi X, et al. A translational study of urine miRNAs in acute myocardial infarction. Journal of Molecular and Cellular Cardiology. 2012;**53**(5):668-676

[174] Ziółkowska NE, Christiano R, Walther TC. Organized living: Formation mechanisms and functions of plasma membrane domains in yeast. Trends in Cell Biology. 2012;**22**(3):151-158

[175] Parolini I, Federici C, Raggi C, Lugini L, Palleschi S, De Milito A, et al. Microenvironmental pH is a key factor for exosome traffic in tumor cells. The Journal of Biological Chemistry. 2009;**284**(49):34211-34222

[176] Joshi BS, de Beer MA, Giepmans BNG, Zuhorn IS. Endocytosis of extracellular vesicles and release of their cargo from endosomes. ACS Nano. 2020;**14**(4):4444-4455

[177] Costa Verdera H, Gitz-Francois JJ, Schiffelers RM, Vader P. Cellular uptake of extracellular vesicles is mediated by clathrin-independent endocytosis and macropinocytosis. Journal of Controlled Release. 2017;**266**:100-108

[178] Tian T, Zhu Y-L, Zhou Y-Y, Liang G-F, Wang Y-Y, Hu F-H, et al. Exosome uptake through clathrin-mediated endocytosis and macropinocytosis and mediating miR-21 delivery. The Journal of Biological Chemistry. 2014;**289**(32): 22258-22267

[179] Gonda A, Kabagwira J, Senthil GN, Wall NR. Internalization of exosomes through receptor-mediated endocytosis. Molecular Cancer Research. 2019;**17**(2): 337-347

[180] Hao S, Bai O, Li F, Yuan J, Laferte S, Xiang J. Mature dendritic cells pulsed with exosomes stimulate efficient cytotoxic T-lymphocyte responses and antitumour immunity. Immunology. 2007;**120**(1):90-102

[181] Yuan D, Zhao Y, Banks WA, Bullock KM, Haney M, Batrakova E, et al. Macrophage exosomes as natural nanocarriers for protein delivery to inflamed brain. Biomaterials. 2017;**142**:1-12

[182] Christianson HC, Svensson KJ, van Kuppevelt TH, Li J-P, Belting M. Cancer cell exosomes depend on cell-surface heparan sulfate proteoglycans for their internalization and functional activity. Proceedings of the National Academy of Sciences of the United States of America. 2013;**110**(43):17380-17385

[183] Sims B, Farrow A, Williams S, Bansal A, Krendelchtchikov A, Gu L, et al. Role of TIM-4 in exosome-dependent entry of HIV-1 into human immune cells. International Journal of Nanomedicine. 2017;**12**:4823-4833

[184] de Abreu RC, Fernandes H, da Costa Martins PA, Sahoo S, Emanueli C, Ferreira L. Native and bioengineered extracellular vesicles for cardiovascular therapeutics. Nature Reviews. Cardiology. 2020;**17**(11):685-697

[185] Conner SD, Schmid SL. Regulated portals of entry into the cell. Nature. 2003;**422**(6927):37-44

[186] Izquierdo-Useros N, Naranjo-Gómez M, Archer J, Hatch SC, Erkizia I, Blanco J, et al. Capture and transfer of HIV-1 particles by mature dendritic cells converges with the exosome-dissemination pathway. Blood. 2009;**113**(12):2732-2741

[187] Heusermann W, Hean J, Trojer D, Steib E, von Bueren S, Graff-Meyer A,

et al. Exosomes surf on filopodia to enter cells at endocytic hot spots, traffic within endosomes, and are targeted to the ER. Journal of Cell Biology. 2016;**213**(2):173-184

[188] Grant BD, Donaldson JG. Pathways and mechanisms of endocytic recycling. Nature Reviews Molecular Cell Biology. 2009;**10**(9):597-608

[189] Spooner RA, Smith DC, Easton AJ, Roberts LM, Lord JM. Retrograde transport pathways utilised by viruses and protein toxins. Virology Journal. 2006;**3**:26

[190] Prada I, Meldolesi J. Binding and fusion of extracellular vesicles to the plasma membrane of their cell targets. International Journal of Molecular Sciences. 2016;**17**(8):1296

[191] Pötgens AJG, Drewlo S, Kokozidou M, Kaufmann P. Syncytin: The major regulator of trophoblast fusion? Recent developments and hypotheses on its action. Human Reproduction Update. 2004;**10**(6):487-496

[192] Tolosa JM, Schjenken JE, Clifton VL, Vargas A, Barbeau B, Lowry P, et al. The endogenous retroviral envelope protein syncytin-1 inhibits LPS/PHA-stimulated cytokine responses in human blood and is sorted into placental exosomes. Placenta. 2012;**33**(11):933-941

[193] Bjerregaard B, Lemmen JG, Petersen MR, Østrup E, Iversen LH, Almstrup K, et al. Syncytin-1 and its receptor is present in human gametes. Journal of Assisted Reproduction and Genetics. 2014;**31**(5):533-539

[194] Bjerregard B, Ziomkiewicz I, Schulz A, Larsson L-I. Syncytin-1 in differentiating human myoblasts: Relationship to caveolin-3 and myogenin. Cell and Tissue Research. 2014;**357**(1):355-362

[195] Buslei R, Strissel PL, Henke C, Schey R, Lang N, Ruebner M, et al. Activation and regulation of endogenous retroviral genes in the human pituitary gland and related endocrine tumours: ERV and pituitary adenomas. Neuropathology and Applied Neurobiology. 2015;**41**(2):180-200

[196] Maliniemi P, Vincendeau M, Mayer J, Frank O, Hahtola S, Karenko L, et al. Expression of human endogenous retrovirus-w including syncytin-1 in cutaneous T-cell lymphoma. PLoS One. 2013;**8**(10):e76281

[197] Søe K, Andersen TL, Hobolt-Pedersen A-S, Bjerregaard B, Larsson L-I, Delaissé J-M. Involvement of human endogenous retroviral syncytin-1 in human osteoclast fusion. Bone. 2011;**48**(4):837-846

[198] Lin Y, Wu J, Gu W, Huang Y, Tong Z, Huang L, et al. Exosome-liposome hybrid nanoparticles deliver CRISPR/Cas9 system in MSCs. Advances in Science. 2018;**5**(4):1700611

[199] Sato YT, Umezaki K, Sawada S, Mukai S-A, Sasaki Y, Harada N, et al. Engineering hybrid exosomes by membrane fusion with liposomes. Scientific Reports. 2016;**6**(1):21933

[200] Usman WM, Pham TC, Kwok YY, Vu LT, Ma V, Peng B, et al. Efficient RNA drug delivery using red blood cell extracellular vesicles. Nature Communications. 2018;**9**(1). DOI: 10.1038/s41467-018-04791-8. https://www.ncbi.nlm.nih.gov/pmc/articles/PMC6004015/

Chapter 4

Morphology and Formation Mechanisms of Cellular Vesicles Harvested from Blood

Veronika Kralj-Iglič, Gabriella Pocsfalvi and Aleš Iglič

Abstract

Theoretical and experimental evidence on cellular vesicles (CVs) isolated from blood is presented. It is suggested that comparison of the observed shapes with theoretical shapes obtained by minimization of membrane-free energy in combination with electron microscopy is key in the assessment of CV identity. We found that shapes of CVs isolated from blood by repetitive centrifugation (up to 20.000 g) and washing, and observed by scanning electron microscopy (SEM) agreed well with theoretically observed shapes. It is indicated that these CVs are colloids deriving from residual blood cells, mostly platelets. SEM images of washed erythrocytes undergoing budding and transmission electron microscopy (TEM) images of isolated erythrocyte microvesicles likewise showed smooth shapes that we described as characteristic for colloidal CVs. Besides these, the CV isolates may contain other small particles, such as exosomes and viruses, as observed in isolates from tomato homogenate, however, we could not identify such particles in isolates from healthy human blood. Theory of deviatoric elasticity underlaying minimization of the membrane free energy and simulated two-component vesicles with the orientational ordering of anisotropic constituents are presented to indicate the interdependence of curvature—sorting of membrane constituents and their orientational ordering in strongly anisotropically curved regions.

Keywords: extracellular vesicles, erythrocyte microvesicles, extracellular vesicle shape, scanning electron images of extracellular vesicles; cryo-TEM images of extracellular vesicles

1. Introduction

The great success of the double helix molecular model of DNA [1] has been based mostly on revealing specific, chemical mechanisms. This model has enabled profound advances in technology; however, it has not completely revealed the causes of common and debilitating physiological and pathophysiological mechanisms. However, the dynamics of the genetic material depend on its interaction with membranous systems which has hitherto not been given adequate attention. Moreover, nanostructures composed of- and enclosed by- biological membranes (e.g., vesicles and nanotubes) were long overlooked [2] due to their small size and fragility. Recently, submicron-sized membrane-enclosed cellular vesicles (CVs) that have been formed in a process in which membrane plays a key role have become a subject of increasing attention. By being released into the cell exterior, they can move more or less freely in the surrounding medium. Extensive studies and empirical knowledge indicate that these tiny particles may have a great impact on living systems, in particular, because they present an intercellular communication system that connects different kingdoms of life. CVs include microexovesicles, exosomes, enveloped viruses, and cellular membrane endovesicles.

The structure of a membrane-enclosed entity that carries a specific cargo presents a foundation stone of life and dwells also on its border. Namely, the physical properties of membrane-enclosed entities are shared with any small particles that attain their configuration according to the minimal energy of the system (referred here as colloid systems). While consideration of biological nanostructures needs the support of a rigorous physical description, new evidence regarding nanoscale features needs interpretation by the development of new physical models, specific to these materials. We believe that the research of nanoscale systems at the cellular level requires the intertwining of existing fields of physics, chemistry, biology, and medicine in the course of their growth. By addressing the physical properties, methods of theoretical physics can be used to describe the system, interpret experimental data, and predict the behavior of the system.

In order to be studied, CVs should be harvested from their natural environment. Presently, integration of different methods is recommended [3], however, new technically advanced solutions are sought. The most commonly used method for CV harvesting involves differential centrifugation [4], which can be followed by using for example sucrose or iodixanol gradient [5]. As this technique is time-consuming and of limited access, alternative techniques were proposed. Ultrafiltration, flow field-flow fractionation, dialysis, size exclusion chromatography (SEC), microchip-based techniques, and precipitation-based methods are being developed to harvest CVs, alone or in combination with ultracentrifugation-based methods. Immunoaffinity-based isolation can also be applied to harvest CVs with particular surface protein composition [6]. Recently, a number of commercial kits are made available and have been widely reported in the literature for CV isolation. For example, ExoQuick (System Biosciences) and Total Exosome Isolation kits (TEI, Invitrogen) rely on polymer precipitation; qEV (Izon) is based on SEC; Millipore uses centrifugal filter devices for ultrafiltration; and exoEasy (Qiagen) is based on membrane affinity binding [7]. But different isolation methods were found to lead to different CV populations [8] due to mechanical and thermal stress and chemical reactions. Although the suggested methods are faster and easier to apply, a recent thorough comparison between isolates obtained from these methods and ultracentrifugation showed that ultracentrifugation is still the most appropriate method among those tested as regards purity [7].

According to their features, CVs are ideal candidates to serve as biomarkers, nanosized drug-delivery vehicles, and mediators for a variety of therapeutics in oncology, immune therapy, and regenerative medicine [9–11]. Thus, CVs have the potential for great clinical impact in nanomedicine. The dual potential of CVs as diagnostic tools and as therapeutic agents supports their use in "theranostics" [11–13]. This area of nanomedicine focuses on multidisciplinary research to set up new systems for various nanobiomedical applications, ranging from the medical use of nanoplatform-based diagnostic agents to therapeutic agents for possible future applications [14]. Furthermore, the theranostic "all-in-one approach" has great potential in the field of personalized medicine, as it enables the detection and monitoring of disease in individual patients, possibly in early clinical stages, as well as targeted drug delivery at the site of the disease.

In order to manipulate cellular vesicles (CVs), the process of their formation should be better understood. Vesiculation of biological membranes was studied theoretically and experimentally. CVs were isolated from different biological samples, including blood [15–20]. In phospholipid vesicles, shape transformations involving evaginations were studied [20–23]. Budding and vesiculation of erythrocytes were also considered [24–29]. While erythrocytes shed vesicles in the final stage of the membrane budding, platelets undergo fragmentation in the shear stress [30].

Visualization of the samples is a prerequisite to identify CVs in samples. As they are very small, CVs are hidden within the organisms or cell assemblies and they cannot be directly observed in their natural environment. The methods used for their harvesting, observation, and assessment are to a large extent invasive enough to transform them to such extent that identification of their original nature is obscured.

We found that a large pool of submicron-sized particles in the isolates from blood was formed during the processing of samples [31–33]. This indicates that the formation of CVs in isolates can be influenced upon by changing the parameters of processing which is of advantage for the production of therapeutic preparations from biological samples.

In this review, we would like to point to some common properties of CVs that we characterize by the key role of the membrane in shape determination and refer to as colloid CVs. Also, we will address other types of cellular particles that can be present in the samples, for example, viruses.

2. Theoretical prediction of the CV shape

Theoretical and experimental studies on (artificial) phospholipid vesicles have shown that the shape of the bilayer membrane-enclosed compartments can be theoretically well described by the properties of the membrane [34]. By considering that membrane is composed of many constituents, methods of statistical physics were used for its description [35, 36]. The key feature in the expression of the energy of a single constituent is the mismatch of the local curvature and the constituent intrinsic curvature (the one fitting the shape of the membrane constituent) [36]. In order to compose the membrane, the constituent attains the local membrane shape that usually differs from its intrinsic shape. Moreover, if the constituent is not symmetric with respect to the axis perpendicular to the membrane surface, the principal axes of the membrane and the constituent are in general rotated by an in-plane angle, meaning that the constituents can attain different in-plane orientations in the membrane which correspond to different energies. The consequence of the mismatch in curvature and orientation is that certain energy is required to insert the constituent at the site, this energy being higher if the mismatch is greater. The thermal motion opposes the complete orientational ordering in the direction with the lowest energy but the constituents will spend on average more time in the orientation with lower energy. As constituents are more or less free to move laterally in the membrane, they can redistribute in a way to minimize the mismatch between the intrinsic and the actual shape. Summing up the contributions of all the constituents and considering the entropic effects due to lateral and orientational ordering yields the expression for the free energy of the membrane F in terms of the mean curvature of the membrane surface H and the curvature deviator D, both composed of the two principal curvatures C_1 and C_2 [34],

$$H = (C_1 + C_2)/2, \tag{1}$$

$$D = (C_1 - C_2)/2, \tag{2}$$

$$F = -kT \int \sum_i m_i \ln \left(q_i^0\, 2\cosh\left(d_{i,\text{eff}}\right)\right) \mathrm{d}A + k_\mathrm{B}T \int \sum_i m_i \ln\left(m_i/m\right)\mathrm{d}A, \tag{3}$$

$$q_i^0 = \exp\left(-\frac{\xi_i (H - H_{i,\text{m}})^2}{2k_\mathrm{B}T} - \frac{\left(\xi_i + \xi_i^*\right)\left(D^2 + D_{i,\text{m}}^2\right)}{4k_\mathrm{B}T}\right), \tag{4}$$

$$d_{i,\text{eff}} = \left(\xi_i + \xi_i^*\right) D D_{i,\text{m}}/k_\mathrm{B}T, \tag{5}$$

where ξ_i and ξ_i^* are constants, k_B is the Boltzmann constant, T is temperature, m_i is the local area number density of the i-th kind of constituents and m is the number density taking all constituents over all membrane area. Integration is performed over membrane surface A. The summation accounts for all types of constituents that are characterized by index i. The intrinsic mean curvature of the membrane surface H_m and the intrinsic curvature deviator D_m refer to respective principal curvatures intrinsic to the shape of the constituent [34].

Free energy given by Eq. (3) consists of two terms—the first one deriving from the single-constituent energy and entropy of orientational ordering, and the second one deriving from lateral distribution of constituents. Eqs. (3) and (4) can be rewritten in the form [34].

$$F = W_\mathrm{B} - \int 2mk_\mathrm{B}T \cosh\left(d_{i,\text{eff}}\right) \mathrm{d}A + k_\mathrm{B}T \int \sum_i m_i \ln\left(m_i/m\right)\mathrm{d}A \tag{6}$$

where W_B has the form of the bending energy of a laterally isotropic membrane [34].

A basic physical principle that the system will attain the shape corresponding to minimal free energy [34]

$$F = \min, \tag{7}$$

at relevant constraints to the system is taken into account such as the requirement of fixed membrane area A and fixed enclosed volume V [37]. In the absence of net external forces acting on the membrane, a convenient geometrical parameter is the relative volume v which represents the volume to area ratio largely determined by the osmotic equilibrium [34]

$$v = \sqrt{V^2/36\pi A^3}. \tag{8}$$

Also, other constraints can be imposed upon the system, for example, constant average mean curvature [34]

$$<H> = \int H\, \mathrm{d}A \Big/ \int \mathrm{d}A. \tag{9}$$

To find the free energy minimum, the above formalism [Eqs. (1)–(5)] is used to state and solve the so-called variational problem in which the principal curvatures are expressed in terms of convenient coordinates. Dimensionless parameters are used for clarity: $c_j = C_j R$, j = 1,2 are the principal curvatures normalized with respect to the radius of the sphere with the surface area A, $R = \sqrt{A/4\pi}$, $h = HR$, and $d = DR$ are the normalized mean curvature and curvature deviator of the membrane, respectively, and $h_{i,\mathrm{m}} = H_{i,\mathrm{m}}R$ and $d_{i,\mathrm{m}} = D_{i,\mathrm{m}}R$ are the intrinsic mean curvature and the intrinsic curvature deviator of the i-th type of constituents, respectively. The area element is normalized with respect to the area A, $\mathrm{d}a = \mathrm{d}A/4\pi R^2$. The constants ξ_i and ξ_i^* are taken to be equal for simplicity. The free energy is normalized with respect to the free energy of the sphere composed of chosen constituents, $8\pi m\xi_i$.

Consistently related distributions of the constituents, their in-plane orientations (if relevant), and the membrane shape are determined simultaneously in a mathematical procedure. Several methods have been developed for this purpose—for example, ansatz [37], numeric solution of differential equation [38], surface evolver [39], or finite element method [40]. The simplest method is the ansatz approach in which the space of possible solutions is assumed within a family of shapes with adjustable parameters. Such a solution can be analytical throughout (depending on the sophistication of the ansatz) and therefore transparent. The advantage of transparent methods is that with some basic mathematical skills the procedures can be repeated and used accordingly. Differential equations expressing minimization of free energy are derived by applying the Euler–Lagrange method [36] in a convenient parametrization. Considering a multicomponent system, the constituents are free to move laterally over the membrane which may present singularities that have not yet been fully explored [36]. Such solutions require numerical procedures that are implemented in respective customized software which normally requires manipulation by a skilled researcher. However, the set of possible solutions is considerably larger than in the ansatz case, as for some classes of shapes the relevant ansatz does not exist. With more freedom in finding a solution, the achieved energy attains lower values. The advantage of the rigorous solution of the system of differential equations is that the lowest possible energy can in principle be found, however, the method is rather demanding for shapes that are not axisymmetric which limits the set of possible solutions.

The theory could be refined by considering particular experimental evidence. For example, it was found that cellular vesicles may go through a process of active and passive solutes' permeation which may cause cyclical expansion and contraction [41].

3. Visualization of CVs

Isolation of CVs from blood first requires removal of blood cells, in particular erythrocytes. Namely, to their prevailing abundance in blood of healthy humans, they present an obstacle in observation of the effects of other blood components. Usually, this is performed by centrifugation at a relatively low speed. Upon centrifugal force particles in the blood are inclined to move toward the bottom of the tube. The motion is roughly determined by the centrifugal force, buoyancy, and resistance force approximated by the Stokes law. It can be seen by equilibrating the forces that the speed of particles is proportional to the square of the particle radius and proportional to its density which means that larger particles will move faster. Besides being the most abundant, erythrocytes are also relatively large. If approaching each other close enough they form rolleaux which effectively speeds up their sedimentation and also creates channels in which smaller particles (platelets and CVs) are pushed out. This creates a counter-flow due to which platelets and CVs accumulate in the plasma above erythrocytes [32, 42]. However, this effect is temporary, as platelets and CVs in the erythrocyte-poor plasma reverse their flow and sediment as well. Plasma obtained by relatively low-speed centrifugation contains platelets as well as residual erythrocytes

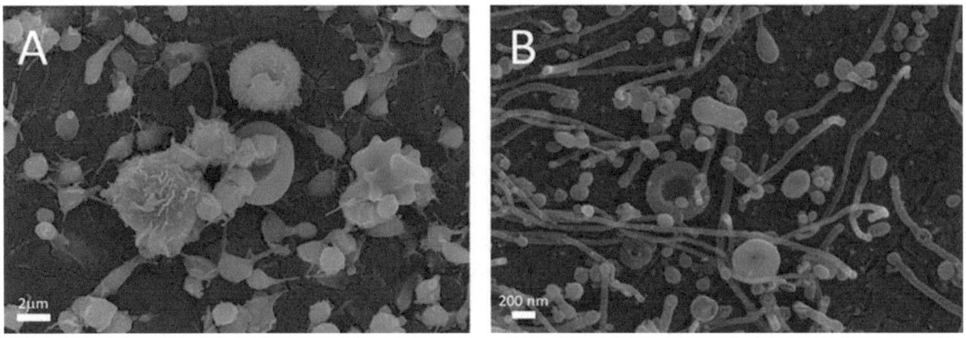

Figure 1.
A: Blood plasma observed by SEM. B: CVs isolated from blood as observed on the inner wall of the tube by SEM. A: From [31]. B: From [47].

and leukocytes (**Figure 1A**). Furthermore, some blood cells shed CVs during the processing. For further elaboration of the sample, different protocols have been proposed, differing in the amount of the required isolate and its purpose. Centrifugation is still the most widely used method as it does not induce changes in the chemical composition of the sample, is relatively simple, time effective and low cost, and enables the simultaneous elaboration of multiple samples. Different methods were suggested for the isolation of CVs from blood [43]. Centrifugation protocols may differ in time and speed of centrifugation as well as in other parameters (e.g., temperature, the type of laboratory material used, up-gradation by technologically advanced procedures) [44–46].

Figure 1B shows the interface of the isolate with the tube wall. The tube was cut and the sample was prepared directly on the piece of the tube for imaging with the scanning electron microscope (SEM). Smooth shapes of the particles in the isolate can be noted.

CV isolates from blood shown in this chapter were obtained by repetitive centrifugation and washing of samples with phosphate- and citrate- buffered saline (PBS). We used different centrifugation protocols. Unilamellar phospholipid vesicles were prepared by electroformation in sugar solution and rinsed into the observation chamber by solution of a lighter sugar with the same osmolarity.

Giant phospholipid vesicles exceed in size several micrometers and can therefore be observed live by optical microscope. Therefore, the comparison between theoretically predicted and experimentally observed shapes is straightforward. For submicron-sized CVs light microscopy does not provide sufficient resolution. The samples can be observed by electron microscope which requires more or less aggressive processing. To observe them by scanning electron microscope, samples are dried and sputtered with heavy metal. For cryo-electron microscopy, they are frozen in thin ice (about 100 nm of thickness) which deforms soft particles larger than this dimension and may cause their degradation. Interpretation of the images of processed samples is not always straightforward.

Figure 2 shows SEM images of CVs found in isolates from blood (a-d), an erythrocyte of a healthy human in physiological *ex vivo* conditions (e), and optical microscope images of a giant phospholipid vesicle (f-i). The corresponding theoretically obtained contours that were obtained by the solution of the variational problem are also given (a-i). A rigorous solution of the system of differential equations was sought. Two sequences of shapes are given, representing the transformation of vesicles composed of a single type of constituents with fixed relative volume v and changing $< h >$. The sequence a-d starts with a discocytic shape which with decreasing $< h >$ transforms into a stomatocyte with a wide dimple. In continuation of the process, the dimple grows inwards while the neck at the top shrinks. The sequence f-i starts with a prolate shape which with increasing $< h >$ transforms into a pear shape. The neck shrinks up to a point in which it becomes infinitesimal.

It can be seen that the shapes of the erythrocyte (**Figure 2e**) and the CV (**Figure 2a**) are the same although the size scale of the CVs is 10 times smaller. Mammalian erythrocytes have no nucleus and also no internal cytoskeleton and their shapes are likewise determined by the minimum of the free energy of the membrane (underlayed with membrane skeleton). Consideration of the membrane skeleton however requires additional assumptions which are important also in describing the formation of CVs.

Figure 3 shows calculated shapes representing the swelling and budding of a membrane-enclosed structure. The swelling was simulated by an increase of v and budding was simulated by an increase of average mean curvature $<h>$. Vesicles were composed of one type of constituents. Calculation of the sequence is conveniently performed within a chosen class of shapes. For a membrane composed from a single type of constituents that favor flat shape, the pear shape sequence (**Figure 2f-i**) is energetically more favorable than the lemon shape sequence (**Figure 3B-D and E-G**) since the formation of the neck is energetically unfavorable for constituents that favor positive (evaginated) and flat regions. The geometrical constraints limit the power of the set of possible shapes. For relative volume 1, there is only one possible shape (e.g., sphere) and this is the shape that attains the biggest possible volume at the given surface area (v cannot exceed 1). With the decrease of v, the set of possible shapes increases. It can be seen that already small decrease of v (e.g., to 0.98 (**Figure 3B-D**))

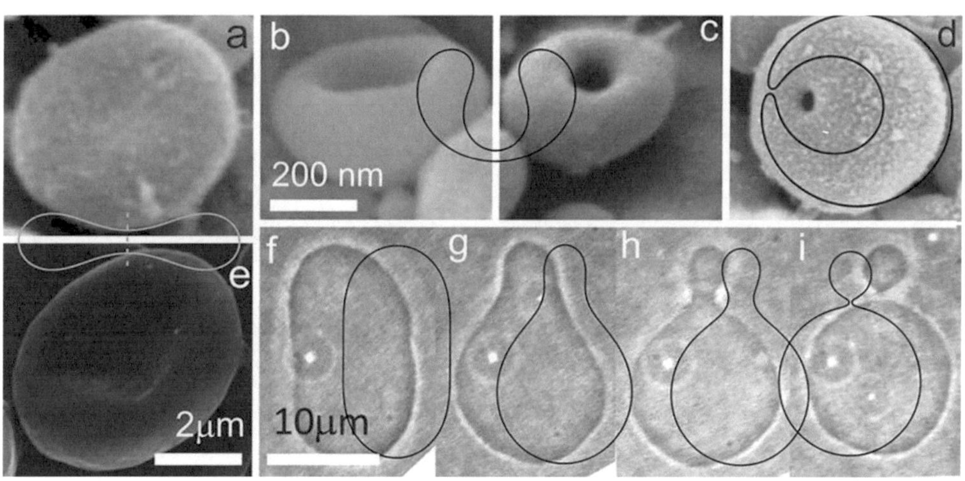

Figure 2.
Experimental: a-d: SEM of CVs found in isolates from blood, e: SEM of a discocyte at physiological ex vivo conditions, f-i: Optical microscope images of giant phospholipid vesicles. Theoretical contours derive from the solution of the variational problem by rigorously solving a system of differential equations. Within sequences a-d and f-i, A and V were fixed while $< h >$ was changing: The parameters of the calculated shapes were $h_m = d_m = 0$, (a, e): $v = 0.6$, $<h> = 1.040$, $<d> = 1.812$, (b, c): $v = 0.6$, $<h> = 0.650$, $<d> = 1.167$, (d): $v = 0.6$, $<h> = 0{,}435$, $<d> = 0.235$, (f): $v = 0.9$, $<h> = 1.050$, $<d> = 0.729$, (g): $v = 0.9$, $<h> = 1.105$, $<d> = 0.697$, (h): $v = 0.9$, $<h> = 1.155$, $<d> = 0.577$, (i): $v = 0.9$, $<h> = 1.240$, $<d> = 0.163$. Adapted from [34].

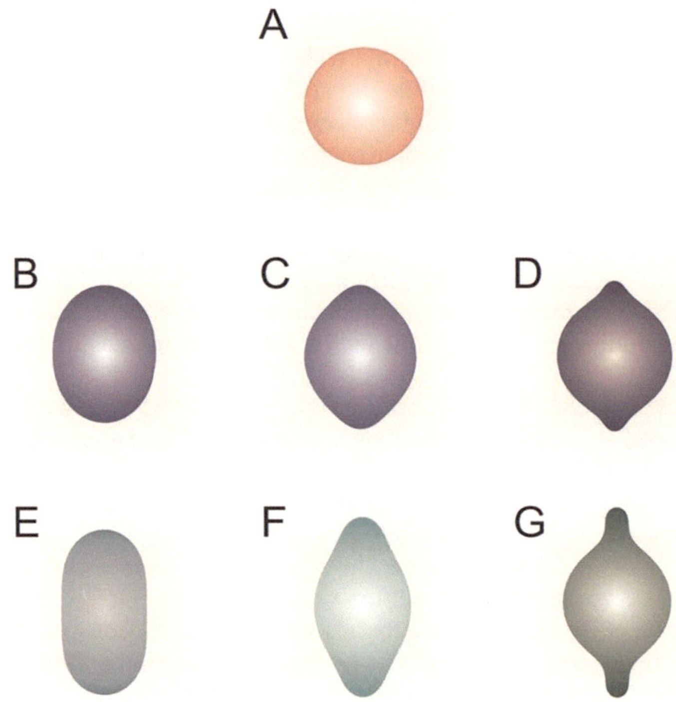

Figure 3.
Shapes corresponding to the minimum of the membrane free energy calculated by rigorously solving a system of differential equations. The parameters of the calculated shapes are $h_m = d_m = 0$, *A:* $v = 1$, $< h > = 1$, *B:* $v = 0.98$, $< h > = 1.0088$, *C:* $v = 0.98$, $< h > = 1.011$, *D:* $v = 0.98$, $< h > = 1.0224$, *E:* $v = 0.90$, $< h > = 1.05035$, *F:* $v = 0.9$, $< h > = 1.0634$, *G:* $v = 0.9$, $< h > = 1.354$.

can induce visible changes in shape with respect to the sphere ($v = 1$). If the vesicle loses 10% of its relative volume, the shape is visibly elongated (**Figure 3E-G**).

The particles that are essentially membrane-enclosed fluid interior deform and eventually undergo fragmentation at the thin necks. As the tearing area is minute the membrane is likely to seal. Smaller fragments are thus created. Fragmentation of residual cells takes place in particular at the interface with the tube wall where the shear force is the highest. Centrifugation at high centripetal accelerations of the rotor was shown to induce the formation of CV aggregates composed of a mixture of CVs highly heterogeneous in size and number of associated CVs [32].

Characteristics of membrane-enclosed vesicles composed of constituents that are not directly interacting are the smoothness of the shape. Good agreement between the calculated and the observed shapes indicates that the particles in the samples are vesicles (membrane-enclosed fluid interior). There are no additional methods needed. According to this principle, **Figure 3B** shows particles that can be identified as CVs deriving from blood cells. It is, however, not clear from this point what is the origin of the CVs, as the material may undergo formation and re-formation of vesicles during the processing, and the constituents of the CVs may come from different cells as well as from the surrounding solution. Such vesicles are colloidal in their nature and their identity depends on the properties of the cells as well as on the processing of the samples (e.g., centrifugation parameters, the composition of the suspension, temperature). They can be considered an artifact; however, this artifact can have clinical significance [30].

4. Shape determination of CVs with complex composition

If we take into account that the membrane is composed of more than one kind of constituents, the set of possible shapes with minimal energy is further expanded as the system has additional freedom by which it can adjust its configuration. The constituents may move to the regions of favorable curvature and additionally adjust their orientation. Allowing the system more freedom effectively increases its stability. To demonstrate the effect of the lateral distribution of constituents, **Figure 4B** shows two views on the shape with minimal free energy determined by the Monte Carlo simulation method. Two different isotropic constituents (favoring strong positive curvature and minute curvature, respectively) were considered. Red color denotes prevalence of the constituents that favor positive curvature and blue color denotes prevalence of the constituents that favor minute curvature; white color denotes the mixture of both kinds. Buds and undulated protrusions have formed largely from the constituents that favor strong positive curvature while the constituents that favor minute curvature organized themselves into a globule where the membrane is almost flat with respect to its thickness. Multiple buds recruit the constituents that favor strong positive curvature and span the almost flat parts to further minimize the energy. Agreement between the shapes observed in budding erythrocytes (**Figure 4A**) and shapes composed of two kinds of constituents does not reach the level evident in **Figure 1**. Budding erythrocyte undergoes detachment of the membrane skeleton at the top of the echinocyte spicules which is not taken into account in the theoretical consideration used to determine the shape shown in **Figure 4B**. Nevertheless, some aspects of the shape (i.e., multiple protrusions and rounded shape of the bud tips) can be noted.

Orientational ordering of the anisotropic constituents becomes evident in strongly anisotropically curved membrane parts such as in the necks and on the tubular parts (**Figure 5**). The variational problem of free energy minimization was solved by the expansion of the shape contour into the Fourier series [48]. Two kinds of constituents were considered—strongly anisotropically curved constituents and isotropic constituents that favor minute curvature. To enhance the formation of strongly anisotropically curved regions, a rod-like structure was imagined inside the vesicle by requiring a fixed length of the shape while the minimization took place. Longer rods induced

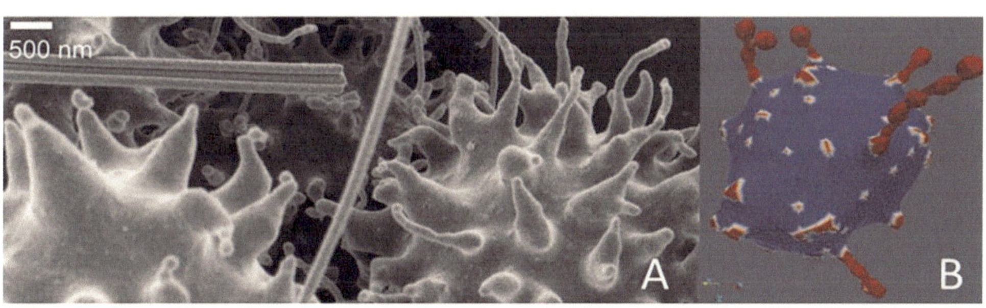

Figure 4.
A: Budding erythrocytes. B: Shape of a vesicle with a two-component membrane, calculated by Monte Carlo simulation. Membrane nanodomains with high intrinsic curvature (red color) are accumulated in undulated membrane protrusions. A: From [34]. B: From [48].

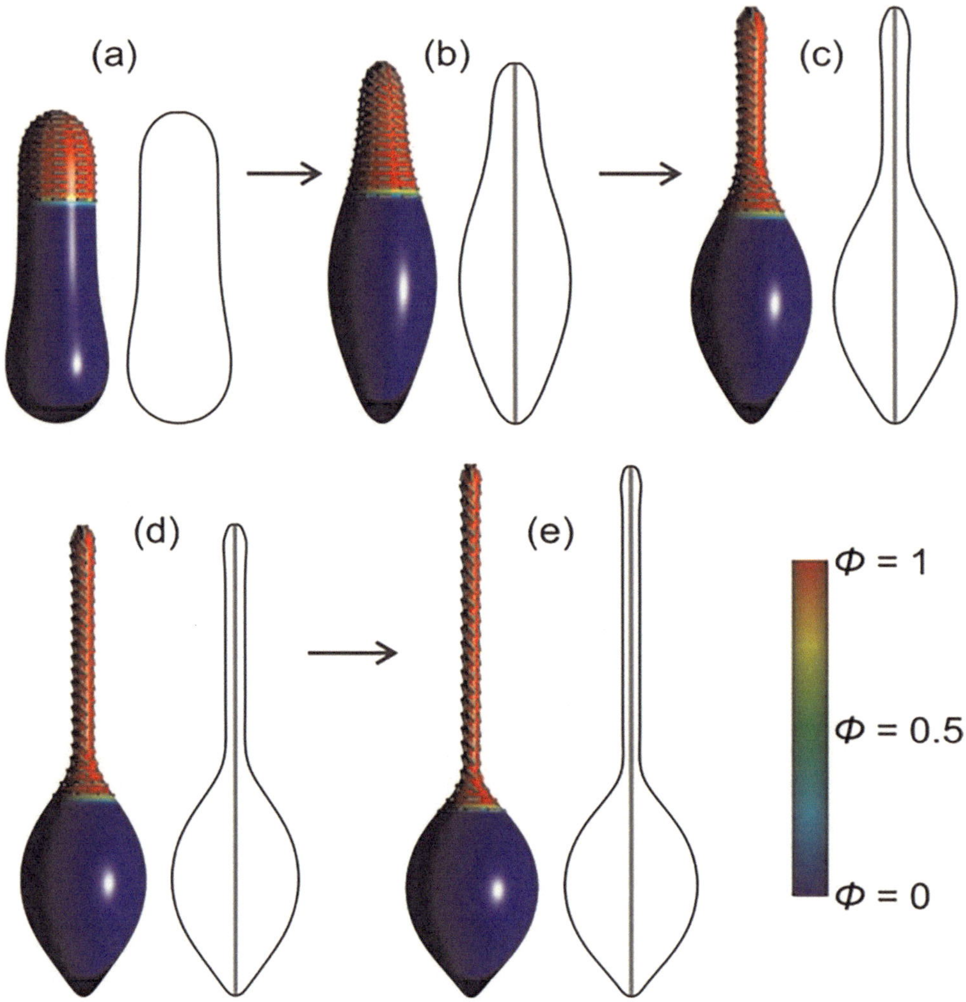

Figure 5.
Shapes of vesicles that are composed of two types of constituents (anisotropic (red) and isotropic (blue)), calculated for different lengths of the rod-like structure inside the vesicle. Panels (a) - (e) show the effect of different lenght of the rod-like structure inside the vesicle as indicated in the figure. The color scale shows the proportion of the two kinds of constituents Φ. The average Φ over all vesicle membrane was 0.25. Orientation of the anisotropic constituents is denoted by short gray lines on the protrusions. From [48].

longer and thinner tubular protrusions with well-ordered anisotropic constituents (**Figure 5**). The constituents redistributed into almost separate parts. Red color denotes prevalence of anisotropic constituents, blue color denotes prevalence of isotropic constituents while the rainbow between these two colors denotes a mixture of both kinds of constituents. It can be seen in **Figure 5** that the anisotropic constituents accumulated in the protrusion and underwent orientational ordering. While in thicker (conical or cylindrical) parts, the constituents ordered circumferentially (**Figure 5a-c**) the direction became tilted in thinner cylindrical parts toward the top of the shape (**Figure 5b**) and was almost uniform on longer tubular parts (**Figure 5d-f**).

Results shown in **Figures 4** and **5** indicate that the curvature induces sorting of membrane constituents to different extents determined by factors such as geometrical constraints and the intrinsic shape of the constituents.

5. Microvesiculation of erythrocytes and fragmentation of platelets

It was found that in *ex vivo* conditions washed erythrocytes undergo transformation into echinocytes (**Figure 6A**) and eventually budding takes place on the top of echinocyte spicules (**Figure 6 B** and **C**) [25, 26, 34]. Vesiculation was accelerated by the addition of amphiphilic molecules into the suspension of washed erythrocytes [25]. The shape of the buds as well as of the vesicles found in the isolates depended on the type of amphiphilic molecules added (**Figure 6B–E**). Vesicles matching in size and shape could be found in the isolate (**Figure 6D** and **E**) [34]. It was assumed that the amphiphilic molecules intercalate into the erythrocyte membrane and change the identity of the membrane constituents which in turn causes shape transformation. While dodecylzwittergent induced budding of globular structures and globular shape of isolated CVs (**Figure 6B** and **D**), dodecylmaltoside induced tubular shape of the buds and the CVs (**Figure 6C** and **E**). Dodecylmaltoside is composed of a carbohydrate tail and a bulky multipolar headgroup. The orientational ordering of the constituents involving dodecylmaltoside can explain the stable shape of tubular buds and vesicles that were observed in experiments. Budding erythrocytes were found also in CV isolates from blood [30] indicating that a part of CVs harvested from blood could be erythrocyte microvesicles.

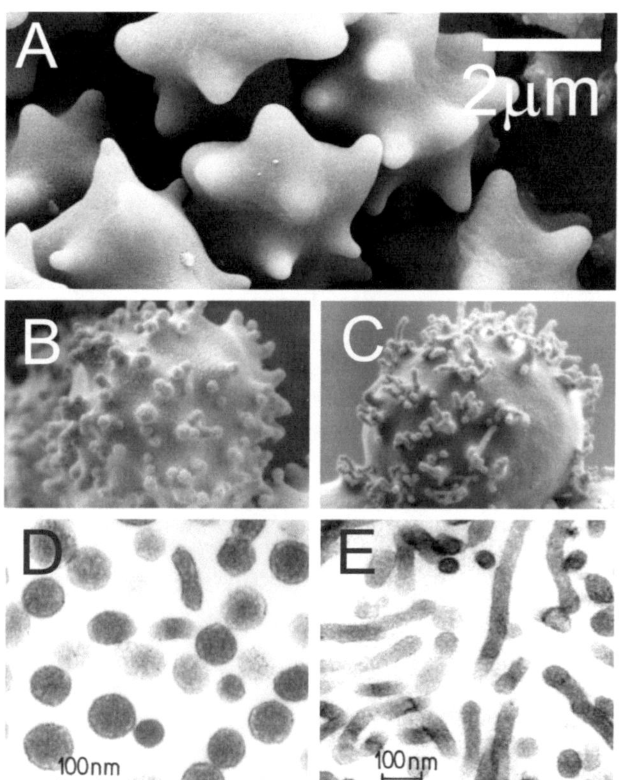

Figure 6.
A: Echinocytes, B: Spheroechinocyte with glubular buds induced by dodecyl-zwittergent, C: Spheroechinocyte with tubular buds induced by dodecylmaltoside, as observed with SEM; D and E: Respective isolated spherical and tubular CVs as observed by TEM. From [34].

The relevance of the model lies in agreement with experiments. The model of isotropic bending describes well the discocyte-stomatocyte transformation of erythrocytes, but cannot distinguish between tubular and spherical budding of the vesicle and therefore the formation of tubular/spherical vesicles. To our best knowledge, the presented model is by now the only model that explained the stability of different nanostructures with strongly anisotropically curved membranes (tubes, thin necks, hexagonal and cubic stacks).

Observations of isolates from blood indicate the presence of a major pool of CVs which shape corresponds to the membrane free energy and can be described as colloidal CVs. It was found [30] that the size of the colloidal CVs in blood isolates was different for different isolation protocols indicating a transient identity of colloidal CVs. In contrast, erythrocyte microvesicles shed from washed erythrocytes were uniform in size and were sensitive to intrinsic curvatures of the membrane constituents. The identity of CVs depends on the processing of samples. As stated above, the most commonly used method for CV isolation involves (differential) centrifugation/ultracentrifugation, which can be followed by ultracentrifugation on a sucrose or iodixanol gradient [4, 5, 49]. Ultrafiltration, dialysis, and size exclusion chromatography are used to harvest fractions of EVs, and immunoaffinity isolation and precipitation methods are used to harvest CVs with particular compositions [50, 51]. But transformation of the material may occur during any of these harvesting procedures, due to chemical/mechanical/thermal stress, as the nanostructures are fragile and prone to interact. Furthermore, the same principle applies to assessment methods. Commonly used methods are flow cytometry, SEM, TEM, atomic force microscopy, light scattering, fluorescence microscopy with analysis of Brownian motion (nanoparticle tracking analysis), and immunoblotting. Also, CV contents are analyzed for high-resolution molecular profiling of protein, microRNA, and lipid content (reviewed in [51]).

Besides colloidal CVs, other types of particles can be expected in biological samples, such as lipoproteins [52] and viruses [53]. Recently, besides the colloidal CVs, rod-like viruses were identified in CV isolates from tomato homogenate. The CVs were isolated by differential centrifugation and by size exclusion chromatography and observed by SEM [53] and by cryo-TEM (**Figure 7**). Three different fractions of the isolate obtained by using iodixanol gradient are shown (denoted by A, B, and C, respectively). The fraction observed in Panel A contained mostly colloidal CVs (black arrow); the fraction observed in Panel B contained colloidal CVs (black arrow) and rod-like virions (white arrow); the fraction observed in Panel C contained mostly

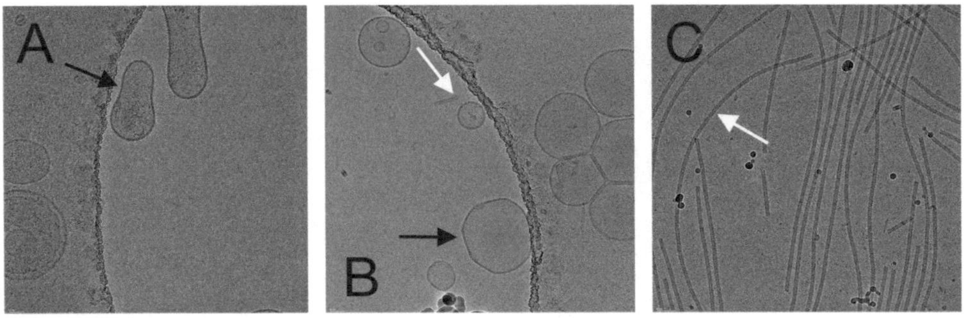

Figure 7.
Cryo-TEM images of three fractions of CV isolate from homogenate of tomato infected by the viruses. Panels A - C show three different fractions of the EV isolate that were separated by iodixanol gradient ultracentrifugation. Black arrows point to CVs, white arrows point to virions. Adapted from [53].

virions (white arrow). It was found by analyzing proteome with mass spectrometry that the samples contained capsid proteins of three tomato viruses [53].

6. Conclusions

Methods for CV characterization such as nanovesicle tracking analysis, flow cytometry, and light scattering can estimate the size and abundance of small particles in samples but are not suitable for their identification. To distinguish viruses or other cell-engineered particles from colloidal vesicles, imaging of samples is crucial, in particular when complemented with the identification of the molecular composition and with the results of the modeling. In constructing a model for CVs, we have implemented a theory based on statistical physics, that was previously used to describe the electric double layer [54]. We made a link to the theory of elasticity and found that cylindrical or saddle shapes that were observed in experiments can be stabilized by constituent redistribution and orientational ordering of anisotropic constituents. Inclusion of the orientational ordering of membrane constituents on strongly anisotropically curved regions is necessary for the description of the formation of CVs within the nanoscale. The so-called deviatoric elasticity of the membrane has been previously introduced [55–57], albeit not originating from statistical physics. We refer to the CVs that are formed due to minimization of the membrane free energy as the colloidal CVs, to distinguish them from cell-engineered CVs, such as viruses. Mechanisms of CV formation and transformation are fundamental and vital and there are prospects that they will in the future contribute to improved solutions in surface functionalization, diagnosis, and theranostics.

Acknowledgements

The authors acknowledge the support of the EU Commission grant Ves4us, and ARRS grants P2-0232, P3-0388, J3-3066, and L3-2621.

Conflict of interest

The authors declare no conflict of interest.

Author details

Veronika Kralj-Iglič[1,2]*, Gabriella Pocsfalvi[2] and Aleš Iglič[3,4]

1 University of Ljubljana, Faculty of Health Sciences, Laboratory of Clinical Biophysics, Ljubljana, Slovenia

2 Extracellular Vesicles and Mass Spetrometry Group, Institute of Biosciences and Bioresources, National Research Council of Italy, Napoli, Italy

3 University of Ljubljana, Faculty of Electrical Engineering, Laboratory of Physics, Ljubljana, Slovenia

4 University of Ljubljana, Faculty of Medicine, Chair of Orthopaedics, Laboratory of Clinical Biophysics, Ljubljana, Slovenia

*Address all correspondence to: veronika.kralj-iglic@fe.uni-lj.si

References

[1] Watson JD, Crick FHC. Molecular structure of nucleic acids; a structure for deoxyribose nucleic acid. Nature. 1953; **171**:737-738. DOI: 10.1038/171737a0

[2] Callier V. Cells talk and help one another via tiny tube networks. Scientific American Quanta Magazine; 2018 Available from: https://www.scientificamerican.com/article/cells-talk-and-help-one-another-via-tiny-tube-networks/

[3] Thery C, Amigorena S, Raposo G, Clayton A. Isolation and characterization of exosomes from cell culture supernatants. Current Protocols in Cell Biology. 2006;**3**:1-29. DOI: 10.1002/0471143030.cb0322s30

[4] Thery C, Witwer KW, Aikawa E, et al. Minimal information for studies of extracellular vesicles 2018 (MISEV2018): A position statement of the International Society for Extracellular Vesicles and update of the MISEV2014 guidelines. Journal of Extracellular Vesicles. 2018;**7**: 1535750. DOI: 10.1080/20013078.2018.1535750

[5] Iwai K, Minamisawa T, Suga K, Yajima Y, Shiba K. Isolation of human salivary extracellular vesicles by iodixanol density gradient ultracentrifugation and their characterizations. Journal of Extracellular Vesicles. 2016;**5**:1-17. DOI: 10.3402/jev.v5.30829

[6] Shtam T, Evtushenko V, Samsonov R, Zabrodskaya Y, Kamyshinsky R, et al. Evaluation of immune and chemical precipitation methods for plasma exosome isolation. PLoS One. 2020; **15**(11):e0242732. DOI: 10.1371/journal.pone.0242732

[7] Tian Y, Gong M, Hu Y, Zhang W, Zhang M, Hu X, et al. Quality and efficiency assessment of six extracellular vesicle isolation methods by nano-flow cytometry. Journal of Extracellular Vesicles. 2020;**9**:1. DOI: 10.1080/20013078.2019.1697028

[8] Freitas D, Balmaña M, Poças J, Campos D, et al. Different isolation approaches lead to diverse glycosylated extracellular vesicle populations. Journal of Extracellular Vesicles. 2019;**8**(1): 1621131. DOI: 10.1080/20013078.2019.1621131

[9] Lener T, Gimona M, Aigner L, Börger V, et al. Applying extracellular vesicles based therapeutics in clinical trials, An ISEV position paper. Journal of Extracellular Vesicles. 2015;**4**:300. DOI: 10.3402/jev.v4.30087

[10] Doyle LM, Wang MZ. Overview of extracellular vesicles, their origin, composition, purpose, and methods for exosome isolation and analysis. Cell. 2019;**8**(7):727. DOI: 10.3390/cells8070727

[11] Fais S, O'Driscoll L, Borras FE, et al. Evidence-based clinical use of nanoscale extracellular vesicles in nanomedicine. ACS Nano. 2016;**104**:3886-3899. DOI: 10.1021/acsnano.5b08015

[12] Soekmadji C, Li B, Huang Y, et al. The future of Extracellular Vesicles as Theranostics - an ISEV meeting report. J Extracell Vesicles. 2020;**9**(1):1809766. DOI: 10.1080/20013078.2020.1809766

[13] Lammers T, Aime S, Hennink WE, et al. Theranostic nanomedicine. Accounts of Chemical Research. 2011; **44**:1029-1038. DOI: 10.1021/ar200019c

[14] Sahoo S, Adamiak M, Mathiyalagan P, et al. Therapeutic and diagnostic translation of extracellular vesicles in cardiovascular diseases. Circulation. 2021;

143:1426-1449. DOI: 10.1161/CIRCULATIONAHA.120.049254

[15] Mrvar-Brečko A, Šuštar V, Janša V, et al. Isolated microvesicles from peripheral blood and body fluids as observed by scanning electron microscope. Blood Cells, Molecules & Diseases. 2010;**44**:307-312. DOI: 10.1016/j.bcmd.2010.02.003

[16] Szatanek R, Baran J, Siedlar M, Baj-Krzyworzeka M. Isolation of extracellular vesicles: Determining the correct approach (Review). International Journal of Molecular Medicine. 2015;**36**:11-17. DOI: 10.3892/ijmm.2015.2194

[17] Konoshenko MY, Lekchnov EA, Vlassov AV, Laktionov PP. Isolation of extracellular vesicles: General methodologies and latest trends. BioMed Research International. 2018;**2018**: 9545347. DOI: 10.1155/2018/8545347

[18] Tulkens J, De Wever O, Hendrix A. Analyzing bacterial extracellular vesicles in human body fluids by orthogonal biophysical separation and biochemical characterization. Nature Protocols. 2020; **15**:40-67. DOI: 10.1038/s41596-019-0236-5

[19] Zhang P, Yeo JC, Lim CT. Advances in thechnologies for purification and enrichment of extracellular vesicles. SLAS Technology. 2019;**24**:477-488. DOI: 10.1177/2472630319846877

[20] Kas J, Sackmann E. Shape transitions and shape stability of giant phospholipid vesicles in pure water induced by area-to-volume changes. Biophysical Journal. 1991;**60**:825-844. DOI: 10.1016/S0006-3495(91)82117-8

[21] Kralj-Iglič V, Gomiščček G, Majhenc J, et al. Myelin-like protrusions of giant phospholipid vesicles prepared by electroformation. Colloids and Surfaces A. 2001;**181**:315-318

[22] Kralj-Iglič V, Iglič A, Gomiščček G, et al. Microtubes and nanotubes of a phospholipid bilayer membrane. Journal of Physics A: Mathematical and General. 2002;**35**:1533-1549. DOI: 10.1088/0305-4470/35/7/305

[23] Iglič A, Kralj-Iglič V. Budding of liposomes – role of intrinsic shape of membrane constituents. In: Ottova-Leitmannova A, editor. Advances in Planar Lipid Bilayers and Liposomes. Vol. 4. Amsterdam: Elsevier; 2006. pp. 253-279. DOI: 10.1016/S1554-4516(06)04008-7

[24] Hägerstrand H, Isomaa B. Vesiculation induced by amphiphiles in erythrocytes. Biochimica et Biophysica Acta. 1989;**982**:179-186. DOI: 10.1016/0005-2736(89)90053-9

[25] Hägerstrand H, Isomaa B. Morphological characterization of exovesicles and endovesicles released from human erythrocytes following treatment with amphiphiles. Biochimica et Biophysica Acta. 1992;**1109**:117-126. DOI: 10.1016/0005-2736(92)90074-v

[26] Hägerstrand H, Kralj-Iglič V, Bobrowska-Hägerstrand M, Iglič A. Membrane skeleton detachment in spherical and cylindrical microexovesicles. Bulletin of Mathematical Biology. 1999;**61**: 1019-1030. DOI: 10.1006/bulm.1999.0128

[27] Kralj-Iglič V, Iglič A, Hägerstrand H, Peterlin P. Stable tubular microexovesicles of the erythrocyte membrane induced by dimeric amphiphiles. Physical Review E. 2000; **61**:4230-4234. DOI: 10.1103/physreve.61.4230

[28] Iglič A, Veranič P, Jezernik K, et al. Spherocyte shape transformation and release of tubular nanovesicles in human erythrocytes. Bioelectrochemistry. 2004; **62**:159-161. DOI: 10.1016/j.bioelechem.2003.07.002

[29] Greenwalt TJ. The how and why of exocytic vesicles. Transfusion. 2006;**46**: 143-152. DOI: 10.1111/j.1537-2995.2006.00692.x

[30] Šuštar V, Bedina Zavec A, Štukelj R, et al. Nanoparticles isolated from blood–a reflection of vesiculability of blood cells during the isolation process. International Journal of Nanomedicine. 2011;**6**: 2737-2748. DOI: 10.2147/IJN.S24537

[31] Božič D, Vozel D, Hočevar M, et al. Enrichment of plasma in platelets and extracellular vesicles by the counterflow to erythrocyte settling. Platelets. 2021;Aug 13: 1-11. DOI: 10.1080/09537104.2021.1961716

[32] Božič D, Hočevar M, Kononenko V, et al. Pursuing mechanisms of extracellular vesicle formation. Effects of sample processing. Advances in Biomembranes and Lipid Self-Assembly. 2020;**32**:113-155. DOI: 10.1016/bs.abl.2020.09.003

[33] Božič D, Hočevar M, Kisovec M, et al. Stability of erythrocyte-derived nanovesicles assessed by light scattering and electron microscpy. International Journal of Molecular Sciences. 2021;**22** (23):12772. DOI: 10.3390/ijms222312772

[34] Kralj-Iglič V, Pocsfalvi G, Mesarec L, et al. Minimizing isotropic and deviatoric membrane energy – An unifying formation mechanism of different cellular membrane nanovesicle types. PLoS One. 2020;**15**(12):e0244796. DOI: 10.1371/journal.pone.0244796

[35] Kralj-Iglič V, Remškar M, Iglič A. Deviatoric elasticity as a mechanism describing stable shapes of nanotubes. In: Reimer A, editor. Horizonts in World Physics. Vol. 244. Nova Science Publisher; 2004. pp. 111-156

[36] Kralj-Iglič V, Babnik B, Gauger DR, et al. Quadrupolar ordering of phospholipid molecules in narrow necks of phospholipid vesicles. Journal of Statistical Physics. 2006;**125**:727-752. DOI: 10.1007/s10955-006-9051-9

[37] Canham PB. The minimum energy of bending as a possible explanation of the biconcave shape of the human red blood cell. Journal of Theoretical Biology. 1970; **26**:61-81. DOI: 10.1016/s0022-5193(70)80032-7

[38] Deuling HL, Helfrich W. The curvature elasticity of fluid membranes. A catalogue of vesicle shapes. Journal of Physics (Paris). 1976;**37**:1335-1345

[39] Yan J, Liu QH, Liu JX, Ou-Yang ZC. Numerical observation of nonaxisymmetric vesicles in fluid membranes. Physical Review E. 1998;**58**:4730-4736. DOI: 10.1103/PhysRevE.58.4730

[40] Feng F, Klug WS. Finite element modeling of lipid bilayer membranes. Journal of Computational Physics. 2006; **220**(1):394-408. DOI: 10.1016/j.jcp.2006.05.023

[41] Oglęcka K, Rangamani P, Liedberg B, Kraut RS, Parikh AN. Oscillatory phase separation in giant lipid vesicles induced by transmembrane osmotic differentials. eLife. 2014;**3**: e03695. DOI: 10.7554/eLife.03695

[42] Pribush A, Meyerstein D, Meyerstein N. The mechanism of erythrocyte sedimentation. Part 1: Channeling in sedimenting blood. Colloids and Surfaces. B, Biointerfaces. 2010;**75**(1):214-223. DOI: 10.1016/j.colsurfb.2009.08.036

[43] Hill AF. Isolation of platelet-derived extracellular vesicles. Chapter 12. Methods in Molecular Biology: Exosomes and Microvesicles. 2017;**1545**:177-188. DOI: 10.1007/978-1-4939-6728-5_12

[44] Onódi Z, Csilla P, Nagy T, et al. Isolation of high-purity extracellular vesicles by the combination of iodixanol density gradient ultracentrifugation and bind-elute chromatography from blood plasma. Frontiers in Physiology. 2018;**9**: 1479. DOI: 10.3389/fphys.2018.01479

[45] Yeo JC, Kenry, Zhao Z, et al. Label-free extraction of extracellular vesicles using centrifugal microfluidics. Biomicrofluidics. 2018;**12**(2):024103. DOI: 10.1063/1.5019983

[46] Linares R, Tan S, Gounou C, et al. High-speed centrifugation induces aggregation of extracellular vesicles. Journal of Extracellular Vesicles. 2015; **4**(1):29509. DOI: 10.3402/jev.v4.29509

[47] Jan Z, Drab M, Drobne D, et al. Decrease in cellular nanovesicles concentration in blood of athletes more than 15 hours after marathon. International Journal of Nanomedicine. 2021;**16**:443-456. DOI: 10.2147/IJN. S282200

[48] Mesarec L, Drab M, Penič S, et al. On the role of curved membrane nanodomains, and passive and active skeleton forces in the determination of cell shape and membrane budding. International Journal of Molecular Sciences. 2021;**22**:2348. DOI: 10.3390/ ijms22052348

[49] Momen-Heravi F. Isolation of extracellular vesicles by ultracentrifugation. In: Kuo W, Jia S, editors. Extracellular Vesicles. Methods in Molecular Biology. Vol. 1660. New York, NY: Humana Press; 2017. DOI: 10.1007/978-1-4939-7253-1_3

[50] Garcia-Romero N, Madurga R, Rackov G, et al. Polyethylene glycol improves current methods for circulating extracellular vesicle-derived DNA isolation. Journal of Translational Medicine. 2019;**17**(1):75. DOI: 10.1186/ s12967-019-1825-3

[51] Hartjes TA, Mytnyk S, Jenster GW, et al. Extracellular vesicle quantification and characterization: Common methods and emerging approaches. Bioengineering. 2019;**6**:7. DOI: 10.3390/ bioengineering6010007

[52] Sódar BW, Kittel A, Pálóczi K, et al. Low-density lipoprotein mimics blood plasma-derived exosomes and microvesicles during isolation and detection. Scientific Reports. 2016;**6**: 24316. DOI: 10.1038/srep24316

[53] Mammadova R, Fiume I, Bokka R, et al. Identification of tomato infecting viruses that co-isolate with nanovesicles using a combined proteomics and electron-microscopic approach. Nanomaterials. 2021;**11**(8):1922

[54] Kralj-Iglič V, Iglič A. A simple statistical mechanical approach to the free energy of the electric double layer including the excluded volume effect. Journal of Physics II (France). 1996;**6**: 477-491. DOI: 10.1051/jp2:1996193

[55] Fischer T. Bending stiffness of lipid bilayers. II. Spontaneous curvature of the monolayers. Journal of Physics II (France). 1992;**2**:327-336

[56] Fischer T. Bending stiffness of lipid bilayers. III. Gaussian curvature. Journal of Physics II (France). 1993;**3**:337-343

[57] Fischer T. Mechanisms for determining the time scales in vesicle budding. Physical Review E. 1994;**50**: 4156-4166. DOI: 10.1103/physreve. 50.4156

Chapter 5

High-Throughput Single Extracellular Vesicle Profiling

Yanling Cai and Di Wu

Abstract

Extracellular vesicles (EVs) are heterogeneous due to their cell of origins, biogenesis, stimuli in the microenvironment and so on. Single EV analysis is required for the study of EV heterogeneity. Besides the investigation of EV biology, single EV analysis technologies are promising approach for liquid biopsy, which relies on the detection of biomarker EVs readily available in body fluids but in trace amount. However, EVs are nano-scaled structures, which beyond the resolution of conventional technologies like optical microscopes, flow cytometers and so on. In this chapter, we will discuss advanced strategies for studying single EVs, including single EV imaging systems, flow cytometers, nano-sensing technologies and single EV barcoding assay.

Keywords: extracellular vesicles, single EV analysis, imaging, flow cytometer, nano-sensing, Proximity Barcoding Assay

1. Introduction

Extracellular vesicles (EVs) are heterogeneous because of their diverse cell of origins, the process of biogenesis, the specific stimuli in their microenvironment and so on. EVs are produced by cells of bacteria, fungi, plants and animals. Inside human bodies, EVs carry molecular signatures of their parent cells and diffuse freely among blood stream and tissues. EVs can be classified into various subpopulations according to their origin, size, density, biogenesis, compositions etc. The origin of the vesicles gives us terminology of prostasome, oncosome etc. According to the size of EVs, investigators utilized terms of exomere, small EVs, large EVs. Depending on the biogenesis, we defined apoptic bodies, microvesicles and exosomes. Apoptic bodies are large vesicles formed due to apoptosis. Microvesicles are vesicles budding directly from cell membrane. Exosomes are the released intraluminal vesicles (ILVs) from multivesicular bodies (MVBs) through fusion of MVB with cell membrane.

To address the heterogeneity of EVs, scientists dedicated to the development of novel techniques for single EV detection. Besides investigation of EV biology, single EV analysis technologies are promising approach for liquid biopsy, which relies on the detection of biomarker EVs readily available in body fluids. Here in this chapter, we will discuss four main strategies of studying single EVs, including single EV imaging systems, flow cytometers, nano-sensing technologies and single EV barcoding assay.

2. Single EV imaging

Due to the sub-200 nm diameter of most EVs, conventional optical microscopies are precluded for direct observation or imaging of EVs. Nano-scaled imaging methods which are applicable to the scale of EVs include electron microscopies (EM), atomic force microscopy (AFM), total internal reflection fluorescence microscopy (TIRFM) and stochastic optical reconstruction microscopy (STORM).

2.1 Electron microscopies (EM)

Electron microscopies (EM) can be used to study the morphology and size of EVs. Scanning electron microscopy utilizes beam of electrons to bombard the surface of samples and detect the backscattered electrons and secondary electrons to construct the image of the detected surface. EVs under SEM imaging are usually round shaped, but could be collapsed and irregular in shape, possibly due to the deformation of EV structure during sample preparation. Transmission electron microscopy (TEM) and cryo-TEM has become one of the gold standards for EV characterization according to MISEV2018 guideline suggested by International Society of Extracellular Vesicles (ISEV). Under TEM observation, EV are round spheres or cup-shaped with lipid membrane visible as bilayers. Immunogold labelling could be combined with EM to get the composition distribution on EVs (**Figure 1**) [1–3].

2.2 Atomic force microscopy (AFM)

With AFM, an oscillating cantilever scans over a substrate with adhered EVs and simultaneously image the spatial dimension of single EVs. Therefore, the morphology

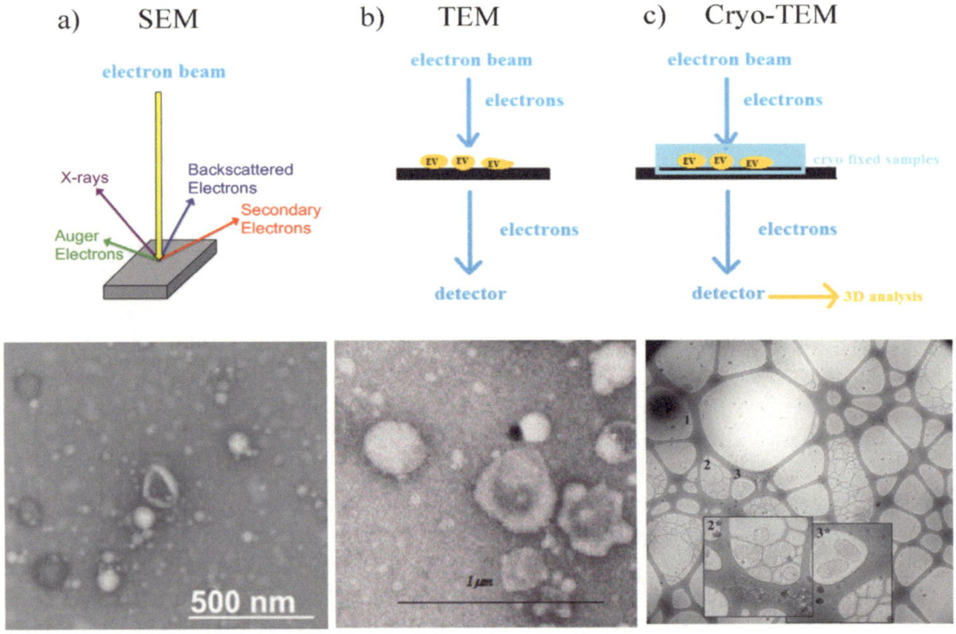

Figure 1.
The principle of a) SEM, b) TEM and c) cryo-TEM with an example of imaged EVs, respectively.

and particle size distribution are analyzed in AFM [2]. AFM tip indents the EVs and records the force-distance curve to provide insight into the mechanical properties, including mechanical stiffness, Young's modulus and bending modulus [4]. Tumor-exosomes of bladder cancer cell lines were proven to show differential mechanical properties dependent on the malignant state [5]. Although constrained in the throughput of EV analysis, the precise analysis of structural and biophysical properties via AFM provides unique measures in EV engineering for targeted delivery and therapeutic (**Figure 2**) [4].

2.3 Microfluidic device for single EV imaging

Combining with fluorescent microscopies, microfluidic devices are widely designed for single EV analysis. The EVs were biotinylated and captured on the streptavidin coated glass surface of a microfluidic device. The repeated cycles of immune staining-imaging-quenching steps allowed the investigators to analyze up to 11 biomarkers on immobilized individual EVs with an inverted microscope (Nikon Eclipse TE2000S) equipped with a sCMOS camera. After image processing, about 600 individual EVs were mapped onto a 2D plot via t-distributed stochastic neighbor embedding (tSNE) [6]. However, Biomarkers expressed on single EVs are at very

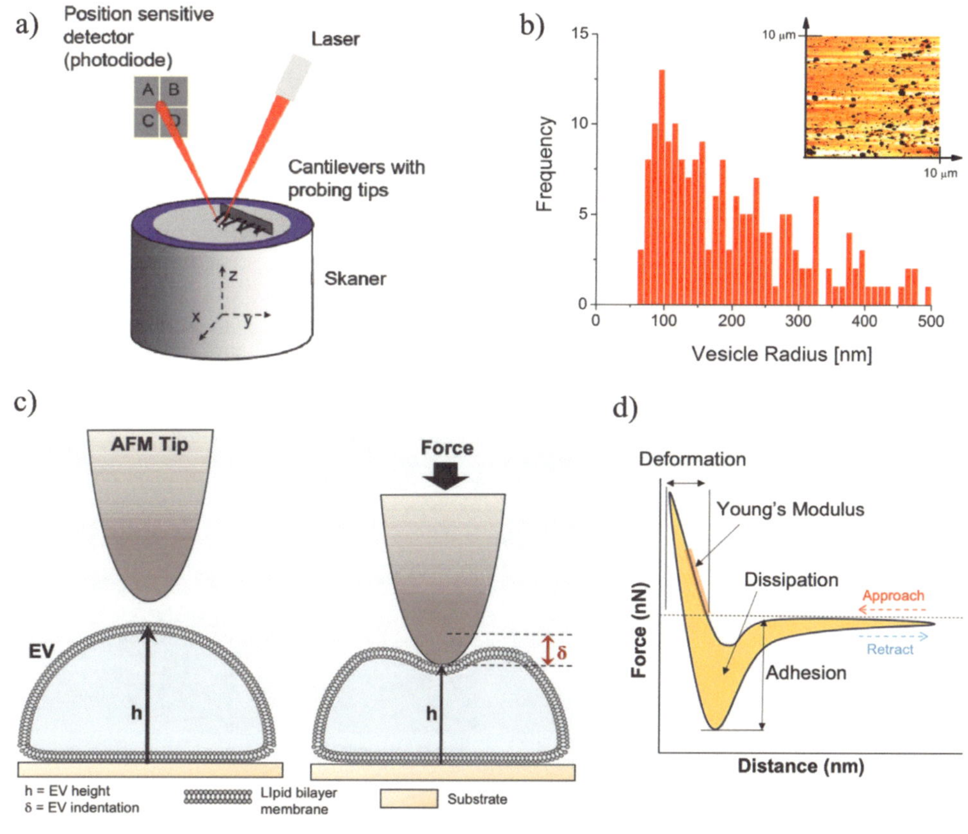

Figure 2.
AFM for analysis of single EVs. a) Principles of AFM detection. Panel b) shows an typical AFM readout of EV morphology detection and particle size distribution plot. c) and d) AFM measurement of biomechanical properties of EVs.

low abundant or has limited number of epitopes for immunostaining. Therefore, EV biomarker could be undetected due to weak signals. Molecular profiling of single EVs often requires high resolution optical microscopies or signal amplification strategies, such as rolling circle amplification, branched DNA probes or enzymatic reactions.

2.4 Total internal reflection fluorescence microscopy (TIRFM)

Total internal reflection fluorescence microscopy (TIRFM) is an optical technique utilized to observe single molecule fluorescence at surfaces and interfaces. It provides a high axial resolution below 100 nm. Single-vesicle imaging analysis could be done via TIRFM. The investigators can visualize multiple marker expressions of individual EVs by using fluorescent probes, and can also classify EV subpopulations by analyzing co-localization of markers [7]. With TIRFM system, targeted miRNA detection at single EVs could be realized with the co-delivery of inactive split DNAzymes and fluorescence-quenched substrates into EVs and a miRNA-activated catalytic cleavage reaction that amplifies fluorescence signal (**Figures 3** and **4**) [8].

2.5 Stochastic optical reconstruction microscopy (STORM)

Stochastic optical reconstruction microscopy (STORM) and direct-STORM (d-STORM) are single-molecule super-resolution imaging techniques with a practical resolution limit of 20 nm. STORM utilized the photoswitchable fluorescent probes to precisely localize detected molecules at a high spatial resolution. AlexaFluor 647-conjugated anti-CD63 antibodies were employed to detect cancer cell-derived EVs [9]. Using photo-switchable lipid dyes such as Dil, d-STORM imaging enabled rapid detection of EVs down to 20–30 nm in size and imaging of EV uptake by live cells in culture [10]. With STORM, droplet-based single-exosome-counting enzyme-linked immunoassay (droplet digital ExoELISA) approach enables absolute counting of EVs with cancer-specific biomarkers (**Figure 5**) [11].

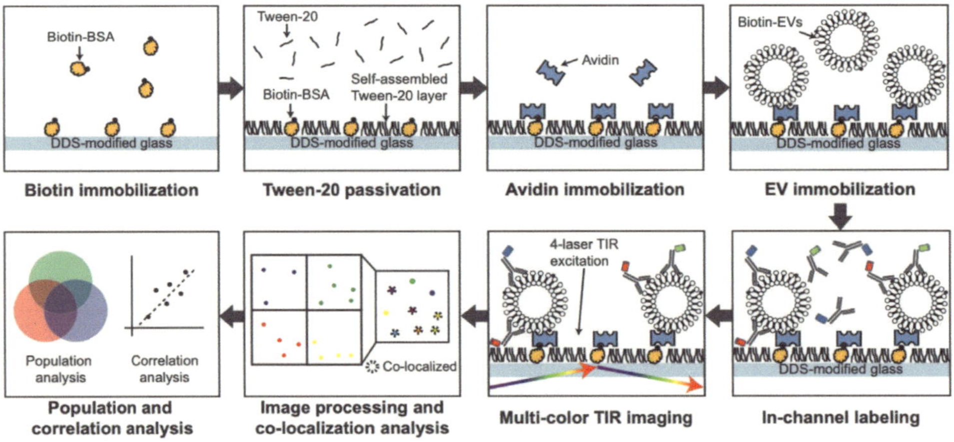

Figure 3.
TIRFM for single EV 3-plexed proteomic biomarker analysis. The colocalization of multiple biomarkers could be simultaneously recorded via multi-color TIR imaging.

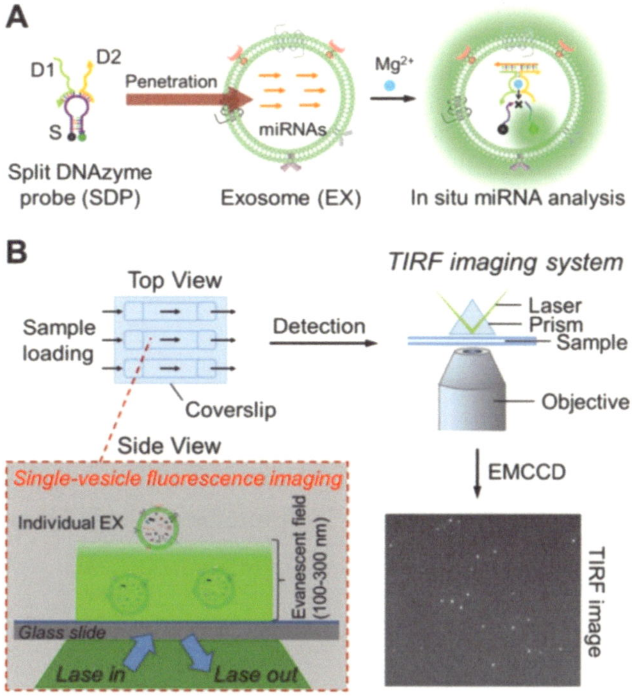

Figure 4.
TIRFM for single EV miRNA biomarker analysis.

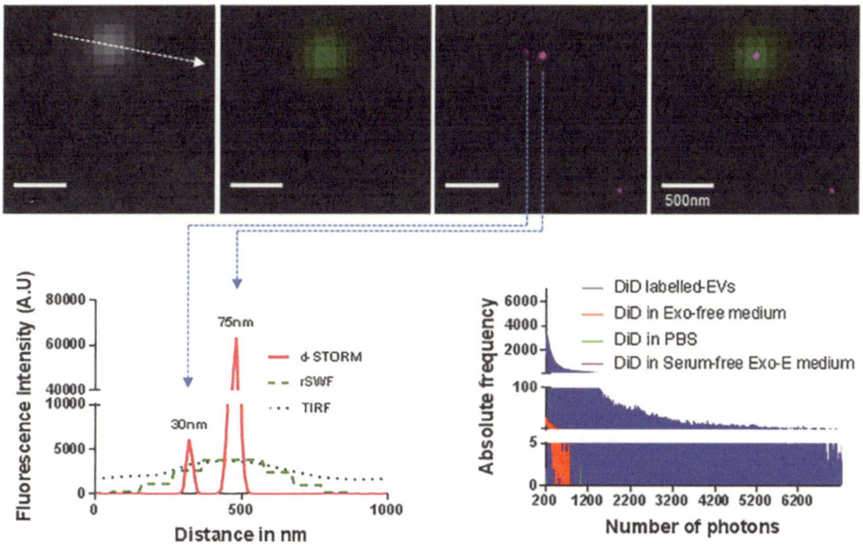

Figure 5.
Imaging of two adjacent exosomes with conventional TIRFM image, PALM/STORM image. Cross-sectional profiles of the two adjacent exosomes shown.

3. Flow cytometer (FCM)

Despite of the advanced technologies superior to the conventional microscopic systems, imaging methods are still not suitable for high-throughput or rapid detection. Therefore, scientists are committed to developing simple, sensitive and high-throughput exosome detection methods. Inspired by single cell technologies, flow cytometer attracted interests of scientists for multiplex large-scale single EV characterization. However, it is difficult to detect particles below 500 nm in size with the conventional FCM. Two approaches could be applied to overcome the problem. First, recognition of proteins on the single EVs was followed by a signal amplification step that could produce detectable structures through FCM. Second, the FCM were customized and optimized for detection in nanoscale.

3.1 Conventional FCM combined with amplified single EV signals

In the first approach, several methods are feasible for signal amplification to produce detectable structures via FCM. The proteins on single EVs could be detected

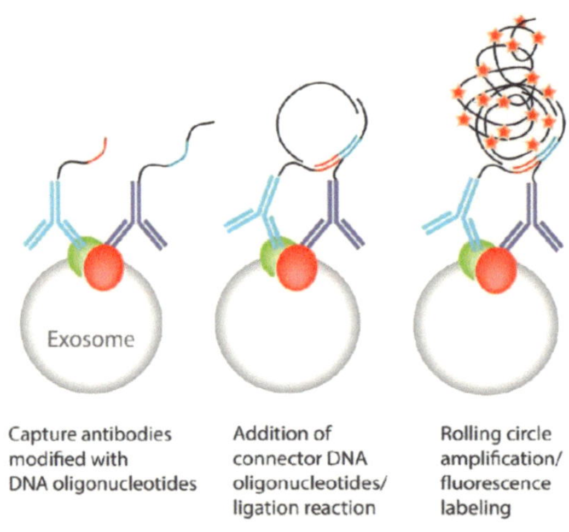

Figure 6.
Colocalization of two proteins on the same individual EV were detected via proximity ligation assay (PLA) and the signal was amplified via rolling circle amplification (RCA) to form a detectable structure under FCM.

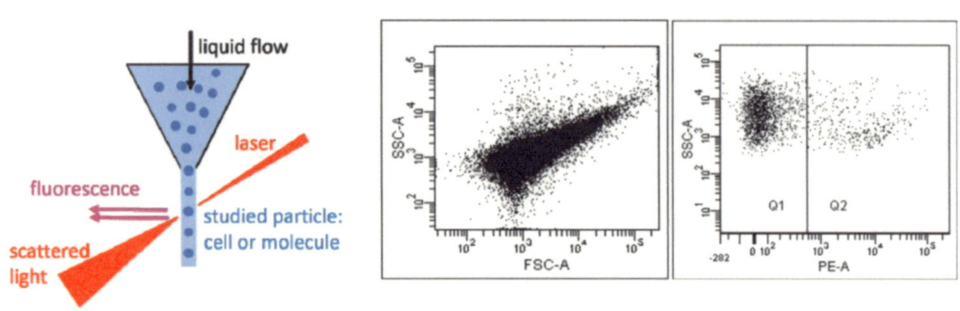

Figure 7.
The principle of flow cytometer (FCM) and an example of detected EV samples with scattered signals and fluorescently labeled Her-2 antigen.

with aptamers with a trigger sequence at the end, which then trigger hybridization chain reaction via DNA hairpins with biotin labels. Addition of avidin-fluorescent labels lead to the formation of an 500 nm fluorescently labeled EV-centered complex to be detect in FCM. Signal amplification via hybridization chain reaction occurred after single EV biomarker recognition by a conformation switchable aptamer. Two biomarkers could be targeted and analyzed simultaneously [12]. In another study, investigators utilized DNA oligonucleotides labeled antibodies to detect two proteins on the same EV. Proximity ligation assay occurs and rolling circle amplification (RCA) reaction produce thousands of copies of DNA sequences for fluorescent labels to bind and consequently forms a structure detectable in FCM (**Figure 6**) [13].

3.2 High resolution FCM and nano FCM

In the second approach, FCM were configured to overcome the limitations of detectable particle size and adapt to the analysis of nanoscale EVs. The improved sensitivity for EV detection includes higher laser power, slower flow and longer signal integration times and so on [14, 15]. Fluorescence based triggering could be superior over light scatter based triggering [16, 17]. Short wavelength laser can detect smaller particle size, for example 405 nm violet side scatter (VSSC) in Beckman Coulter CytoFLEX S [18]. As particle size decreases, the scattered light decreases rapidly. Data acquisition always needs to be optimized for better SSC resolution and efficacy of the detection of dim fluorescent single EVs. Therefore, novel noise reduction algorithm is applied for the high sensitivity detection of small particles. Calibrating light scatter detection for EV analysis have been proposed [19]. Nowadays, Apogee flow cytometry with micro-PLUS mode announced a detectable size of as low as 80 nm [20]. Yan's lab developed nano flow cytometer (nFCM) employing single molecule fluorescence detection in a sheathed flow. Two single-photon counting avalanche photodiodes (APDs) were used to detect side scatter and fluorescence of individual EVs, respectively, which enables phenotyping of single EVs as small as 40 nm (**Figure 7**) [21].

4. Nano-sensing technology

4.1 Nanoparticle tracking analysis (NTA)

Nanoparticle tracking analysis (NTA) is well acknowledged and commercially available technique for EV characterization. By tracking the Brownian motion of each particle, the hydrodynamic diameter of each particle is calculated with Stockes-Einstein equation. Thereafter, size distribution and concentration of EVs in a fluid sample could be obtained. Combined with fluorescent labeled antibodies or molecular beacons, EV subpopulation with expression of certain protein or miRNAs, respectively, could be quantified (**Figure 8**).

As a modified tracking analysis of EVs, an on-chip microcapillary electrophoresis system was built with a laser dark-field microscope. The tracking analysis of the electrophoretic migration of individual exosomes were performed and the zeta potential distribution of exosomes were able to be analyzed. The system consists of a chip, a pair of platinum electrodes, a DC power supply, a laser source, an inverted microscope, and an EMCCD camera (**Figure 9**) [22].

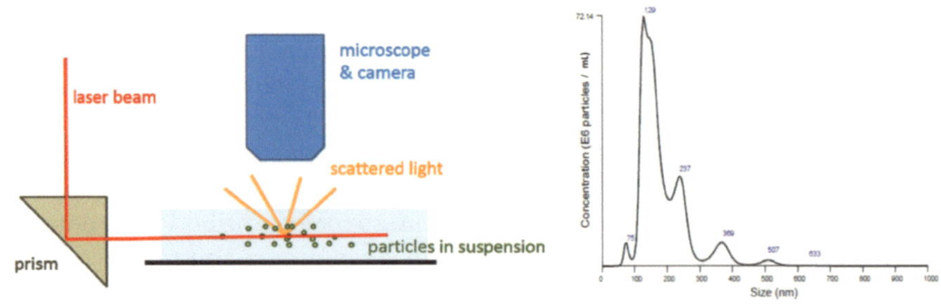

Figure 8.
The principle of nanoparticle tracking analysis (NTA) and its obtained EV size distribution.

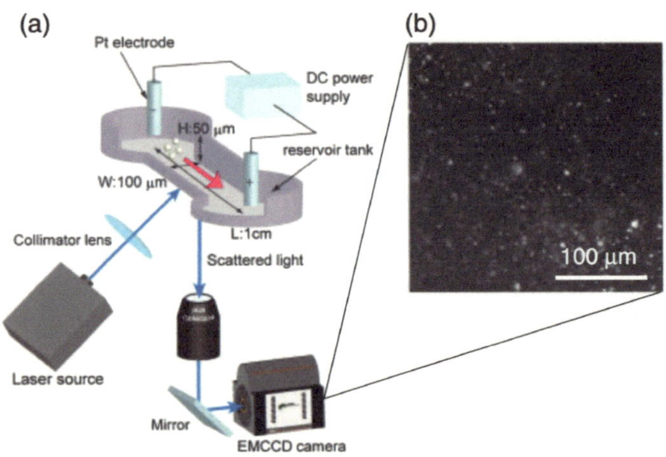

Figure 9.
Tracking analysis combined with zeta potential measurement of single EVs.

4.2 Fluorescence correlation spectroscopy (FCS)

Fluorescence Correlation Spectroscopy (FCS) were set up for characterize the GFP labeled vesicles at the single molecule – single vesicle level. FCS detects temporal fluorescence fluctuations in a defined confocal volume at a single photon sensitivity [23].

4.3 Surface plasmon resonance (SPR)

Surface plasmon resonance (SPR) detects real-time interactions between the receptors and the targeted biomolecules. SPR sensors show high sensitivity for label-free exosome quantification, which could be attributed to a nanoscale sensing range closing to exosome size. The receptors were immobilized on the surface of the biosensor. When a solution of biomolecules flows across, the targeted biomolecules interact with the receptors to change the refractive index of the sensor surface. The response of the sensor is measured in resonance units (RU) and is proportional to the mass of molecules interacting with the immobilized receptor. (**Figure 10**) Nano-plasmonic exosome (nPLEX) system was developed based on SPR principles and utilized 200 nm plasmonic nanoholes for EV capture and detection. nPLEX enables quantitative detection and proteomic profiling of EVs at the throughput of 12 biomarkers

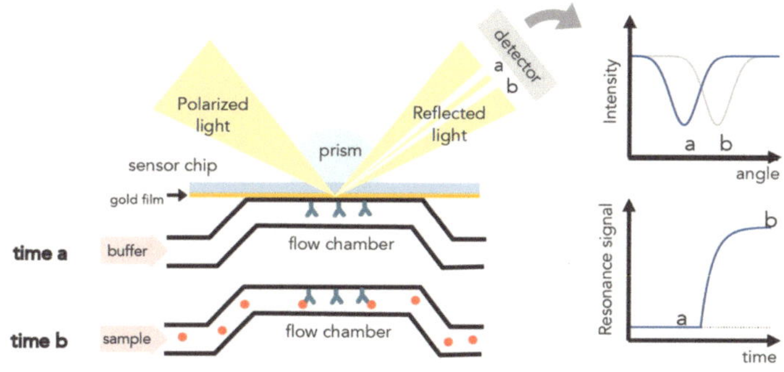

Figure 10.
The priciple of surface plasmon resonance (SPR).

detection with antibodies immobilized at 12 channels of a microfluidic chip [24]. The nPLEX-FL assay enables multiplexed single EV analysis of targeted markers with improved sensitivities. In nPLEX-FL assay, EVs are biotinylated and then captured on the nanohole surface with avidin coating. EVs are immunostained by fluorescent labeled antibodies. Plasmon-enhanced fluorescence detection can amplify fluorescence signals using plasmonic metallic nanostructures. The labeled EVs are imaged, and their fluorescent intensities are analyzed. Therefore, biomarker distribution analysis could be performed on a single-EV level [25]. A localized surface plasmon resonance imaging (LSPRi) platform improves the limit of detection down to the single exosome limit. With a 400 nanopillar array sensor chip, single EVs are captured by nanopillar for both imaging and spectrometer measurement (**Figure 11**) [26].

4.4 Raman spectroscopy

Raman spectroscopy is a spectral analysis method that determines the chemical properties of samples by measuring the vibration mode of covalent bonds caused by a radiation laser. Raman spectroscopy is non-destructive label-free analysis and therefore is suitable for analysis of biological samples. Laser tweezers Raman spectroscopy (LTRS)explore the chemical content of individual EVs, in which a tightly focused laser beam traps small particles at the laser's focal point. A confocal detection of Raman scattering from the precise focal volume allows EVs to be studied individually. Based on the exosomal chemical differences, EVs could be classified into subpopulations (**Figure 12**) [27–29].

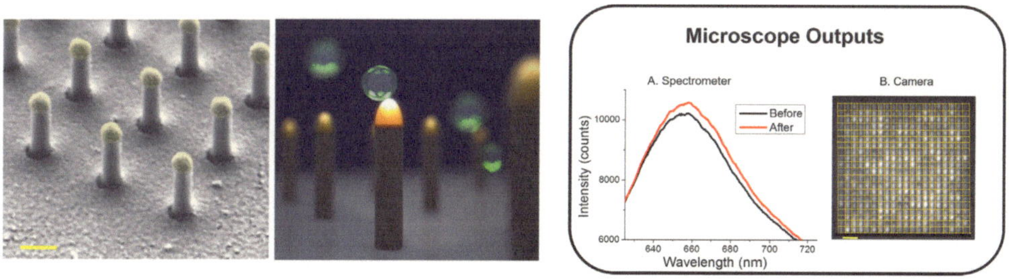

Figure 11.
A localized surface plasmon resonance imaging (LSPRi) nanopillar platform enables both spectrum and image readout.

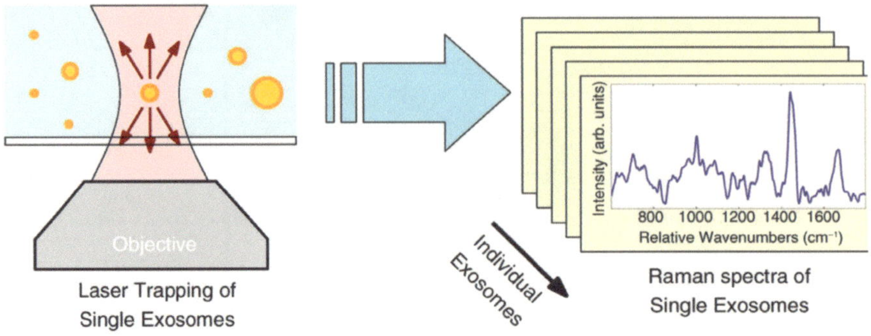

Figure 12.
The principle of laser tweezers Raman spectroscopy (LTRS).

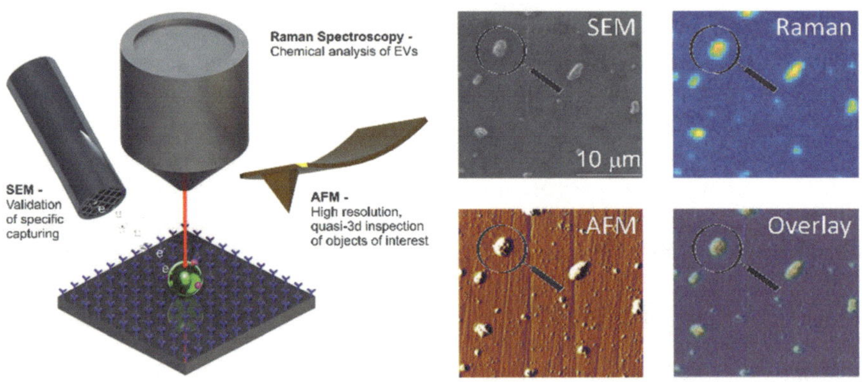

Figure 13.
Multi modal characterization of antibody captured individual EVs via SEM, AFM and Raman spectroscopy.

Surface-enhanced Raman spectroscopy (SERS) is a type of molecular vibration spectrum with enhanced Raman scattering signals from molecules adsorbed on noble metal (Ag, Au) nanostructures. SERS has been applied for the detection of cancer exosomes due to its high sensitivity, specificity, and multiplexing capability. Single EV analysis was performed in several studies. An graphene-coated periodic gold-pyramid were used as SERS substrate to detect single EVs and EV samples from different biological sources were distinguished in their Raman signature [30]. In another study, gold nanoparticles (AuNP) coated single EVs were formed due to electrostatic interaction between cationic AuNP and anionic EV membrane. SERS spectra of individual ELVs with a high density of AuNP were then acquired and variations in the SERS spectra of individual ELVs could be used for EV heterogeneity analysis [31]. Multi modal characterization of individual EVs could be achieved after EV capture on antibody functionalized stainless steel substrate followed by characterization via SEM, AFM and Raman spectroscopy (**Figure 13**) [32].

5. Barcoding based single EV analysis

In the methods described above, the number of simultaneously analyzed biomarkers is limited to types of distinguishable readout signals like fluorescent dyes etc.

To achieve multiplexed high-throughput single EV analysis, researchers unitized molecular tools named Proximity Barcoding Assay (PBA) to detect colocalized proteins on the same EV [33]. The principle is that only proteins on the same individual EV are in proximity and therefore labeled with the same EV tags in PBA.

PBA probes were prepared by labelling of selected antibodies with DNA oligonucleotides containing unique 8-nt protein tag (**Figure 14a**). Rolling circle amplification

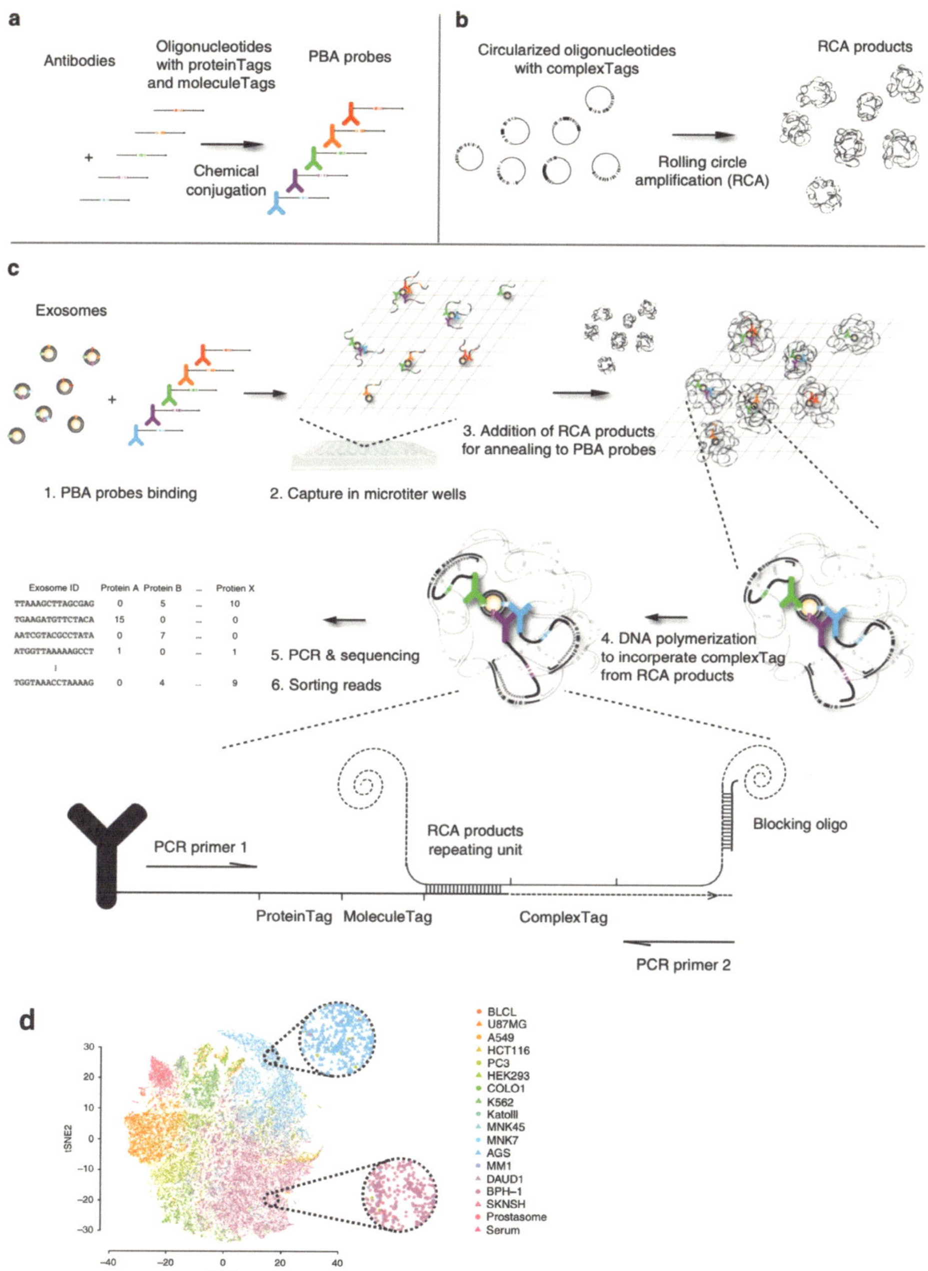

Figure 14.
Schematic illustration of proximity barcoding assay (PBA) and 38-plex analysis of 18 EV samples.

	Analytical methods	capabilities	Throughput	Advantages	Disadvantages	Sample requirement
Imaging	AFM	Dimensional and mechanical properties	1–100 EVs per image	High resolution Unique mechanical property	Device dependent No composition information	Purified EV sample
	SEM	Morphology and size	1–100 EVs per image	High resolution	Device dependent No composition information Collapse of EV during imaging	Purified EV sample
	TEM	Morphology and size	1–100 EVs per image	High resolution	Device dependent No composition information	Purified EV sample 300 μl
	TIRFM	Fluorescently labeled EVs Protein on single EVs	1–100 EVs per image	Biomarker analysis	Resolution of about 100 nm Limited number of biomarkers	Purified EV sample 300 μl
	STORM	Fluorescently labeled EVs Protein on single EVs	1–100 EVs per image	High resolution of about 20 nm Biomarker analysis	Limited number of biomarkers	Purified EV sample
Flow cytometer	Signal amplification for conventional FCM	Proteins on single EVs	10^3–10^4 EVs per minute	Biomarker analysis	Device needed Dependent on antibody affinity Limited number of biomarkers	Purified EV sample, 500 μl 10^8–10^{10} particle/ml
	Nano FCM or high resolution FCM	Biomarkers on single EVs	10^3–10^4 EVs per minute	Biomarker analysis Direct analysis of EVs	Not for small EVs Limited by antibody affinity Limited number of biomarkers	Purified EV sample, 500 μl 10^8–10^{10} particle/ml

	Analytical methods	capabilities	Throughput	Advantages	Disadvantages	Sample requirement
Nano-sensing	NTA	Size distribution Concentration of EV samples	10^6 EVs	Precise size distribution and quantification	Limited possibility in detection of biomarker	Purified EV sample 100 µl 10^8–10^{10} particle/ml
	SPR	Amount of EVs interacting with specific receptor	NA	Label-free Real time High sensitivity	Device dependent, Limited possibility in detection of biomarkers	Purified EV sample
	Raman spectroscopy	Chemical composition	10–100 EVs per analysis	Label-free, real time,	Device dependent, Limited possibility in detection of biomarkers	Purified EV sample
Barcoding	Proximity Barcoding Assay	Proteomic profiling (38-plex protein count for each EV)	>10^6 EVs per test	Highly multiplexed High- throughput Small sample volume No EV enrichment needed	Dependent on antibody affinity	Body fluid (cell-free) or purified EVs 2–20 µl

Table 1.
Comparison of single EV analysis methods.

(RCA) products were prepared from circularized oligonucleotides comprising a 15-nt random sequence. Because15-nt random sequence are capable of encoding 4^{15} (about one billion) unique sequences to work as EV tags, RCA products are utilized as templates to barcode antibody conjugated oligonucleotides on single EVs (**Figure 14b**). EVs were incubated with PBA probes and then captured via immobilized cholera toxin subunit B (CTB). Oligonucleotides on PBA probes brought together by binding the same EV are allowed to hybridize to a unique RCA product and therefore obtain the same EV tag via enzymatic extension. Successfully extended DNA molecules on PBA probes are amplified by PCR. The PCR product were subjected to DNA sequencing to record the combinations of EV tag - protein tag thus revealing the proteomic profiles of individual EVs.

In the study of 38-plex PBA analysis, individual EVs of 16 types of cell lines, prostasome and serum could be profiled. The information researchers could obtain includes protein expression and the pattern of protein combinations on individual EVs. The differences of samples were visualized by t-SNE according to their protein compositions.

6. Conclusion

Based on the technology, we discussed the general strategies of single EVs analysis including imaging, flow cytometer, nano-sensing and single EV barcoding. EVs could be classified into subsets according to their chemical composition, size, expression of specific biomarkers etc. The advancing of technologies to reveal EV heterogeneity are the pursuit of scientists from multiple disciplines.

Each EV has a unique story, the start and the end of their journey, the cargo they carried and the mission they accomplished. Revealing the heterogeneity of EVs and getting closer to the detailed facts of the abundant populations of EV will endow us the knowledge of how cells in our bodies talks to each other and what they say about our health (**Table 1**).

Acknowledgements

This work was supported by the National Natural Science Foundation of China under Grant 81802052; China Postdoctoral Science Foundation under Grant 2018 M643318; the Guangdong Key Laboratory funds of Systems Biology and Synthetic Biology for Urogenital Tumors under Grant 2017B030301015.

Conflict of interest

D.W. has filed a patent application (PCT/SE2014/051133) describing the PBA technique. D.W. is a shareholder of Vesicode and Secretech.

Author details

Yanling Cai[1,2] and Di Wu[3,4*]

1 Guangdong Provincial Key Laboratory of Systems Biology and Synthetic Biology for Urogenital Tumors, Department of Urology, The First Affiliated Hospital of Shenzhen University, Shenzhen Second People's Hospital (Shenzhen Institute of Translational Medicine), Shenzhen, China

2 Shenzhen Key Laboratory of Genitourinary Tumor, Department of Urology, The First Affiliated Hospital of Shenzhen University, Shenzhen Second People's Hospital (Shenzhen Institute of Translational Medicine), Shenzhen, China

3 Shenzhen SecreTech Co., Ltd., Shenzhen, China

4 Vesicode AB, Solna, Sweden

*Address all correspondence to: di.wu@vesicode.com

References

[1] Y. Wu, W. Deng, and D. J. Klinke, "Exosomes: Improved methods to characterize their morphology, RNA content, and surface protein biomarkers," Analyst, vol. 140, no. 19, pp. 6631-6642, 2015, doi: 10.1039/c5an00688k.

[2] R. Szatanek, M. Baj-Krzyworzeka, J. Zimoch, M. Lekka, M. Siedlar, and J. Baran, "The methods of choice for extracellular vesicles (EVs) characterization," Int. J. Mol. Sci., vol. 18, no. 6, 2017, doi: 10.3390/ijms18061153.

[3] L. Pascucci et al., "Paclitaxel is incorporated by mesenchymal stromal cells and released in exosomes that inhibit in vitro tumor growth: A new approach for drug delivery," J. Control. Release, vol. 192, pp. 262-270, 2014, doi: 10.1016/j.jconrel.2014.07.042.

[4] M. LeClaire, J. Gimzewski, and S. Sharma, "A review of the biomechanical properties of single extracellular vesicles," Nano Sel., vol. 2, no. 1, pp. 1-15, 2021, doi: 10.1002/nano.202000129.

[5] B. Whitehead et al., "Tumour exosomes display differential mechanical and complement activation properties dependent on malignant state: Implications in endothelial leakiness," J. Extracell. Vesicles, vol. 4, no. 1, 2015, doi: 10.3402/jev.v4.29685.

[6] K. Lee et al., "Multiplexed profiling of single extracellular vesicles," ACS Nano, vol. 12, no. 1, pp. 494-503, 2018, doi: 10.1021/acsnano.7b07060.

[7] C. Han et al., "Single-vesicle imaging and co-localization analysis for tetraspanin profiling of individual extracellular vesicles," J. Extracell. Vesicles, vol. 10, no. 3, 2021, doi: 10.1002/jev2.12047.

[8] D. He et al., "Total internal reflection-based single-vesicle in situ quantitative and stoichiometric analysis of tumor-derived exosomal microRNAs for diagnosis and treatment monitoring," Theranostics, vol. 9, no. 15, pp. 4494-4507, 2019, doi: 10.7150/thno.33683.

[9] C. Chen et al., "Imaging and intracellular tracking of cancer-derived exosomes using single-molecule localization-based super-resolution microscope," ACS Appl. Mater. Interfaces, vol. 8, no. 39, pp. 25825-25833, 2016, doi: 10.1021/acsami.6b09442.

[10] Z. Nizamudeen et al., "Rapid and accurate analysis of stem cell-derived extracellular vesicles with super resolution microscopy and live imaging," Biochim. Biophys. Acta - Mol. Cell Res., vol. 1865, no. 12, pp. 1891-1900, 2018, doi: 10.1016/j.bbamcr.2018.09.008.

[11] C. Liu et al., "Single-exosome-counting immunoassays for cancer diagnostics," Nano Lett., vol. 18, no. 7, pp. 4226-4232, 2018, doi: 10.1021/acs.nanolett.8b01184.

[12] W.Shen et al., "A single extracellular vesicle (EV) flow cytometry approach to reveal EV heterogeneity," Angewandte Chemie International Edition. doi: 10.1002/anie.201806901.

[13] L. Löf et al., "Detecting individual extracellular vesicles using a multicolor in situ proximity ligation assay with flow cytometric readout," Sci. Rep., vol. 6, no. September, pp. 1-9, 2016, doi: 10.1038/srep34358.

[14] E. J. van der Vlist, E. N. M. Nolte-'t Hoen, W. Stoorvogel, G. J. A. Arkesteijn, and M. H. M. Wauben, "Fluorescent labeling of nano-sized vesicles released

by cells and subsequent quantitative and qualitative analysis by high-resolution flow cytometry.," Nat. Protoc., vol. 7, no. 7, pp. 1311-1326, 2012, doi: 10.1038/nprot.2012.065.

[15] S. A. Stoner et al., "High sensitivity flow cytometry of membrane vesicles," Cytom. Part A, vol. 89, no. 2, pp. 196-206, 2016, doi: 10.1002/cyto.a.22787.

[16] T. G. Kormelink, G. J. A. Arkesteijn, F. A. Nauwelaers, G. van den Engh, E. N. M. Nolte-'t Hoen, and M. H. M. Wauben, "Prerequisites for the analysis and sorting of extracellular vesicle subpopulations by high-resolution flow cytometry," Cytom. Part A, vol. 89, no. 2, pp. 135-147, 2016, doi: 10.1002/cyto.a.22644.

[17] N. Arraud, C. Gounou, R. Linares, and A. R. Brisson, "A simple flow cytometry method improves the detection of phosphatidylserine-exposing extracellular vesicles," J. Thromb. Haemost., vol. 13, no. 2, pp. 237-247, 2015, doi: 10.1111/jth.12767.

[18] N. Salmond, K. Khanna, G. R. Owen, and K. C. Williams, "Nanoscale flow cytometry for immunophenotyping and quantitating extracellular vesicles in blood plasma," Nanoscale, vol. 13, no. 3, pp. 2012-2025, 2021, doi: 10.1039/d0nr05525e.

[19] J. A. Welsh et al., "MIFlowCyt-EV: a framework for standardized reporting of extracellular vesicle flow cytometry experiments," J. Extracell. Vesicles, vol. 9, no. 1, 2020, doi: 10.1080/20013078.2020.1713526.

[20] G. Kibria et al., "A rapid, automated surface protein profiling of single circulating exosomes in human blood," Sci. Rep., vol. 6, no. November, pp. 1-9, 2016, doi: 10.1038/srep36502.

[21] Y. Tian et al., "Quality and efficiency assessment of six extracellular vesicle isolation methods by nano-flow cytometry," J. Extracell. Vesicles, vol. 9, no. 1, 2020, doi: 10.1080/20013078.2019.1697028.

[22] K. Kato et al., "Electrokinetic evaluation of individual exosomes by on-chip microcapillary electrophoresis with laser dark-field microscopy," Jpn. J. Appl. Phys, vol. 52, no. 6 PART 2, 2013, doi: 10.7567/JJAP.52.06GK10.

[23] G. Corso et al., "Systematic characterization of extracellular vesicles sorting domains and quantification at the single molecule–single vesicle level by fluorescence correlation spectroscopy and single particle imaging," J. Extracell. Vesicles, vol. 8, no. 1, 2019, doi: 10.1080/20013078.2019.1663043.

[24] H. Im et al., "Label-free detection and molecular profiling of exosomes with a nano-plasmonic sensor," Nat. Biotechnol., vol. 32, no. 5, pp. 490-495, 2014, doi: 10.1038/nbt.2886.

[25] J. Min et al., "Plasmon-enhanced biosensing for multiplexed profiling of extracellular vesicles," Adv. Biosyst., vol. 4, no. 12, pp. 1-8, 2020, doi: 10.1002/adbi.202000003.

[26] D. Raghu et al., "Nanoplasmonic pillars engineered for single exosome detection," PLoS One, vol. 13, no. 8, pp. 1-13, 2018, doi: 10.1371/journal.pone.0202773.

[27] Z. J. Smith et al., "Single exosome study reveals subpopulations distributed among cell lines with variability related to membrane content," J. Extracell. Vesicles, vol. 4, no. 1, 2015, doi: 10.3402/jev.v4.28533.

[28] K. Ajito and K. Torimitsu, "Single nanoparticle trapping using a Raman

tweezers microscope," Appl. Spectrosc., vol. 56, no. 4, pp. 541-544, 2002, doi: 10.1366/0003702021955015.

[29] K. Ajito and K. Torimitsu, "Laser trapping and Raman spectroscopy of single cellular organelles in the nanometer range," Lab Chip, vol. 2, no. 1, pp. 11-14, 2002, doi: 10.1039/b108744b.

[30] Z. Yan et al., "A label-free platform for identification of exosomes from different sources," ACS Sensors, vol. 4, no. 2, pp. 488-497, 2019, doi: 10.1021/acssensors.8b01564.

[31] S. Stremersch et al., "Identification of individual exosome-like vesicles by surface enhanced raman spectroscopy," Small, vol. 12, no. 24, pp. 3292-3301, 2016, doi: 10.1002/smll.201600393.

[32] P. Beekman et al., "Immuno-capture of extracellular vesicles for individual multi-modal characterization using AFM, SEM and Raman spectroscopy," Lab Chip, vol. 19, no. 15, pp. 2526-2536, 2019, doi: 10.1039/c9lc00081j.

[33] D. Wu et al., "Profiling surface proteins on individual exosomes using a proximity barcoding assay ", Nat Commun 10, 3854 (2019). doi.org/10.1038/s41467-019-11486-1

Section 3

Role of Extracellular Vesicles in Human Disease

Chapter 6

Extracellular Vesicles as Intercellular Communication Vehicles in Regenerative Medicine

Gaspar Bogdan Severus, Ionescu Ruxandra Florentina, Enache Robert Mihai, Dobrică Elena Codruța, Crețoiu Sanda Maria, Crețoiu Dragoș and Voinea Silviu Cristian

Abstract

Extracellular vesicles (EVs) represent cell-specific carriers of bioactive cargos that can be of importance in either physiological or pathological processes. Frequently, EVs are seen as intercellular communication vehicles, but it has become more and more evident that their usefulness can vary from circulating biomarkers for an early disease diagnosis to future therapeutic carriers for slowing down the evolution of different afflictions and their ability to restore damaged tissue/organs. Here, we summarize the latest progress of EVs classification, biogenesis, and characteristics. We also briefly discuss their therapeutic potential, with emphasis on their potential application in regenerative medicine.

Keywords: extracellular vesicles, exosomes, microvesicles, intercellular communication, stem cells, regenerative medicine

1. Introduction

Extracellular vesicles (EVs) are cell-derived membranous structures released by a multitude of cell types into the extracellular environment, from where they can enter body fluids and reach distant tissues, releasing their content [1]. Considered an essential pathway for intercellular communication, EVs are non-traditional lipid membrane-enclosed structures, with nanometric sizes [2]. Many studies have shown that EVs are produced by both prokaryotes and eukaryotes, indicating a persistent evolution of their signaling mechanism during a time, giving EVs an increasingly important role in the future [3, 4]. In general, EVs from the human blood are derived from platelets, but they can also be released from leukocytes, erythrocytes, endothelial cells, smooth muscle cells, and even cancer cells [5, 6].

Internal (platelet activation, pH variations, hypoxia, etc.) and external (irradiation, injury, etc.) factors can stimulate cells to produce EVs, that are secreted in lacrimal fluid, breast milk, broncho-alveolar lavage fluid, blood, ascites, urine, faces, etc. [3, 7, 8].

The content of EVs can vary to a great extent (lipids, proteins, nucleic acid species) and depends on the cell of origin [4, 6].

Their main function is represented by intercellular communication [2]. EVs can influence a variety of biological processes, transferring functional molecules (mRNA, microRNAs, and proteins) between cells [6, 9]. Their content is shuttled between cells, making EVs essential for a multitude of physiological, but also pathological processes (**Figure 1**). The various substances contained in the EVs can be taken up by other cells, both from the proximity of the cells of origin, but also from distant locations where they are transported by biofluids, inducing various phenotypic responses [10]. Apparently, this uptake is pH-dependent and can be of significance, especially in the tumor microenvironment [7].

EVs can also be considered as a possible source of biomarkers for early disease diagnosis [6, 11]. The implication of EVs in several diseases, including cancer, infectious diseases, neurodegenerative diseases, and blood diseases amplified the research interest, aiming to discover new possible therapeutic targets. EVs content can provide important leads about the type and stage of cancer, while during oncological treatment, the composition of EVs can change, which can be beneficial for therapeutic evaluation [5, 9, 12–14].

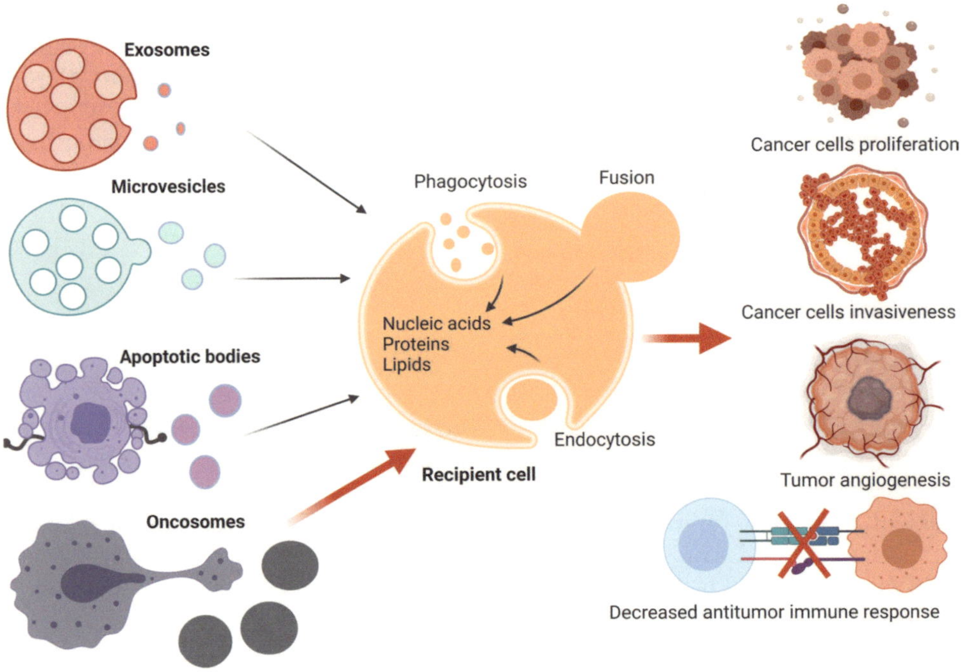

Figure 1.
EVs, produced by different cell types, can be taken up by a recipient cell via phagocytosis, endocytosis, or membrane fusion. Thus, they can determine some biological effects. Oncosomes, a particular type of EVs produced by cancer cells, can stimulate the proliferation and invasiveness of cancer cells and tumor angiogenesis. They can also decrease antitumor immune response. Created with BioRender.com (last accessed on October 26, 2021).

EVs have various physiological and pathological roles. Current evidence points out their involvement in embryonic development, regenerative medicine (tissue regeneration), immunity modulation, angiogenesis, stress response, senescence, cell proliferation and differentiation, the capture of dissipated cancerous cells [4, 15–17].

Moreover, EVs can be regarded as therapeutic solutions and can act like possible alternatives to stem cell (SC) therapy [4].

Their role was and will continue to be exploited until reaching its maximum. Nowadays, EVs are also regarded as potential drug delivery and gene transport devices [18, 19].

Shortly, EVs are common vehicles between different cell types. Nowadays, their importance has attracted considerable scientific attraction due to their involvement in disease pathogenesis, different therapies, and also in many translational pathways. Extracellular vesicles are certainly a breakthrough in the regenerative medicine field, their involvement in many processes such as apoptosis, cell proliferation, differentiation, migration, angiogenesis, oxidative stress, aging, and inflammation being recently described. Lately, extracellular vesicles were also pointed out as important vehicles for multiple therapies due to their multifaceted roles.

The current chapter will summarize the most up-to-date knowledge about the role of EVs in regenerative medicine and will discuss the effects that EVs may have on tissue regeneration, a phenomenon that was initially focused only on cell therapies or tissue engineering. It will also approach the EVs' significance and crucial role in mediating cell-to-cell communication, especially their relationship with SCs and their biodisponibility in damaged tissue.

2. Extracellular vesicles: definition and main characteristics

Over time, the definition and role of EVs have been strongly questioned. Unanimously, considered as ranging from 20 to 200 nm to 10 μm in diameter, EVs can be differentiated into three major classes: exosomes, microvesicles (MVs), or ectosomes and apoptotic bodies. However, there are a limited number of studies on apoptotic bodies, so frequently the term EVs refers to exosomes and microvesicles [17]. Moreover, recent research underlined the possibility of subdividing EVs, for example, mitochondrial protein-enriched EVs or other categories of exosomes, based on their proteins and RNA profile (such as large or small exosome vesicles) [20–22].

EVs are regularly classified based on biogenesis, release pathway, size, content, and function [1, 6, 23]:

1. *Exosomes* are produced and secreted by all cell types and have a characteristic diameter between 30 and 150 nm [24]. The biogenesis and release pathway begin with early endosomes, deriving from inwardly budding of the plasma membrane of the cell. The same process will be applied then to the limiting membrane of the early endosomes, representing the second phase. The maturation of the early endosomes will lead to multivesicular bodies (MVBs) formation [23]. Both early endosomes and MVBs are participating in performing certain functions related to cellular material (especially proteins), like endocytic and trafficking functions [25]. Finally, MVBs present two possible routes of evolution: one refers to degradation of MVBs by lysosomes, including its components, and the second one to attaching MVBs to the plasma membrane of the cells and releasing its constituents, exosomes, in the extracellular space [6, 26, 27]. Even though the specific

factors that regulate these mechanisms are not well known, the most underlined pathway researched by studies is the endosomal sorting complexes required for transport (ESCRT) [28]. Based on the primary mechanism involved in the biogenesis of exosomes, ESCRT proteins, it is obvious that exosomes contain these proteins [29]. However, another mechanism involved independently is based on the sphingomyelinase enzyme, which was studied by researchers because cells without ESCRT mechanism can still produce CD63 positive exosomes, a protein from the tetraspanin family [30]. Exosomes also contain glycoproteins, low levels of proteins associated with endoplasmic reticulum and Golgi apparatus, cholesterol, ceramide, noncoding RNA, mRNA, miRNA, and cytosol [6, 24, 31].

Some well-known functions of exosomes are the facilitation of communication between cells, cell preservation, association with cancer evolution, stimulation of immune response, involvement in the functions of the nervous system (myelination, growth, and survival of the nerve cells, but also the progression of neurological diseases by containing pathogenic proteins, as a beta-amyloid peptide, superoxide dismutase and α-synuclein [24, 32–35]. Because of their constituents, exosomes are becoming more and more attractive for researchers to discover new implications in diseases and potentially new therapeutic methods [24]. For example, as already mentioned, exosomes contain α-synuclein, which is involved in Parkinson's disease [36]. New studies are concentrating on the association with glioblastoma, acute kidney disease, pancreatic or lung cancer, vaccines or other immunological uses, and diminishing tissue injury [37–41].

2. *Microvesicles* are a type of EVs measuring between 100 nm and 1 μm [1]. Their biogenesis and release pathway are still not well known. However, MVs are produced by outward budding of the plasma membrane of the cells, involving cytoskeleton elements (actin and microtubules and other cytoskeletal proteins like ARF6 and RhoA), molecular motors (kinesins and myosins) and fusions machinery (ESCRT, SNAREs, and tethering factors) [1, 42, 43]. The content of the MVs, largely determined by their biogenesis, is represented by proteins associated with cytosol and plasma membrane (especially tetraspanins), cytoskeletal proteins, integrins, glycosylated and phosphorylated proteins, and heat shock proteins [24, 44, 45]. In addition, MVs contain cholesterol, mRNA, miRNA, and cytosol [6]. Other specific markers helping in differentiation between MVs and exosomes need to be further studied [24]. Like exosomes, MVs participate in communication between cells, a particular characteristic being their ability to deliver proteins, lipids, or nucleic acids to another cell [1, 23]. Primarily, this function facilitates communication between healthy cells, but on the other hand, it can be a way to spread cancerous cells in the body, leading to metastasis [46]. That's why future studies must focus on this individuality of MVs, to develop potentially new therapeutic methods in cancer. Other possible purposes of MVs use in the future are, as already noted, the same as with exosomes [24].

A particular type of MVs is represented by *oncosomes*, which are secreted by the shedding of plasma membrane blebs of cancer cells [2, 47]. Even if their main characteristics are still not well known, some experimental studies on glioblastoma and prostate cancer have shown that their biogenesis is linked to serine/threonine kinase 1 (AKT1) and epidermal growth factor receptor (EGFR) pathways [48, 49]. Their size depends on the stage of cancer, reaching up to 1000 nm in the final stages, thus being

	Exosomes	Microvesicles	Oncosomes	Apoptotic bodies
Size	30–150 nm [24];	100 nm–1 μm [1];	100–1000 nm [6];	50–5000 nm [24];
Biogenesis	I. Early endosomes [23]; II. Maturation of early endosomes [23]; III. MVBs formation [23];	Direct outward budding of the plasma membrane of the cells [1];	Shedding of plasma membrane blebs of cancer cells serine/threonine kinase 1 (AKT1) and epidermal growth factor receptor (EGFR) pathways [47–49];	I. Cell contraction [24, 50]; II. Increased hydrostatic pressure [24, 50]; III. Separation between cytoskeleton and plasma membrane of the cell [24, 50];
Release pathway	MVBs attach to the plasma membrane of the cells and release their constituents [6, 26, 27];			Released into extracellular space by apoptotic cells [24, 50];
Content	ESCRT proteins, tetraspanin family proteins, glycoproteins, low levels of proteins associated with endoplasmic reticulum and Golgi apparatus, cholesterol, ceramide, noncoding RNA, mRNA, miRNA, and cytosol [6, 24, 29–31];	Proteins associated with cytosolic and plasma membrane (especially tetraspanins) [24, 44]; cytoskeletal proteins, integrins, glycosylated and phosphorylated proteins, and heat shock proteins, cholesterol, mRNA, miRNA, and cytosol [6, 45];	Oncogenic proteins, miRNA, and enzymes for amino acid, glucose or glutamine metabolism [6, 47];	Chromatin, low levels of glycosylated proteins and intact organelles, including proteins associated with mitochondria, endoplasmic reticulum, Golgi apparatus, and nucleus [24, 51];
Function	Intercellular communication, cell preservation, association with cancer evolution, stimulation of immune response, involvement in the functions of the nervous system (myelination, growth, and survival of the nerve cells, but also the progression of neurological diseases) [24, 32–35];	The same as exosomes, with a particular association with metastatic disease [46];	Cancer evolution and metastasis [6];	Insufficiently known [24];

Table 1.
Classification of EVs and their main characteristics.

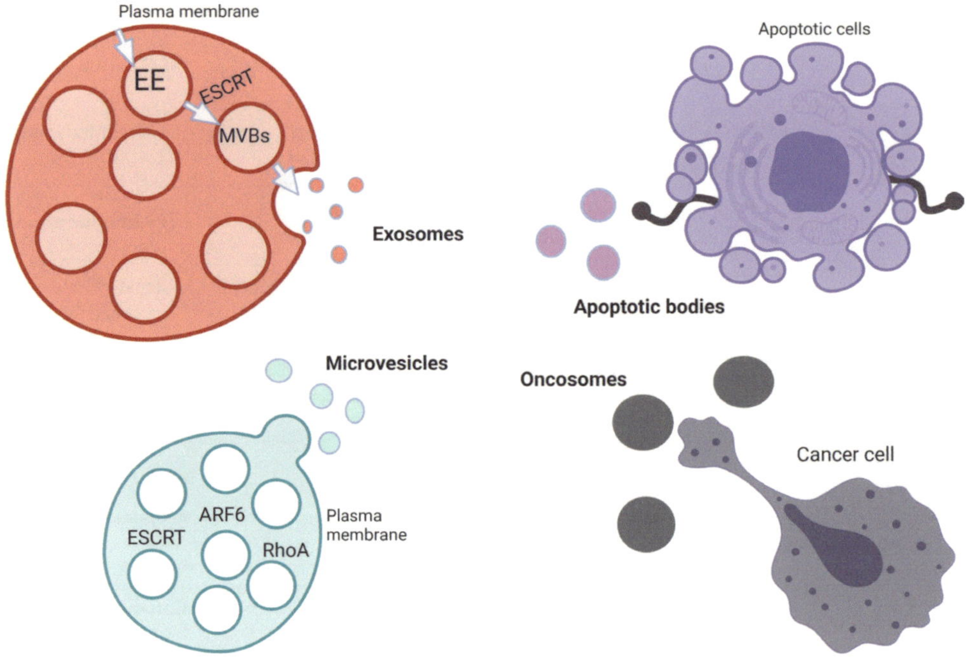

Figure 2.
The biogenesis of EVs. (a) Exosomes are produced from early endosomes (EE) and endosomal pathway (ESCRT) and released into extracellular space by fusion of multivesicular bodies (MVBs) with the plasma membrane; (b) microvesicles are produced by direct outward budding of the plasma membrane, with the involvement of cytoskeleton elements (like ARF6 and RhoA) or ESCRT; (c) apoptotic bodies are released into extracellular space by apoptotic cells; (d) oncosomes are secreted by shedding of plasma membrane blebs of cancer cells. Created with BioRender.com (last accessed on September 21, 2021).

the largest EVs. The content of oncosomes is represented by elements involved in the evolution of cancer and metastasis, like oncogenic proteins, miRNA, and enzymes for amino acid, glucose, or glutamine metabolism [6, 47].

3. *Apoptotic bodies* are a particular type of EVs measuring between 50 nm and 5000 nm. There are few studies related to apoptotic bodies and thus their characteristics are not well known. Their biogenesis is related to the separation between cytoskeleton and plasma membrane of the cell, because of cell contraction and consequently increased hydrostatic pressure, afterward being released into extracellular space by apoptotic cells [6, 50].

The composition of apoptotic bodies consists of chromatin, low levels of glycosylated proteins, and intact organelles, including proteins associated with the mitochondria, endoplasmic reticulum, Golgi apparatus, and nucleus [6, 51].

The biogenesis and main characteristics of the EVs are summarized in **Table 1** and **Figure 2**.

3. Intercellular communication through EVs

EVs can carry a big amount of information within/on their surface to another cell, influencing physiological and pathological pathways [6]. For a better understanding, in this chapter, some of these processes will be exemplified to illustrate the roles of EVs in intercellular communication.

3.1 Implantation and embryonic development

The implantation process refers to the development of the trophoblasts by the embryo, which then will adhere and invade the uterine wall. This is a crucial step in embryonic development, and any inaccuracy can have severe consequences [2]. EVs are secreted by both maternal and embryonic cells. In the first case, studies have shown that endometrial epithelial cells produce EVs that stimulate the activation of focal adhesion kinase (FAK), increasing the adhesion of trophoblasts to the uterine wall [52]. Regarding embryonic production of EVs, recent studies have shown the involvement of MVs. Laminin and fibronectin, two extracellular matrix proteins, found on the surface of MVs, are playing an important role in this case. MVs are transported to trophoblasts, where laminin and fibronectin activate integrins on the surface of the trophoblast, stimulating the activation of c-Jun N-terminal kinase (JNK) and FAK and thus promoting migration of trophoblastic cells and rates of implantation [53]. Embryonic development is influenced by communication between the cells of embryos, through the secretion of factors that are still not well known. Some studies have suggested that EVs could be involved in these processes. For example, a study conducted by P. Qu *et al*. on bovine cells has shown that embryos without replaced culture medium contain CD9 positive exosomes and have a better chance of a healthy pregnancy [54]. Another study conducted by I.M. Saadeldin *et al*. concluded that EVs are influencing the communication between embryos. They combined cloned embryos with embryos from an unfertilized egg cell and showed that the latter are secreting CD 9 positive exosomes and EVs containing RNA transcripts that encoded some pluripotency genes, improving the features of the cloned embryos if co-cultured [55].

The roles of the EVs in implantation and embryonic development are illustrated in **Figure 3**.

3.2 Cancer development

EVs are produced by stromal cells, which can be found, along with cancerous cells, as components of a tumor mass. In this case, EVs act like a bidirectional transferring mechanism between stromal cells and cancerous cells, influencing tumor evolution [6]. The biogenesis, release pathways, and the contents of EVs will be modified by the tumor microenvironment. Circulating DNA, contained by EVs will be transferred between apoptotic bodies (derived from apoptotic tumor cells) and other cells, leading to increased expression of oncogenes [56]. Tumor-derived EVs play a crucial role in all steps of cancer development, being more and more studied, to discover new treatment methods [57].

For a better understanding of the role of EVs in tumoral processes (cell proliferation, apoptosis resistance, angiogenesis, local invasion and metastasis, therapy resistance, etc.) we will discuss this with respect to some cancer types.

Some studies have shown that exosomes produced and released by ovarian cancer cells can carry RNAs and miRNAs, influencing cell transformation and tumor evolution. RNA-binding protein LIN 28, a marker of SCs, is associated with an unfavorable outcome when present in malignancies. Ovarian cancer cells which express high LIN 28 levels can secrete exosomes, which can further interact with noncancerous cells, leading to variations of gene expressions and cell behavior. This can lead

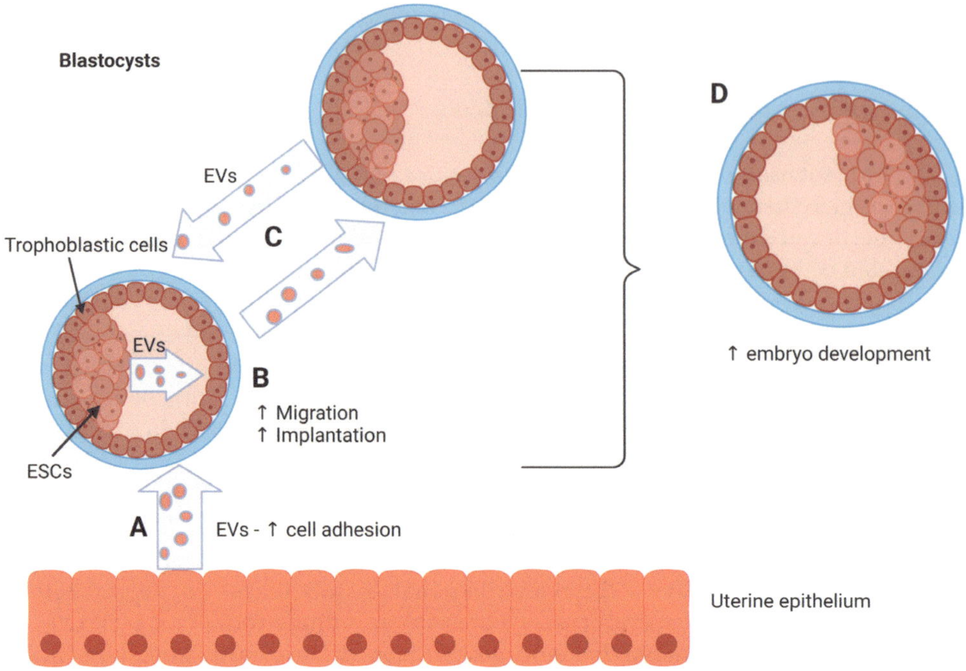

Figure 3.
The roles of the EVs in embryo communication. (A) The uterine epithelium secretes EVs that stimulate the adhesion of trophoblastic cells to the uterus; (B) embryonic stem cells (ESCs) produce EVs that stimulate the trophoblasts to migrate and implant into the uterus; (C) the co-culturing of embryos increases (↑) embryo development (D), mediated by EVs. Created with BioRender.com (last accessed on September 22, 2021).

to consequential amplification of genes responsible for epithelial to mesenchymal transition, human embryonic kidney 293 cells (HEK 293) invasion, and migration [58]. SKOV3, an ovarian cancer cell line, is also involved in cancer development by producing and releasing exosomes that can stimulate the M2 macrophage phenotype and consequently migration and proliferation of cancerous cells [6].

In breast cancer, studies have shown that EVs contain two extracellular matrix proteins, discoidin I-like domains 3 and epidermal growth factor-like repeats, that can activate FAK cascade and, along with an independent mechanism of microRNA biogenesis possessed by EVs, they play a crucial role in cancer development [59, 60].

In glioblastoma, EVs are transferring between cells the protein chloride intracellular channel-1, which stimulates the growth of the recipient cells, and the splicing factor RNA-binding motif protein 11, which increases survival [61, 62]. Moreover, the effect of EVs on angiogenesis, an important process in tumor growth, has been studied on glioma cells and it has been reported that EVs contain factors that promote angiogenesis by stimulating vascular endothelial growth factors [63].

In bladder and gastric cancer and melanoma, EVs are releasing platelet-derived growth factor receptor-beta, which is stimulating PI3K/AKT and MAP/ERK pathways, thus increasing cell proliferation and apoptosis resistance [64, 65].

The role of the EVs in intercellular communication and cancer development occurs not only locally but also remotely, leading to metastatic disease. The most studied components of EVs involved in this process are miRNAs that can influence angiogenesis, local invasion, colonization, immune modulation, etc., and annexin II, a membrane-associated protein, by stimulating angiogenesis [66, 67]. Also, peritoneal metastases of ovarian cancer are accelerated by matrix metalloproteinase-1 from EVs [68].

Therapy response in cancer can be influenced by EVs, until the emergence of multidrug resistance, by transferring some drug resistance traits from cancer cells to recipient cells, like drug efflux pumps (decreasing drug concentrations in the cells by drug efflux), apoptotic regulators (simulating anti-apoptotic pathways), proteins involved in metal ion transportation (decreasing the effect of a metal-based therapy, as cisplatin), but also microRNAs, functional mRNAs and lncRNAs (long non-coding RNAs) [57, 69–71].

3.3 Therapeutic potential of EVs

As already mentioned, EVs have an important role in cell-cell communication and thus in physiological and pathological processes, leading to an increased interest in studying their ability to generate new therapeutic methods. Over time, several studies have tried to demonstrate the involvement of EVs in immunological modulation, tissue regeneration, bioengineering, transportation of therapeutic agents, etc. [4]. One focuses our attention on explaining some other therapeutic potential of EVs, while the role of EVs in tissue regeneration will be separately discussed.

One of the first studied therapeutic potentials of EVs has been in immunotherapy. EVs produced by mesenchymal stromal cells (MSCs), especially exosomes, can induce an M2-like phenotype (anti-inflammatory, regenerative) in monocytes *in vitro* and thus polarization of activated CD4 T-cells to regulatory T-cells [72]. Some experimental studies performed in rats have shown that allograft rejection can be decreased by regulatory T-cells (activated by exosomes) in kidney and intestinal transplantation in rats and by exosomes derived from immature dendritic cells in cardiac transplantation [73–75]. In ischemic events, MSCs are producing exosomes that are decreasing myocardial inflammation after 24 h, by secreting anti-inflammatory cytokines and MVs that are reducing renal inflammation and fibrosis [74, 75].

4. EVs as drug delivery vehicles

Today's medicine is increasingly focused on personalized treatment methods, on targeted therapies that act at the molecular level. One of the concepts aimed at these aspects is that of theranostics, which aims at diagnosis, treatment, and concomitant follow-up of the response by using very specific drug delivery systems [76]. In this sense, EVs are an extremely useful tool for passive diagnosis (especially in neoplastic pathologies, through the ability to identify the tumor type based on the miRNA, mRNA, and mitochondrial RNA profile of EVs) and active (by associating EVs with advanced imaging methods). Thus, numerous platforms based on EVs technology have been developed for theranostic purposes, namely, transition metal-labeled exosomes, nanoparticle-loaded exosomes, bioluminescently labeled exosomes, nanocluster loaded exosomes, metabolically labeled exosomes. The main applications at present are those in the oncology field, but the lack of uniformity of clear classifications, increased immunogenicity of EVs, and the limited number of drugs that can be loaded at EVs highlight the need for future studies for widespread application of these diagnostic and therapeutic tools [77].

EVs are used as transporters for a variety of substances, ranging from small molecules, small interfering RNA, mRNA, and microRNAs to drugs with suboptimal pharmaceutical effects, carrying active constituents through biological barriers [4, 11].

Exosomal transporters present advantages, as they can travel efficiently between cells, smoothly passing their cargo along the cell membrane, keeping it biologically active, and crossing hard-to-penetrate barriers, such as the blood-brain barrier. Important issues regarding exosome-based drug delivery vehicles are the precise method of exosome loading, without altering the biological characteristics, and the scalable repeatable production of exosome categories [18, 78].

Exosomes tend to have special homing targets, influenced by their cell of origin [18]. Their membrane can be modified, to amplify the targeting of specific cells [18, 79].

The content of EVs can be loaded exogenously (integration of small proteins, RNA, or other molecules) or endogenously (assuring that cells possess the ways to integrate small molecules, proteins or RNA into EVs during their formation) [4, 80]. Exogenous changes of EVs can be done after their collection, incorporating the cargo into EVs through different methods: coincubation (with no modification of vesicle size distribution or integrity, electroporation, and sonication) [79, 80]. The endogenous loading can be obtained through artificial adjustment of the parental cell to overexpress certain proteins or RNA, that can be integrated into secreted EVs afterward [81].

Human MSCs, multipotent adult progenitors, could be an adequate source of exosomes for drug delivery. Their transplantation has been investigated in numerous trials and proved to be safe, MSCs also produce immunologically inert exosomes [18]. As a delivery system, studies have shown that EV derived from MSCs can transfer therapeutic drugs to diseased cells [19]. Their efficiency is based on the adhesion proteins on their surface (like integrins, extracellular matrix proteins, tetraspanins), which facilitate the penetration of the cellular membrane and the accumulation of EVs in the diseased cells [82]. Other characteristics that make EVs an ideal candidate for a drug delivery system are their decreased toxicity and immunogenicity, as well as their potential to cross the blood-brain barrier [83, 84]. A study conducted by S. Kamerkar *et al*. has shown that EVs produced by MSCs can transfer small interfering RNA targeting the oncogenic KRas (G12D) mutants to pancreatic cancer cells, increasing cells apoptosis and decreasing the risk of metastatic disease [85].

Exosomes have significant immune properties, modulating immunological responses and facilitating antigen presentation [11, 86]. Exosomes derived from dendritic cells can conduct MHC class I/peptide complexes to other dendritic cells for *in vivo* activation of cytotoxic T lymphocytes and promote T cell-dependent antitumor responses *in vivo* [86, 87]. Dendritic cell exosomes have been previously loaded with antigenic peptides, to activate T cell proliferation, with possible use as vaccines against infectious or neoplastic diseases. Due to the immunogenic nature of dendritic cell exosomes, their use as drug delivery vehicles is not ideal. A more suitable choice would be human ESCs-derived mesenchymal cells [86, 88, 89].

Clinical trials with therapies based on EVs are studied in malignancies, such as melanoma, non-small cell lung cancer, colon cancer, metastatic pancreatic cancer, bronchopulmonary dysplasia, malignant ascites, and pleural effusion, but also chronic kidney disease, type 1 diabetes, insulin resistance and chronic inflammation polycystic ovary syndrome, ulcers, and acute ischemic stroke [4]. Exosomes were shown to transport curcumin and chemotherapeutics, such as doxorubicin and paclitaxel [19, 90].

The implication of MSCs was also evaluated in patients with anthracycline-induced cardiomyopathy. Mitochondrial transfer, mediated by large EVs, diminished injury determined by doxorubicin, in patient-specific induced pluripotent SC-derived cardiomyocytes. MSCs could ameliorate cardiac function in anthracycline-induced cardiomyopathy, regardless of regeneration effects [91].

Liposomes possess many favorable characteristics as drug delivery vehicles, being used in the transportation of anti-cancer drugs, anti-fungal medication, and analgesics [92–98]. Liposomes have a phospholipid membrane that helps with the incorporation of hydrophilic or hydrophobic drugs, and they can also deliver the carried drugs to the targeted points through plasma membrane breaching. To diminish the recognition by opsonins and their clearance, liposomes can be covered with polymers (PEG). Their membranes can be adapted, to present ligands or antibody elements, which can interact with specific cells and amplify the targeted drug delivery. Liposomes, with easy-to-control properties, can be loaded with drugs, DNA, diagnostic instruments, enzymes, or peptides. Drugs included in liposomes have attenuated toxicity and do not provoke unwanted toxic reactions [99]. Liposomal drugs have various routes of administration, such as parenteral, oral, topical, and even through aerosols [99].

Synthetic liposomes, although very useful, are overcome by EVs (naturally derived liposomes), which have lower toxicity [19]. Exosomes are considered superior drug delivery vehicles, as an alternative to liposomes. In contrast to the latter, exosomes are usually adequately tolerated by the human body and do not present intrinsic toxicity. They can deliver their content through the plasmatic membrane and protect against its early transformation and elimination [19]. Since exosomes can be found in a variety of biological fluids, such as blood, urine, breast milk the delivered drugs will be well tolerated, less toxic, and with a longer circulating half-life [19, 100].

5. Regenerative medicine and EVS

Regenerative medicine is a relatively new concept and a complex domain that involves the restoration of damaged tissues using multiple techniques (e.g., SCs, biomaterials, differentiated autologous cells, or combinations of the aforementioned techniques). Regenerative medicine is focusing on repairing, regrowing, or replacing injured, malfunctioning, or missing tissue and addresses many tissular types: skin, heart tissue, cartilage tissue, bone tissue, adipose tissue etc. [101]. Thus, SC research focuses on their properties of repairing damaged tissues, either by producing new tissues by division and differentiation or by their partial repair.

Stem cells represent a highly interesting resource and were considered the ideal choice for regenerative therapies. SCs are defined as non-specialized cells, characterized by an enormous capacity of differentiation, which varies depending on their origin (embryonic, fetal, or adult). SCs are capable of differentiation into adipocytes, osteocytes, chondrocytes, endothelial cells, cardiomyocytes, pericytes, and smooth muscle cells [102–107]. They can also differentiate into neurogenic, cardiovascular, and neovascular pathways [108–113]. Allogeneic transplantation can be used in other applications due to the immunosuppressive properties of SCs [114].

Over the last decades, in an attempt to better understand SCs to use them in the processes of tissue repair and regeneration, multiple classifications have been made, depending on many aspects: the organism of origin (embryo, fetus, infant, adult), the tissue of origin within the adult-origin cells (mesenchymal tissue, hematopoietic, nervous, gastrointestinal, cutaneous, etc.), and the ability to divide (totipotent, pluripotent, multipotent, etc.) [115–117]. The understanding, even partial, of SCs' ability to divide, especially the asymmetric division of adult SCs has opened new horizons in terms of reparative and regenerative medicine [115]. Overpassing the initial idea that considers EVs as cellular debris, nowadays they are seen as tools for intercellular communication and as possible therapeutic vehicles. However, the same cannot be said about SCs. In the last decades, the interest for their properties has gained more and more interest. However, they have been regarded as cells at the origin of many pathologies since 1933, when Sabin *et al.* emphasize the possibility of radioactive damage to lymphoid tissue by affecting SCs [115, 118].

Although at the beginning researchers, scientists, and clinical doctors considered that the success of stem cell transplantation depends on the purity of the transplanted cells, not always the purer means also the better. Over time, it has become increasingly clear that the success of SC therapy depends on EVs and the soluble secreted factors because they play important paracrine roles. Our recent work also demonstrated that some other cellular types, such as the newly discovered telocytes, can act as cellular adjutants participating in the regenerative processes possibly through the released EVs influencing the microenvironment of the stem cell niche [118, 119]. Other additional evidence suggests that EVs can have not only regenerative properties, but also immunomodulatory roles, consequently summing up the therapeutic effects of stem cells. EVs, by contrast to stem cells, are nonimmunogenic and are not able to self-replicate [120]. In addition, EVs display powerful therapeutic potential, with positive outcomes regarding regeneration in many tissues (**Table 2**).

The central point of the pathophysiological mechanisms by which SCs contribute to the tissular repair are EVs that function as carriers of many biomolecules, such as miRNA, mRNA, cytokines, growth factors, differentiating factors with a key role in the main processes involved in tissue regeneration: immunomodulation, angiogenesis, differentiation [2]. Thus, multiple preclinical studies performed *in vitro* or *in vivo* on animal subjects have tried to identify the molecules involved and their role, important advances being made in diseases with high mortality and morbidity such as myocardial infarction, neuronal degeneration, osteoarticular diseases, skin ulcers, corneal damage etc. [121, 133, 134].

The use of SCs, despite promising results, has many disadvantages that require careful control of this procedure and the formation of very specific microenvironments to induce the differentiation of these cells [135]. Among the disadvantages mentioned before, those given by the ethical considerations of embryo use, the risk of uncontrolled differentiation, and the appearance of teratomas and genetic instability (especially those of embryonic origin) are the most important. To these disadvantages, a lower capacity for division and differentiation is added, as well as a laborious procedure for adult SCs acquirement [136, 137].

Among SCs, MSCs secrete growth factors and cytokines, with autocrine and paracrine properties. These substances inhibit the local immune system, fibrosis, and apoptosis, amplifying mitosis and differentiation of tissue-intrinsic reparative cells. These phenomena are known as trophic effects and differ from the direct differentiation of MSCs for tissue repair [138].

The numerous functions of MSCs in tissue regeneration and implicitly in the possible treatment of many diseases are mainly achieved by the secretion of exosomes loaded with key molecules (cytokines, growth factors, miRNA) and by molecules secreted directly into the extracellular environment with paracrine action [138–141].

However, they cannot produce infinite numbers of exosomes, repetitive isolation of cells being needed. The advantages of MSCs exosomes are their non-immunogenic

Involved tissue	Type of stem cells from which EVs derived	Involved molecules within the EVs	Type of effect on tissue repair	References
Myocardial (myocardial infarction)	MSCs	miRNA 19a, 132, 146-3p, 220, 221 mRNA	Antiapoptotic Proangiogenic Cardioprotective	[121–123]
	CSCs	miRNA 132, 146a, 210 TCA-3 SDF-1 VEGF Eritropoetin bFGF Osteopontin SCF Activin A DKK homolog 1 TGF beta	Antiapoptotic Proangiogenic (stimulates tubules formation in endothelial cells) Improve ejection fraction	[121, 124, 125]
	BM-MSCs	miRNA 22, 126, 130a, 182	Antiapoptotic Proangiogenic Antifibrotic Immunomodulatory Anti-inflammatory	[121–123, 126, 127]
Bone and cartilage (osteoarthritis)	MSCs	miRNAs (92a, 125b, 320) MMP-13	Modulates the immune response Protects the chondrocyte Stimulates regeneration, matrix, and chondrocytes proliferation	[128–130]
Skin (ulcers)	ADSc	IL 2,6,7,9,21 etc. TNF FGF CCL 2,4, 38 etc. BMP 5,7 etc. More than 70 specific miRNAs (204, 210-3p etc.)	Proangiogenic Improve epithelialization Improve epithelial width Decrease scar formation Decrease wound diameter	[131, 132]

MSCs—mesenchymal stem cells; CSCs—cardiac stem cells; BM-MSCs—bone marrow mesenchymal stem cells; ADSc—Adipose tissue stem cells; miRNA—microRNA; TCA-3—T-cell activation gene-3; SDF-1—stromal derived factor 1; VEGF—vascular endothelial growth factor; bFGF—basic fibroblast growth factor; SCF—F box containing complex; DKK 1—Dickkopf-related protein 1; TGF beta—transforming growth factor 1; MMP-13—matrix metalloproteinase; IL—interleukin; TNF—tumor necrosis factor; CCL—CC chemokine ligand; BMP—bone morphogenic protein.

Table 2.
The role of stem cell-derived EVs depending on their content and tissue type.

property, the intrinsic therapeutic capacity of reducing tissue damage, large *ex vivo* expansion, conveniently reachable source, and clinically tested cell source [18].

MSCs are a source of small EVs which can favor angiogenesis and cell proliferation in infarcted myocardium, can inhibit cardiac remodeling, and improve ventricular functions [121, 142, 143].

Moreover, MSCs can repair infarcted myocardium through paracrine interactions. EVs derived from MSCs have a better therapeutic effect than simple MSCs therapy. In animal subjects, suffering from myocardial infarction, exosomes derived from MSCs diminished inflammation, improved cardiac function, stimulated cardiomyocyte H9C2 cell proliferation, inhibited apoptosis induced by H_2O_2 and cardiac fibrosis, and slowed down the transformation of fibroblasts into myofibroblasts mediated by TGF-β [144].

Although macrovascular reperfusion is the gold standard therapy for acute myocardial infarction, heart failure developed due to deficient cardiac remodeling is still a major issue for long-term therapeutic management. Angiogenesis is crucial for tissular regeneration, therefore, interest for therapeutic enhancement of angiogenesis has increased. Preclinical and human clinical trials showed conflicting results, the use of one growth factor not being enough to promote adequate angiogenesis [144–147]. Cell transplantation could be an alternative/another solution [148, 149]. Stem cells were utilized as sources for new cardiac cells production (endothelial progenitor cells, MSCs, cardiac progenitor cells). Paracrine factors secreted by transplanted cells seem to influence endogenous repair of damaged tissues [121].

In vivo studies indicate that small EVs from MSCs that overexpress Akt can amplify neovascularization, ameliorating the left ventricle ejection fraction [150]. The angiogenic involvement is supported by the treatment of renal ischemic reperfusion injury with small EVs derived from umbilical cord MSCs ameliorated capillary density through promoting VEGF up-regulation, independently from HIF-1α [151]. Small EVs can also deliver miRNAs (miR-125a) to endothelial cells, favoring angiogenesis [152]. Comparable outcomes were also obtained from the use of MSCs which overexpress hypoxia-inducible factors (HIF-1α). Injection of exosomes from MSCs, containing Jagged1, and hypoxia-inducible factor—MSCs cultures led to angiogenesis *in vivo* and *in vitro*. Exosomes derived from HIF-1α-overexpressing MSCs have a strong angiogenic function, through an expansion in the packaging of Jagged1 [153]. In addition, the immune system has a big role in the repair of the ischemic myocardium, in the inflammatory and angiogenesis phases. Chemokines, cytokines, and the release of EVs with paracrine actions sustain this restoration. EVs favor tissular regeneration and angiogenesis, therefore, research in this area is of high interest for patients suffering from acute myocardial infarction.

A study evaluating the effect of intracoronary administration of cardiac-derived SCs-secreted small EVs showed a lower number and altered polarization state of CD68+ macrophages in the infarcted myocardium, with elevated expression of anti-inflammatory genes (Arg1, IL4ra, Tgfb1, Vegfa). Macrophages primed with EVs from cardiac-derived SCs displayed high levels of miR-181b, which targets protein kinase C δ. Therefore, exosomal transfer of miR-181b into macrophages lowered the levels of protein kinase C δ transcript, underlining the cardioprotective properties of stem cell infusion after reperfusion [154].

According to L. Cambier and colleagues, cardiosphere-derived cells proved to reduce myocardial infarction size through secreted EVs-Y RNA fragment, found in generous concentrations in EVs from cardiosphere-derived cells, correlated with the potency of these cells *in vivo*. This fragment can be transferred from cardiac cells to target macrophages through EVs, inducing transcription and secretion of IL-10, offering cardioprotection. *In vivo* injection of EV-Y RNA fragment after reperfusion reduced the infarct size [121, 155].

One of the major issues in diabetic patients is inadequate myocardial angiogenesis, which is responsible for an elevated risk for ischemic heart disease in these patients. Exosomes loaded with miRNAs (miR-320-3p or 320a) derived from diabetic cardiomyocytes proved, they can influence angiogenesis in endothelial cell cultures. Moreover, miR-320-3p, together with the miR-29 family and miR-7a can regulate insulin secretion and its signaling pathways [156, 157].

6. Conclusions

It is becoming increasingly evident that EVs are involved in a multitude of biological processes and can play modulatory and regulatory roles. If one adds their new potential as biomarkers for various diseases and their therapeutic delivery cargo abilities, it is certain that one cannot over neglect their ability to support stem-cell-based therapies. Future avenues are seen at the horizon when further research will probably allow us to use engineered-EVs to support endogenous repair and thus create a modern regenerative medicine.

Conflict of interest

The authors declare no conflict of interest.

Author details

Gaspar Bogdan Severus[1,2*†], Ionescu Ruxandra Florentina[3†], Enache Robert Mihai[4†], Dobrică Elena Codruța[5,6†], Crețoiu Sanda Maria[7*], Crețoiu Dragoș[8] and Voinea Silviu Cristian[9]

1 Surgery Department, Carol Davila University of Medicine and Pharmacy, Bucharest, Romania

2 Surgery Clinic, Bucharest Emergency Clinical Hospital, Bucharest, Romania

3 Department of Cardiology I, Central Military Emergency Hospital "Dr. Carol Davila", Bucharest, Romania

4 Department of Radiology and Medical Imaging, Fundeni Clinical Institute, Bucharest, Romania

5 Department of Pathophysiology, University of Medicine and Pharmacy of Craiova, Craiova, Romania

6 Department of Dermatology, Elias Emergency University Hospital, Bucharest, Romania

7 Department of Morphological Sciences, Division of Cell and Molecular Biology and Histology, Carol Davila University of Medicine and Pharmacy, Bucharest, Romania

8 Alessandrescu-Rusescu National Institute for Mother and Child Health, Fetal Medicine Excellence Research Center, Bucharest, Romania

9 Department of Surgical Oncology, Prof. Dr. Alexandru Trestioreanu Oncology Institute, Carol Davila University of Medicine and Pharmacy, Bucharest, Romania

*Address all correspondence to: bogdan.gaspar@umfcd.ro and sanda.cretoiu@umfcd.ro

† Equal contribution.

References

[1] Zaborowski MP, Balaj L, Breakefield XO, Lai CP. Extracellular vesicles: Composition, biological relevance, and methods of study. Bioscience. 2015;**65**:783-797

[2] Hur YH, Cerione RA, Antonyak MA. Extracellular vesicles and their roles in stem cell biology: Extracellular vesicles and stem Cells. Stem Cells. 2020; **38**:469-476

[3] Witwer KW, Buzás EI, Bemis LT, Bora A, Lässer C, Lötvall J, et al. Standardization of sample collection, isolation and analysis methods in extracellular vesicle research. Journal of Extracellular Vesicles. 2013;**2**:20360

[4] Wiklander OPB, Brennan MÁ, Lötvall J, Breakefield XO, El Andaloussi S. Advances in therapeutic applications of extracellular vesicles. Science Translational Medicine. 2019;**11**:eaav8521

[5] Nomura S. Extracellular vesicles and blood diseases. International Journal of Hematology. 2017;**105**:392-405

[6] Voichitoiu A-D, Mihaela Radu B, Pavelescu L, Cretoiu D, Teona Deftu A, Suciu N, et al. Extracellular vesicles in cancer. In: Bona AGD, Calderon JAR, editors. Extracellular Vesicles and Their Importance in Human Health. London, England: IntechOpen; 2020

[7] Parolini I, Federici C, Raggi C, Lugini L, Palleschi S, De Milito A, et al. Microenvironmental pH is a key factor for exosome traffic in tumor cells. The Journal of Biological Chemistry. 2009;**284**:34211-34222

[8] Kucharzewska P, Belting M. Emerging roles of extracellular vesicles in the adaptive response of tumour cells to microenvironmental stress. Journal of Extracellular Vesicles. 2013;**2**:20304

[9] Xu R, Rai A, Chen M, Suwakulsiri W, Greening DW, Simpson RJ. Extracellular vesicles in cancer—Implications for future improvements in cancer care. Nature Reviews. Clinical Oncology. 2018;**15**:617-638

[10] O'Brien K, Breyne K, Ughetto S, Laurent LC, Breakefield XO. RNA delivery by extracellular vesicles in mammalian cells and its applications. Nature Reviews. Molecular Cell Biology. 2020;**21**:585-606

[11] Valadi H, Ekström K, Bossios A, Sjöstrand M, Lee JJ, Lötvall JO. Exosome-mediated transfer of mRNAs and microRNAs is a novel mechanism of genetic exchange between cells. Nature Cell Biology. 2007;**9**:654-659

[12] Desrochers LM, Antonyak MA, Cerione RA. Extracellular vesicles: Satellites of information transfer in cancer and stem cell biology. Developmental Cell. 2016;**37**:301-309

[13] EL Andaloussi S, Mäger I, Breakefield XO, Wood MJA. Extracellular vesicles: Biology and emerging therapeutic opportunities. Nature Reviews. Drug Discovery. 2013;**12**: 347-357

[14] Wendler F, Stamp GW, Giamas G. Tumor-stromal cell communication: Small vesicles signal big changes. Trends in Cancer. 2016;**2**:326-329

[15] Kurian NK, Modi D. Extracellular vesicle mediated embryo-endometrial cross talk during implantation and in pregnancy. Journal of Assisted

Reproduction and Genetics. 2019;**36**:189-198

[16] Lamichhane TN, Sokic S, Schardt JS, Raiker RS, Lin JW, Jay SM. Emerging roles for extracellular vesicles in tissue engineering and regenerative medicine. Tissue Engineering. Part B, Reviews. 2015;**21**:45-54

[17] De Jong OG, Van Balkom BWM, Schiffelers RM, Bouten CVC, Verhaar MC. Extracellular vesicles: Potential roles in regenerative medicine. Frontiers in Immunology. 2014;**5**:608

[18] Lai RC, Yeo RWY, Tan KH, Lim SK. Exosomes for drug delivery - a novel application for the mesenchymal stem cell. Biotechnology Advances. 2013; **31**:543-551

[19] Tian Y, Li S, Song J, Ji T, Zhu M, Anderson GJ, et al. A doxorubicin delivery platform using engineered natural membrane vesicle exosomes for targeted tumor therapy. Biomaterials. 2014;**35**:2383-2390

[20] Jang SC, Crescitelli R, Cvjetkovic A, Belgrano V, Bagge RO, Höög JL, et al. A subgroup of mitochondrial extracellular vesicles discovered in human melanoma tissues are detectable in patient blood. BioRxiv. 2017;**174193**:1-14. DOI: 10.1101/174193

[21] Zhang H, Freitas D, Kim HS, Fabijanic K, Li Z, Chen H, et al. Identification of distinct nanoparticles and subsets of extracellular vesicles by asymmetric flow field-flow fractionation. Nature Cell Biology. 2018;**20**:332-343

[22] Tkach M, Kowal J, Théry C. Why the need and how to approach the functional diversity of extracellular vesicles. Philosophical Transactions of the Royal Society of London. Series B, Biological Sciences. 2018;**373**:1-9. DOI: 10.1098/rstb.2016.0479

[23] Yáñez-Mó M, Siljander PR-M, Andreu Z, Zavec AB, Borràs FE, Buzas EI, et al. Biological properties of extracellular vesicles and their physiological functions. Journal of Extracellular Vesicles. 2015;**4**:27066

[24] Doyle LM, Wang MZ. Overview of extracellular vesicles, their origin, composition, purpose, and methods for exosome isolation and analysis. Cell. 2019;**8**:727

[25] Mathivanan S, Simpson RJ. ExoCarta: A compendium of exosomal proteins and RNA. Proteomics. 2009;**9**:4997-5000

[26] Villarroya-Beltri C, Baixauli F, Mittelbrunn M, Fernández-Delgado I, Torralba D, Moreno-Gonzalo O, et al. ISGylation controls exosome secretion by promoting lysosomal degradation of MVB proteins. Nature Communications. 2016;**7**:13588

[27] Bebelman MP, Smit MJ, Pegtel DM, Baglio SR. Biogenesis and function of extracellular vesicles in cancer. Pharmacology & Therapeutics. 2018;**188**:1-11

[28] Wollert T, Hurley JH. Molecular mechanism of multivesicular body biogenesis by ESCRT complexes. Nature. 2010;**464**:864-869

[29] Tauro BJ, Greening DW, Mathias RA, Ji H, Mathivanan S, Scott AM, et al. Comparison of ultracentrifugation, density gradient separation, and immunoaffinity capture methods for isolating human colon cancer cell line LIM1863-derived exosomes. Methods. 2012;**56**:293-304

[30] Buschow SI, van Balkom BWM, Aalberts M, Heck AJR, Wauben M,

Stoorvogel W. MHC class II-associated proteins in B-cell exosomes and potential functional implications for exosome biogenesis. Immunology and Cell Biology. 2010;**88**:851-856

[31] Sinha A, Ignatchenko V, Ignatchenko A, Mejia-Guerrero S, Kislinger T. In-depth proteomic analyses of ovarian cancer cell line exosomes reveals differential enrichment of functional categories compared to the NCI 60 proteome. Biochemical and Biophysical Research Communications. 2014;**445**:694-701

[32] Bobrie A, Colombo M, Raposo G, Théry C. Exosome secretion: Molecular mechanisms and roles in immune responses. Traffic. 2011;**12**:1659-1668

[33] Lachenal G, Pernet-Gallay K, Chivet M, Hemming FJ, Belly A, Bodon G, et al. Release of exosomes from differentiated neurons and its regulation by synaptic glutamatergic activity. Molecular and Cellular Neurosciences. 2011;**46**:409-418

[34] Bakhti M, Winter C, Simons M. Inhibition of myelin membrane sheath formation by oligodendrocyte-derived exosome-like vesicles. The Journal of Biological Chemistry. 2011;**286**:787-796

[35] Wang S, Cesca F, Loers G, Schweizer M, Buck F, Benfenati F, et al. Synapsin I is an oligomannose-carrying glycoprotein, acts as an oligomannose-binding lectin, and promotes neurite outgrowth and neuronal survival when released via glia-derived exosomes. The Journal of Neuroscience. 2011; **31**:7275-7290

[36] Guo M, Wang J, Zhao Y, Feng Y, Han S, Dong Q, et al. Microglial exosomes facilitate α-synuclein transmission in Parkinson's disease. Brain. 2020;**143**:1476-1497

[37] Giusti I, Delle Monache S, Di Francesco M, Sanità P, D'Ascenzo S, Gravina GL, et al. From glioblastoma to endothelial cells through extracellular vesicles: Messages for angiogenesis. Tumour Biology. 2016;**37**:12743-12753

[38] Zhou H, Pisitkun T, Aponte A, Yuen PST, Hoffert JD, Yasuda H, et al. Exosomal Fetuin-A identified by proteomics: A novel urinary biomarker for detecting acute kidney injury. Kidney International. 2006;**70**:1847-1857

[39] Melo SA, Luecke LB, Kahlert C, Fernandez AF, Gammon ST, Kaye J, et al. Glypican-1 identifies cancer exosomes and detects early pancreatic cancer. Nature. 2015;**523**:177-182

[40] Sandfeld-Paulsen B, Aggerholm-Pedersen N, Bæk R, Jakobsen KR, Meldgaard P, Folkersen BH, et al. Exosomal proteins as prognostic biomarkers in non-small cell lung cancer. Molecular Oncology. 2016;**10**: 1595-1602

[41] Gatti S, Bruno S, Deregibus MC, Sordi A, Cantaluppi V, Tetta C, et al. Microvesicles derived from human adult mesenchymal stem cells protect against ischaemia-reperfusion-induced acute and chronic kidney injury. Nephrology, Dialysis, Transplantation. 2011;**26**: 1474-1483

[42] Cai H, Reinisch K, Ferro-Novick S. Coats, tethers, Rabs, and SNAREs work together to mediate the intracellular destination of a transport vesicle. Developmental Cell. 2007;**12**:671-682

[43] Monguió-Tortajada M, Gálvez-Montón C, Bayes-Genis A, Roura S, Borràs FE. Extracellular vesicle isolation methods: Rising impact of size-exclusion chromatography. Cellular and Molecular Life Sciences. 2019;**76**(12):2369-2382

[44] Escola JM, Kleijmeer MJ, Stoorvogel W, Griffith JM, Yoshie O, Geuze HJ. Selective enrichment of tetraspan proteins on the internal vesicles of multivesicular endosomes and on exosomes secreted by human B-lymphocytes. The Journal of Biological Chemistry. 1998;**273**:20121-20127

[45] Heijnen HF, Schiel AE, Fijnheer R, Geuze HJ, Sixma JJ. Activated platelets release two types of membrane vesicles: Microvesicles by surface shedding and exosomes derived from exocytosis of multivesicular bodies and alpha-granules. Blood. 1999;**94**:3791-3799

[46] Altonsy MO, Ganguly A, Amrein M, Surmanowicz P, Li SS, Lauzon GJ, et al. Beta3-tubulin is critical for microtubule dynamics, cell cycle regulation, and spontaneous release of microvesicles in human malignant melanoma cells (A375). International Journal of Molecular Sciences. 2020;**21**:1656

[47] Morello M, Minciacchi VR, de Candia P, Yang J, Posadas E, Kim H, et al. Large oncosomes mediate intercellular transfer of functional microRNA. Cell Cycle. 2013;**12**: 3526-3536

[48] Al-Nedawi K, Meehan B, Micallef J, Lhotak V, May L, Guha A, et al. Intercellular transfer of the oncogenic receptor EGFRvIII by microvesicles derived from tumour cells. Nature Cell Biology. 2008;**10**:619-624

[49] Di Vizio D, Kim J, Hager MH, Morello M, Yang W, Lafargue CJ, et al. Oncosome formation in prostate cancer: Association with a region of frequent chromosomal deletion in metastatic disease. Cancer Research. 2009;**69**: 5601-5609

[50] Orlando KA, Stone NL, Pittman RN. Rho kinase regulates fragmentation and phagocytosis of apoptotic cells. Experimental Cell Research. 2006;**312**:5-15

[51] Théry C, Boussac M, Véron P, Ricciardi-Castagnoli P, Raposo G, Garin J, et al. Proteomic analysis of dendritic cell-derived exosomes: A secreted subcellular compartment distinct from apoptotic vesicles. Journal of Immunology. 2001;**166**: 7309-7318

[52] Greening DW, Nguyen HPT, Elgass K, Simpson RJ, Salamonsen LA. Human endometrial exosomes contain hormone-specific cargo modulating trophoblast adhesive capacity: Insights into endometrial-embryo interactions. Biology of Reproduction. 2016;**94**:38

[53] Desrochers LM, Bordeleau F, Reinhart-King CA, Cerione RA, Antonyak MA. Microvesicles provide a mechanism for intercellular communication by embryonic stem cells during embryo implantation. Nature Communications. 2016;**7**:11958

[54] Qu P, Qing S, Liu R, Qin H, Wang W, Qiao F, et al. Effects of embryo-derived exosomes on the development of bovine cloned embryos. PLoS One. 2017;**12**:e0174535

[55] Saadeldin IM, Kim SJ, Choi YB, Lee BC. Improvement of cloned embryos development by co-culturing with parthenotes: A possible role of exosomes/microvesicles for embryos paracrine communication. Cellular Reprogramming. 2014;**16**:223-234

[56] Bergsmedh A, Szeles A, Henriksson M, Bratt A, Folkman MJ, Spetz AL, et al. Horizontal transfer of oncogenes by uptake of apoptotic bodies. Proceedings of the National Academy of Sciences of the United States of America. 2001;**98**:6407-6411

[57] Xavier CPR, Caires HR, Barbosa MAG, Bergantim R, Guimarães JE, Vasconcelos MH. The role of extracellular vesicles in the hallmarks of cancer and drug resistance. Cell. 2020;**9**:1141

[58] Enriquez VA, Cleys ER, Da Silveira JC, Spillman MA, Winger QA, Bouma GJ. High LIN28A expressing ovarian cancer cells secrete exosomes that induce invasion and migration in HEK293 cells. BioMed Research International. 2015;**2015**:701390

[59] Lee J-E, Moon P-G, Cho Y-E, Kim Y-B, Kim I-S, Park H, et al. Identification of EDIL3 on extracellular vesicles involved in breast cancer cell invasion. Journal of Proteomics. 2016;**131**:17-28

[60] Melo SA, Sugimoto H, O'Connell JT, Kato N, Villanueva A, Vidal A, et al. Cancer exosomes perform cell-independent microRNA biogenesis and promote tumorigenesis. Cancer Cell. 2014;**26**:707-721

[61] Setti M, Osti D, Richichi C, Ortensi B, Del Bene M, Fornasari L, et al. Extracellular vesicle-mediated transfer of CLIC1 protein is a novel mechanism for the regulation of glioblastoma growth. Oncotarget. 2015;**6**:31413-31427

[62] Pavlyukov MS, Yu H, Bastola S, Minata M, Shender VO, Lee Y, et al. Apoptotic cell-derived extracellular vesicles promote malignancy of glioblastoma via intercellular transfer of splicing factors. Cancer Cell. 2018;**34**:119-135, e10

[63] Sun X, Ma X, Wang J, Zhao Y, Wang Y, Bihl JC, et al. Glioma stem cells-derived exosomes promote the angiogenic ability of endothelial cells through miR-21/VEGF signal. Oncotarget. 2017;**8**:36137-36148

[64] Yang L, Wu X-H, Wang D, Luo C-L, Chen L-X. Bladder cancer cell-derived exosomes inhibit tumor cell apoptosis and induce cell proliferation in vitro. Molecular Medicine Reports. 2013;**8**:1272-1278

[65] Vella LJ, Behren A, Coleman B, Greening DW, Hill AF, Cebon J. Intercellular resistance to BRAF inhibition can be mediated by extracellular vesicle–associated PDGFRβ. Neoplasia. 2017;**19**:932-940

[66] Dhondt B, Rousseau Q, De Wever O, Hendrix A. Function of extracellular vesicle-associated miRNAs in metastasis. Cell and Tissue Research. 2016;**365**:621-641

[67] Maji S, Chaudhary P, Akopova I, Nguyen PM, Hare RJ, Gryczynski I, et al. Exosomal Annexin II promotes angiogenesis and breast cancer metastasis. Molecular Cancer Research. 2017;**15**:93-105

[68] Yokoi A, Yoshioka Y, Yamamoto Y, Ishikawa M, Ikeda S-I, Kato T, et al. Malignant extracellular vesicles carrying MMP1 mRNA facilitate peritoneal dissemination in ovarian cancer. Nature Communications. 2017;**8**:14470

[69] Gong J, Luk F, Jaiswal R, George AM, Grau GER, Bebawy M. Microparticle drug sequestration provides a parallel pathway in the acquisition of cancer drug resistance. European Journal of Pharmacology. 2013;**721**:116-125

[70] Dong H, Wang W, Chen R, Zhang Y, Zou K, Ye M, et al. Exosome-mediated transfer of lncRNA-SNHG14 promotes trastuzumab chemoresistance in breast cancer. International Journal of Oncology. 2018;**53**:1013-1026

[71] Khoo X-H, Paterson IC, Goh B-H, Lee W-L. Cisplatin-resistance in oral

squamous cell carcinoma: Regulation by tumor cell-derived extracellular vesicles. Cancers (Basel). 2019;**11**:1166

[72] Zhang B, Yin Y, Lai RC, Tan SS, Choo ABH, Lim SK. Mesenchymal stem cells secrete immunologically active exosomes. Stem Cells and Development. 2014;**23**:1233-1244

[73] Yu X, Huang C, Song B, Xiao Y, Fang M, Feng J, et al. CD4+CD25+ regulatory T cells-derived exosomes prolonged kidney allograft survival in a rat model. Cellular Immunology. 2013;**285**:62-68

[74] Arslan F, Lai RC, Smeets MB, Akeroyd L, Choo A, Aguor ENE, et al. Mesenchymal stem cell-derived exosomes increase ATP levels, decrease oxidative stress and activate PI3K/Akt pathway to enhance myocardial viability and prevent adverse remodeling after myocardial ischemia/reperfusion injury. Stem Cell Research. 2013;**10**:301-312

[75] Zou X, Zhang G, Cheng Z, Yin D, Du T, Ju G, et al. Microvesicles derived from human Wharton's jelly mesenchymal stromal cells ameliorate renal ischemia-reperfusion injury in rats by suppressing CX3CL1. Stem Cell Research & Therapy. 2014;**5**:40

[76] Soekmadji C, Li B, Huang Y, et al. The future of extracellular vesicles as theranostics—An ISEV meeting report. Journal of Extracellular Vesicles. 2020;**9**(1):1809766

[77] Ailuno G, Baldassari S, Lai F, Florio T, Caviglioli G. Exosomes and extracellular vesicles as emerging theranostic platforms in cancer research. Cell. 2020;**9**(12):2569

[78] Batrakova EV, Kim MS. Development and regulation of exosome-based therapy products. Wiley Interdisciplinary Reviews. Nanomedicine and Nanobiotechnology. 2016;**8**:744-757

[79] Alvarez-Erviti L, Seow Y, Yin H, Betts C, Lakhal S, Wood MJA. Delivery of siRNA to the mouse brain by systemic injection of targeted exosomes. Nature Biotechnology. 2011;**29**:341-345

[80] Didiot M-C, Hall LM, Coles AH, Haraszti RA, Godinho BM, Chase K, et al. Exosome-mediated delivery of hydrophobically modified siRNA for Huntingtin mRNA silencing. Molecular Therapy. 2016;**24**:1836-1847

[81] Lamichhane TN, Jeyaram A, Patel DB, Parajuli B, Livingston NK, Arumugasaamy N, et al. Oncogene knockdown via active loading of small RNAs into extracellular vesicles by sonication. Cellular and Molecular Bioengineering. 2016;**9**:315-324

[82] Johnsen KB, Gudbergsson JM, Skov MN, Pilgaard L, Moos T, Duroux M. A comprehensive overview of exosomes as drug delivery vehicles—Endogenous nanocarriers for targeted cancer therapy. Biochimica et Biophysica Acta. 1846;**2014**:75-87

[83] El Andaloussi S, Lakhal S, Mäger I, Wood MJA. Exosomes for targeted siRNA delivery across biological barriers. Advanced Drug Delivery Reviews. 2013;**65**:391-397

[84] Chen CC, Liu L, Ma F, Wong CW, Guo XE, Chacko JV, et al. Elucidation of exosome migration across the blood-brain barrier model in vitro. Cellular and Molecular Bioengineering. 2016;**9**:509-529

[85] Kamerkar S, LeBleu VS, Sugimoto H, Yang S, Ruivo CF, Melo SA, et al. Exosomes facilitate therapeutic targeting of oncogenic KRAS in pancreatic cancer. Nature. 2017;**546**:498-503

[86] Qazi KR, Gehrmann U, Domange Jordö E, Karlsson MCI, Gabrielsson S. Antigen-loaded exosomes alone induce Th1-type memory through a B-cell-dependent mechanism. Blood. 2009;**113**:2673-2683

[87] André F, Chaput N, Schartz NEC, Flament C, Aubert N, Bernard J, et al. Exosomes as potent cell-free peptide-based vaccine. I. Dendritic cell-derived exosomes transfer functional MHC class I/peptide complexes to dendritic cells. Journal of Immunology. 2004;**172**:2126-2136

[88] Hsu D-H, Paz P, Villaflor G, Rivas A, Mehta-Damani A, Angevin E, et al. Exosomes as a tumor vaccine: Enhancing potency through direct loading of antigenic peptides. Journal of Immunotherapy. 2003;**26**:440-450

[89] Yeo RWY, Lai RC, Zhang B, Tan SS, Yin Y, Teh BJ, et al. Mesenchymal stem cell: An efficient mass producer of exosomes for drug delivery. Advanced Drug Delivery Reviews. 2013;**65**:336-341

[90] Sun D, Zhuang X, Xiang X, Liu Y, Zhang S, Liu C, et al. A novel nanoparticle drug delivery system: The anti-inflammatory activity of curcumin is enhanced when encapsulated in exosomes. Molecular Therapy. 2010;**18**:1606-1614

[91] O'Brien CG, Ozen MO, Ikeda G, Vaskova E, Jung JH, Bayardo N, et al. Mitochondria-rich extracellular vesicles rescue patient-specific cardiomyocytes from doxorubicin injury: Insights into the SENECA trial. JACC: CardioOncology. 2021;**3**:428-440

[92] Gonçalves A, Braud AC, Viret F, Genre D, Gravis G, Tarpin C, et al. Phase I study of pegylated liposomal doxorubicin (Caelyx) in combination with carboplatin in patients with advanced solid tumors. Anticancer Research. 2003;**23**:3543-3548

[93] Harrington KJ, Lewanski C, Northcote AD, Whittaker J, Peters AM, Vile RG, et al. Phase II study of pegylated liposomal doxorubicin (Caelyx) as induction chemotherapy for patients with squamous cell cancer of the head and neck. European Journal of Cancer. 2001;**37**:2015-2022

[94] Schmidinger M, Wenzel C, Locker GJ, Muehlbacher F, Steininger R, Gnant M, et al. Pilot study with pegylated liposomal doxorubicin for advanced or unresectable hepatocellular carcinoma. British Journal of Cancer. 2001;**85**:1850-1852

[95] Perez AT, Domenech GH, Frankel C, Vogel CL. Pegylated liposomal doxorubicin (Doxil®) for metastatic breast cancer: The cancer research network, Inc., experience. Cancer Investigation. 2002;**20**:22-29

[96] Seiden MV, Muggia F, Astrow A, Matulonis U, Campos S, Roche M, et al. A phase II study of liposomal lurtotecan (OSI-211) in patients with topotecan resistant ovarian cancer. Gynecologic Oncology. 2004;**93**:229-232

[97] Schwonzen M, Kurbacher CM, Mallmann P. Liposomal doxorubicin and weekly paclitaxel in the treatment of metastatic breast cancer. Anti-Cancer Drugs. 2000;**11**:681-685

[98] Sundar S, Jha TK, Thakur CP, Mishra M, Singh VP, Buffels R. Single-dose liposomal amphotericin B in the treatment of visceral leishmaniasis in India: A multicenter study. Clinical Infectious Diseases. 2003;**37**:800-804

[99] Torchilin VP. Recent advances with liposomes as pharmaceutical carriers. Nature Reviews. Drug Discovery. 2005;**4**:145-160

[100] Caby M-P, Lankar D, Vincendeau-Scherrer C, Raposo G, Bonnerot C. Exosomal-like vesicles are present in human blood plasma. International Immunology. 2005;**17**:879-887

[101] Mao AS, Mooney DJ. Regenerative medicine: Current therapies and future directions. Proceedings of the National Academy of Sciences of the United States of America. 2015;**112**:14452-14459

[102] Bruder SP, Kurth AA, Shea M, Hayes WC, Jaiswal N, Kadiyala S. Bone regeneration by implantation of purified, culture-expanded human mesenchymal stem cells. Journal of Orthopaedic Research. 1998;**16**:155-162

[103] Dennis JE, Merriam A, Awadallah A, Yoo JU, Johnstone B, Caplan AI. A quadripotential mesenchymal progenitor cell isolated from the marrow of an adult mouse. Journal of Bone and Mineral Research. 1999;**14**:700-709

[104] Haynesworth SE, Goshima J, Goldberg VM, Caplan AI. Characterization of cells with osteogenic potential from human marrow. Bone. 1992;**13**:81-88

[105] Johnstone B, Hering TM, Caplan AI, Goldberg VM, Yoo JU. In vitro chondrogenesis of bone marrow-derived mesenchymal progenitor cells. Experimental Cell Research. 1998;**238**:265-272

[106] Pittenger MF, Mackay AM, Beck SC, Jaiswal RK, Douglas R, Mosca JD, et al. Multilineage potential of adult human mesenchymal stem cells. Science. 1999;**284**:143-147

[107] Gojo S, Gojo N, Takeda Y, Mori T, Abe H, Kyo S, et al. In vivo cardiovasculogenesis by direct injection of isolated adult mesenchymal stem cells. Experimental Cell Research. 2003;**288**:51-59

[108] Lin Y-C, Ko T-L, Shih Y-H, Lin M-YA FT-W, Hsiao H-S, et al. Human umbilical mesenchymal stem cells promote recovery after ischemic stroke. Stroke. 2011;**42**:2045-2053

[109] Kohyama J, Abe H, Shimazaki T, Koizumi A, Nakashima K, Gojo S, et al. Brain from bone: Efficient "meta-differentiation" of marrow stroma-derived mature osteoblasts to neurons with Noggin or a demethylating agent. Differentiation. 2001;**68**:235-244

[110] Sanchez-Ramos J, Song S, Cardozo-Pelaez F, Hazzi C, Stedeford T, Willing A, et al. Adult bone marrow stromal cells differentiate into neural cells in vitro. Experimental Neurology. 2000;**164**:247-256

[111] Woodbury D, Schwarz EJ, Prockop DJ, Black IB. Adult rat and human bone marrow stromal cells differentiate into neurons. Journal of Neuroscience Research. 2000;**61**:364-370

[112] Kobayashi T, Hamano K, Li TS, Katoh T, Kobayashi S, Matsuzaki M, et al. Enhancement of angiogenesis by the implantation of self bone marrow cells in a rat ischemic heart model. The Journal of Surgical Research. 2000;**89**:189-195

[113] Sato T, Iso Y, Uyama T, Kawachi K, Wakabayashi K, Omori Y, et al. Coronary vein infusion of multipotent stromal cells from bone marrow preserves cardiac function in swine ischemic cardiomyopathy via enhanced neovascularization. Laboratory Investigation. 2011;**91**:553-564

[114] Le Blanc K, Pittenger M. Mesenchymal stem cells: Progress toward promise. Cytotherapy. 2005;**7**:36-45

[115] Chagastelles PC, Nardi NB. Biology of stem cells: An overview. Kidney International. Supplement. 2011;**1**:63-67

[116] Ullah I, Subbarao RB, Rho GJ. Human mesenchymal stem cells - current trends and future prospective. Bioscience Reports. 2015;**35**:1-18

[117] Laplane L, Solary E. Towards a classification of stem cells. eLife. 2019;**8**:1-5. DOI: 10.7554/eLife.46563

[118] Sabin FR, Doan CA, Forkner CE. The production of osteogenic sarcomata and the effects on lymph nodes and bone marrow of intravenous injections of radium chloride and mesothorium in rabbits. The Journal of Experimental Medicine. 1932;**56**:267-289

[119] Albulescu R, Tanase C, Codrici E, Popescu DI, Cretoiu SM, Popescu LM. The secretome of myocardial telocytes modulates the activity of cardiac stem cells. Journal of Cellular and Molecular Medicine. 2015;**19**:1783-1794

[120] Romano M, Zendrini A, Paolini L, Busatto S, Berardi AC, Bergese P, et al. Extracellular vesicles in regenerative medicine. In: Rossi F, Rainer A, editors. Nanomaterials for Theranostics and Tissue Engineering. United States: Elsevier; 2020. pp. 29-58

[121] Sánchez-Alonso S, Alcaraz-Serna A, Sánchez-Madrid F, Alfranca A. Extracellular vesicle-mediated immune regulation of tissue remodeling and angiogenesis after myocardial infarction. Frontiers in Immunology. 2018;**9**:2799

[122] Saludas L, Oliveira CC, Roncal C, Ruiz-Villalba A, Prósper F, Garbayo E, et al. Extracellular vesicle-based therapeutics for heart repair. Nanomaterials (Basel). 2021;**11**:570

[123] Rogers RG, Ciullo A, Marbán E, Ibrahim AG. Extracellular vesicles as therapeutic agents for cardiac fibrosis. Frontiers in Physiology. 2020;**11**:479

[124] Barile L, Lionetti V, Cervio E, Matteucci M, Gherghiceanu M, Popescu LM, et al. Extracellular vesicles from human cardiac progenitor cells inhibit cardiomyocyte apoptosis and improve cardiac function after myocardial infarction. Cardiovascular Research. 2014;**103**:530-541

[125] Maring JA, Lodder K, Mol E, Verhage V, Wiesmeijer KC, Dingenouts CKE, et al. Cardiac progenitor cell-derived extracellular vesicles reduce infarct size and associate with increased cardiovascular cell proliferation. Journal of Cardiovascular Translational Research. 2019;**12**:5-17

[126] Feng Y, Huang W, Wani M, Yu X, Ashraf M. Ischemic preconditioning potentiates the protective effect of stem cells through secretion of exosomes by targeting Mecp2 via miR-22. PLoS One. 2014;**9**:e88685

[127] Bian S, Zhang L, Duan L, Wang X, Min Y, Yu H. Extracellular vesicles derived from human bone marrow mesenchymal stem cells promote angiogenesis in a rat myocardial infarction model. Journal of Molecular Medicine. 2014;**92**:387-397

[128] Velot É, Madry H, Venkatesan JK, Bianchi A, Cucchiarini M. Is Extracellular vesicle-based therapy the next answer for cartilage regeneration? Frontiers in Bioengineering and Biotechnology. 2021;**9**:645039

[129] Qiao K, Chen Q, Cao Y, Li J, Xu G, Liu J, et al. Diagnostic and therapeutic role of extracellular vesicles in articular cartilage lesions and degenerative joint diseases. Frontiers in Bioengineering and Biotechnology. 2021;**9**:698614

[130] Wang X, Thomsen P. Mesenchymal stem cell-derived small extracellular vesicles and bone regeneration. Basic & Clinical Pharmacology & Toxicology. 2021;**128**:18-36

[131] Jin T, Gu J, Li Z, Xu Z, Gui Y. Recent advances on extracellular vesicles in central nervous system diseases. Clinical Interventions in Aging. 2021;**16**:257-274

[132] Pomatto M, Gai C, Negro F, Cedrino M, Grange C, Ceccotti E, et al. Differential therapeutic effect of extracellular vesicles derived by bone marrow and adipose mesenchymal stem cells on wound healing of diabetic ulcers and correlation to their cargoes. International Journal of Molecular Sciences. 2021;**22**:3851

[133] Boere J, Malda J, van de Lest CHA, van Weeren PR, Wauben MHM. Extracellular vesicles in joint disease and therapy. Frontiers in Immunology. 2018;**9**:2575

[134] Branscome H, Paul S, Yin D, El-Hage N, Agbottah ET, Zadeh MA, et al. Use of stem cell extracellular vesicles as a "holistic" approach to CNS repair. Frontiers in Cell and Development Biology. 2020;**8**:455

[135] Sun Q, Zhang Z, Sun Z. The potential and challenges of using stem cells for cardiovascular repair and regeneration. Genes and Diseases. 2014;**1**:113-119

[136] Evans MJ, Kaufman MH. Establishment in culture of pluripotential cells from mouse embryos. Nature. 1981;**292**:154-156

[137] Chambers I, Smith A. Self-renewal of teratocarcinoma and embryonic stem cells. Oncogene. 2004;**23**:7150-7160

[138] Caplan AI, Dennis JE. Mesenchymal stem cells as trophic mediators. Journal of Cellular Biochemistry. 2006;**98**:1076-1084

[139] Hare JM, Traverse JH, Henry TD, Dib N, Strumpf RK, Schulman SP, et al. A randomized, double-blind, placebo-controlled, dose-escalation study of intravenous adult human mesenchymal stem cells (prochymal) after acute myocardial infarction. Journal of the American College of Cardiology. 2009;**54**:2277-2286

[140] Le Blanc K, Frassoni F, Ball L, Locatelli F, Roelofs H, Lewis I, et al. Mesenchymal stem cells for treatment of steroid-resistant, severe, acute graft-versus-host disease: A phase II study. Lancet. 2008;**371**:1579-1586

[141] Chen L, Tredget EE, Wu PYG, Wu Y. Paracrine factors of mesenchymal stem cells recruit macrophages and endothelial lineage cells and enhance wound healing. PLoS One. 2008;**3**:e1886

[142] Zhang Z, Yang J, Yan W, Li Y, Shen Z, Asahara T. Pretreatment of cardiac stem cells with exosomes derived from mesenchymal stem cells enhances myocardial repair. Journal of the American Heart Association. 2016;**5**:1-16. DOI: 10.1161/JAHA.115.002856

[143] Gnecchi M, He H, Noiseux N, Liang OD, Zhang L, Morello F, et al. Evidence supporting paracrine hypothesis for Akt-modified mesenchymal stem cell-mediated cardiac protection and functional improvement. The FASEB Journal. 2006;**20**:661-669

[144] Shao L, Zhang Y, Lan B, Wang J, Zhang Z, Zhang L, et al. MiRNA-sequence indicates that mesenchymal stem cells and exosomes have similar mechanism to enhance cardiac repair. BioMed Research International. 2017;**2017**:4150705

[145] Schwarz ER, Speakman MT, Patterson M, Hale SS, Isner JM, Kedes LH, et al. Evaluation of the effects of intramyocardial injection of DNA expressing vascular endothelial growth factor (VEGF) in a myocardial infarction model in the rat--angiogenesis and angioma formation. Journal of the American College of Cardiology. 2000;**35**:1323-1330

[146] Pearlman JD, Hibberd MG, Chuang ML, Harada K, Lopez JJ, Gladstone SR, et al. Magnetic resonance mapping demonstrates benefits of VEGF-induced myocardial angiogenesis. Nature Medicine. 1995;**1**:1085-1089

[147] Simons M, Annex BH, Laham RJ, Kleiman N, Henry T, Dauerman H, et al. Pharmacological treatment of coronary artery disease with recombinant fibroblast growth factor-2: Double-blind, randomized, controlled clinical trial: Double-blind, randomized, controlled clinical trial. Circulation. 2002;**105**: 788-793

[148] Gerbin KA, Murry CE. The winding road to regenerating the human heart. Cardiovascular Pathology. 2015;**24**: 133-140

[149] Orlic D, Kajstura J, Chimenti S, Jakoniuk I, Anderson SM, Li B, et al. Bone marrow cells regenerate infarcted myocardium. Nature. 2001;**410**:701-705

[150] Ma J, Zhao Y, Sun L, Sun X, Zhao X, Sun X, et al. Exosomes derived from Akt-modified human umbilical cord mesenchymal stem cells improve cardiac regeneration and promote angiogenesis via activating platelet-derived growth factor D. Stem Cells Translational Medicine. 2017;**6**:51-59

[151] Zou X, Gu D, Xing X, Cheng Z, Gong D, Zhang G, et al. Human mesenchymal stromal cell-derived extracellular vesicles alleviate renal ischemic reperfusion injury and enhance angiogenesis in rats. American Journal of Translational Research. 2016;**8**:4289-4299

[152] Liang X, Zhang L, Wang S, Han Q, Zhao RC. Exosomes secreted by mesenchymal stem cells promote endothelial cell angiogenesis by transferring miR-125a. Journal of Cell Science. 2016;**129**:2182-2189

[153] Gonzalez-King H, García NA, Ontoria-Oviedo I, Ciria M, Montero JA, Sepúlveda P. Hypoxia inducible factor-1α potentiates jagged 1-mediated angiogenesis by mesenchymal stem cell-derived exosomes: Angiogenesis mediated by Jagged1 in MSC exosomes. Stem Cells. 2017;**35**:1747-1759

[154] de Couto G, Gallet R, Cambier L, Jaghatspanyan E, Makkar N, Dawkins JF, et al. Exosomal MicroRNA transfer into macrophages mediates cellular postconditioning. Circulation. 2017;**136**:200-214

[155] Cambier L, Couto G, Ibrahim A, Echavez AK, Valle J, Liu W, et al. Y RNA fragment in extracellular vesicles confers cardioprotection via modulation of IL −10 expression and secretion. EMBO Molecular Medicine. 2017;**9**:337-352

[156] Beuzelin D, Kaeffer B. Exosomes and miRNA-loaded biomimetic nanovehicles, a focus on their potentials preventing type-2 diabetes linked to metabolic syndrome. Frontiers in Immunology. 2018;**9**:2711

[157] Dooley J, Garcia-Perez JE, Sreenivasan J, Schlenner SM, Vangoitsenhoven R, Papadopoulou AS, et al. The microRNA-29 family dictates the balance between homeostatic and pathological glucose handling in diabetes and obesity. Diabetes. 2016;**65**:53-61

Chapter 7

Roles of Extracellular Vesicles in Human Reproduction

Weidong Zhou, Teng Zhang, Yikai Lian and Wenjing Zhang

Abstract

Extracellular vesicles (EVs) are newly identified as cell-to-cell communication mediators that carry and transfer various regulatory molecules. Recent studies have shown that EVs play important roles in normal physiology and pathological conditions of human reproduction. In the female reproductive system, EVs in follicular fluid, oviduct fluid, and uterine luminal fluid are considered as vehicles to regulate follicular development, oocyte maturation and mediate embryo–maternal crosstalk to affect embryo implantation and pregnancy. In the male reproductive system, prostasomes and epididymosomes are involved in regulating sperm maturation, motility, capacitation, acrosome reaction, and fertilization. EVs transmitted cargos also play important roles in reproduction-related pathologies, such as polycystic ovarian syndrome, endometriosis, pregnancy complications, male infertility, and gynecological malignant tumors. In view of the important roles in the reproductive system, EVs may be used as biomarkers or therapeutic targets for reproductive abnormalities and related diseases. In this chapter, we sorted EVs in human reproduction through their physical/pathological functions and mechanisms, and listed several EVs as biomarkers and clinical therapeutic applications in the future.

Keywords: extracellular vesicles, exosome, reproductive system, polycystic ovarian syndrome, endometriosis, pregnancy disorders, male infertility

1. Introduction

Cell communication is vital for all living organisms, whether between environment or host with single-celled organisms, or between cells in multicellular organisms [1]. As new-found mediators, extracellular vesicles (EVs) mediate a cell communication mechanism that is different from classic ways as long/short range of secretory signaling, receptor-dependent contact signaling, gap junction, etc. According to cell origin, biogenetic mechanisms, and physical features (size and density), EVs can be classified into three major categories: apoptotic body (ABs), microvesicles (MVs), and exosomes (EXOs) [2]. These vesicles act as shuttles, transporting "cargo" like protein, lipids, nucleic acids, and other regulatory chemicals from one tissue or cell to another through body fluid [3]. When EVs and their targets fuse, their cargos can influence the functioning of target cells, allowing them to engage in a variety of physiological

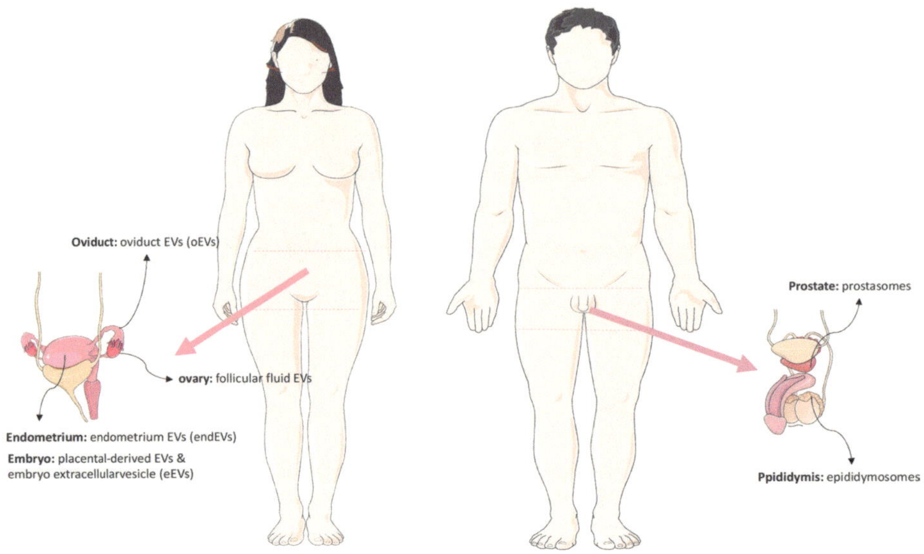

Figure 1.
EVs from human reproductive systems.

and pathological processes. That is why EVs have a significant role in multiple biological processes and diseases, such as immune function, tissue homeostasis, cancer, and neurodegenerative diseases. Recently, EVs have also been described in human female and male reproductive systems, including the oviduct, ovary, endometrium, embryo, prostate, and epididymis (**Figure 1**). EVs are key mediators in human reproduction and take part in physiological processes like gametogenesis (spermatogenesis and oogenesis), fertilization (coordination of sperm capacitation and acrosome reaction), embryogenesis, and implantation (crosstalk between mother and embryo). EVs and trigger and/or maintenance some conditions such as infertility, impotence, polycystic ovary syndrome (PCOS), endometriosis (EMs), premature ovarian failure (POF), and preeclampsia (toxemia) [2, 4].

2. EVs in the female reproductive system

The female reproductive system contains external and internal parts. Internal genital organs are mainly composed of the gonad (ovary), oviduct, and uterus (namely womb). As an organ whose basic functions are ovulation and the production of hormones, ovary influences a woman's feminine physical characteristics and affects the process of reproduction. In fertilization preparation, an ovulated oocyte enters the oviduct to meet spermatozoa (SPZ). Then fertilization occurs in oviduct when a sperm and an oocyte combine and fuse to form a zygote, which develops into a blastocyte and then implant uterus endometrium to further embryonic development. Finally, in the perinatal period, the fetus is delivered from the dilated cervix [5, 6]. Among these progresses, EVs coordinate the maturation of female gamete and mediate crosstalk to help successful embryo implantation (known as blastocyst implantation) and pregnancy in follicular fluid, oviduct, endometrium, vagina, semen, and embryo.

2.1 EVs derived from the ovary

The ovarian follicle is the basic unit of female ovary, which contains an oocyte that is surrounded by two-layer granulosa cells (GCs, including membrane/mural granulosa and cumulus granulosa), stromal cells (theca cells), and a liquid called follicular fluid (FF). FF offers a suitable micro-environment for follicle growth and oocyte development (oogenesis) in the ovary, so their relative stability of composition is indispensable. Similar to gap junctions between oocytes and somatic cells which mediated cell-to-cell communication, EVs in GCs, somatic cells, and FF mediate signal transmission and cell communication [7, 8], suggesting that EVs are carriers for molecular transfer in the ovary.

The first EVs in FF was reported in equine, which contains microRNAs (miRNAs) and proteins, and they could be taken by surrounding GCs *in vitro* and *in vivo* [9]. MiRNAs are involved in RNA silence as post-transcriptional regulators [10]. There are numerous miRNAs indirectly regulating follicular growth and oocyte maturation in human FF [7]: (1) miR-10b, miR-21-5p, miR-31, miR-95, miR-99b-3p, miR-134, miR-135b, miR-140-3p, and miR-190b participate in WNT signaling through glycoprotein signaling molecular WNTs potentially regulate follicle formation, growth, and ovulation/luteinization [11]. (2) miR-203, miR-218 regulate MAPK (ERK1/2) pathway which activation is crucial for oocyte meiotic resumption, GCs proliferation and cumulus expansion [12]. (3) miR-489, miR-493, miR-503, miR-542-5p, miR-654-3p, miR-874, miR-886-5p, miR-887 as mediators of TGFβ. (4) miR-337-5p, miR-339-3p, miR-370, miR-449a, miR-455-5p, miR-483-5p as mediators of ErbB, etc. In addition, as we know, the expansion of cumulus-oocyte complex (COC) is a key process of ovulation. A study showed that bovine FF EVs taken in cumulus could increase gene expression of cumulus expansion (PTGS2, PTX3, and TNFAIP6) and support the expansion process *in vitro* [13].

2.2 EVs derived from the oviduct

Oviduct is the place where occur oocyte transportation, fertilization, and initial embryo development. Oviductal extracellular vesicles (oEVs), as the main components of oviduct fluid, can orchestrate gamete/embryo-oviduct and embryo-maternal interactions, hence supporting oocyte mature (in canine), embryonic development, and motility, capacitation, and fertilizing ability of sperm [14, 15]. The contents of oEVs (mRNAs, miRNAs, and proteins) are dynamic at different stages of the estrous cycle, suggesting that cargos of oEVs are under hormonal control [16]. The first oEVs in the oviductal fluid was described in mouse, and then they were identified in other mammals including human. In humans, oEVs (oviductosomes) arising from the apocrine pathway carry and deliver plasma membrane Ca^2 + -ATPase (PMCA) which are fertility-modulating proteins. PMCA delivered to sperm can prevent premature capacitation and maintain Ca^2 + levels homeostasis; lack of PMCA leads to sperm motility loss and male infertility in mice [17]. Bovine oEVs can pass through the zona pellucida and then be internalized by embryonic cells to increase blastocyst rate and improve the quality of embryo. The possible mechanism may involve: (1) The proteins of bovine oEVs involved in cell communication, cell metabolism, localization, and reproduction [18]; (2) The miRNAs of bovine oEVs contribute to a successful pregnancy through inducing changes of embryonic transcriptome [15]; (3) Exosomes derived from bovine oviduct epithelial cells (BOECs) improve the mitochondrial health of bovine embryo because they re-establish the tricarboxylic acid cycle (TCA-cycle) flux [19].

2.3 EVs derived from the endometrium

Successful pregnancy not only needs intercommunication in ovary and oviduct, but also the crosstalk between endometrium and blastocyst especially the trophoblasts. Indeed, several studies have found that the secretions in endometrium affect the embryo. The first reported EVs of the female reproductive system are endometrium EVs (endEVs) in mouse uterine cavity [20]. EndEVs were found in endometrial fluid during the menstrual/estrous cycle, and were also released from endometrial epithelial cells cultured *in vitro* [21]. Several studies reported that the EVs in the endometrium have autocrine/paracrine effects on regulation of receptivity and implantation of uterine [22].

In EndEVs of primary endometrial epithelial cells (ECCs), 35% of the proteins had not been reported before, indicating that the contents of EndEVs are unique [23]. EndEVs influence blastocyst implantation through their hormone-specific protein cargos to modulate trophoblast capacity of adhesion, migration, invasion, and so on. Among these protein cargos, metalloproteinases (MMP-14 and ADAM10) regulated by the endometrial receptivity-related hormone are essential to trophoblast invasion and pregnancy outcome because they regulate (activate and degrade) other factors in endEVs. *In vitro*, when EndEVs are internalized by trophoblast cells, they can activate Focal Adhesion Kinase (FAK) signaling, thereby enhance the adhesive capacity of trophoblast cells [24–26]. The proteins of bovine endEVs in ECCs change during the peri-implantation period: endEVs enhance the expression of cell apoptosis genes in the preimplantation stage, but cell adhesion genes in the post-implantation stage [27].

MiRNAs of EndEVs act as modifiers for implantation. EndEVs contain specific miRNAs cargo which targets predictably genes involved in blastocyst implantation and crucial signaling pathways such as the VEGF, the Jak–STAT, and the Toll-like receptor in primary ECC and ECC1 (an endometrial epithelial cell line) [21]. During the window of implantation (the period allows blastocyte invasion), maternal miRNAs are differently expressed in human EndEVs. One of the miRNAs, miR-30d upregulates Itgb3, Itga7, and Cdh5, which are involved in embryo adhesion in murine; treating miR-30d *in vitro* to murine embryo can increase embryo adhesion [28].

Sperm adhesion molecule 1(SPAM1)/PH-20 is a hyaluronidase which enhances sperm fertility. EVs in murine uterine luminal fluid can deliver SPAM1 to sperm membranes to increase sperm penetration through cumulus cell layers around oocyte, and adhesive to zona pellucida, same as SPAM1 transferred from epididymosome to SPZ play a significant role in sperm maturation and motility [2, 20]. While as mentioned earlier, in oEVs SPAM1 interaction with sperm to inhibit premature acrosomal reaction, showing that the EVs with the same cargo in different organs have multiple functions [17].

2.4 EVs derived from the placenta and embryo

Pregnancy is a unique immunomodulatory state in which the maternal immune system temporarily tolerates the paternal antigen so as to protect the fetus from allogeneic rejection, at the same time maintaining immune surveillance to protect the mother from external pathogens. EVs have been identified in trophoblastic cells, placenta, and maternal circulation [22]. Maternal immune response is effected by the EVs from endometrium, embryo, and trophoblast cells in early pregnancy, while by the EVs from the human placenta in late pregnancy [29].

Placenta is an important organ and performs a variety of functions to support pregnancy. Placental-derived EVs are important media for intercellular communication, considered to be potential mediators in regulating maternal immune response to achieve a successful pregnancy and maternal and infant health outcomes [30]. These EVs inhibit the immune response to the developing fetus and establish and maintain a systemic inflammatory response against infectious invaders [31]. The underlying mechanisms include: (1) Compared with non-pregnancy, during pregnancy, the cytokines such as transforming growth factor-β 1 and IL-10 increase, and the ability to induce caspase-3 activity in cytotoxic natural killer (NK) cells enhance in peripheral blood EVs, which promoting immunosuppressive phenotype by inducing apoptosis to help regulate the maternal immune response to the fetus [32]. (2) In amnion (the innermost layer of placenta), miR-21 helps in embryo growth; histone 3 (H3), heat shock protein 70 (HSP70), activated form of prosenescenceP-p38, MAPK participate in stress response and are related to pregnancy outcome [33].

In addition to the placenta, embryos can also secrete EVs (embryo extracellular vesicle, eEVs) to participate in the regulation of implantation and other pregnancy processes. Embryos transferred to uterine are known as blastocyst, which consists of a fluid-filled cavity, inner cell mass (ICM), and trophectoderm/trophoblast (T) cells. Mouse embryonic stem cells from the ICM can produce MVs which reach the trophoblast ectoderm, thereupon enhancing the migration ability of trophoblast cells, both as isolated cells and in the whole embryo. Laminin and fibronectin exist in the eEVs of ICM orchestrate ICM to attach to the integrin on the surface of trophoblast cells and stimulates the cascade of c-Jun N-terminal kinase and FAK, increasing the migration of trophoblast cells. Injecting these eEVs into the blastocyst cavity of day 3.5 blastocysts can improve successful implantation rate [34]. EEVs-derived PIBF alters the maternal immune system by increasing IL-3, IL-4, and IL-10 to achieve a Th2-dominant cytokine balance. The aberrant expressions of PIBF may lead to pregnancy failure [35]. In eEVs of trophoblast also carry a variety of factors regulating maternal immunity to protect the fetus, including: tissue factors (TFs), soluble vascular endothelial growth factor receptor1 (sFlt-1), immunosuppressive factors, and so on [33].

3. EVs in the male reproductive system

The male reproductive system consists of internal organs - gonads (testis), reproductive ducts (epididymis, vas deferens, ejaculatory ducts, male urethra), and accessory glands (seminal vesicle, prostate, urethral bulbar gland) and external organs. Testes produce sperm and secrete male sex hormones. SPZ/sperm produced by the testis are first stored in the epididymis; during ejaculation, SPZ are excreted through the vas deferens, ejaculatory ducts, and urethra. Semen is the protect fluid around SPZ containing seminal plasma (SP), which is derived from testis (5%), epididymis and prostate (20%), seminal vesicles (65%), and seminal vesicles (65%). Seminal plasma lipids and/or EVs are rich in lipids, sugars, growth factors, TF and proteins, which play important roles in sperm survival, membrane integrity, maturation, motility, capacitation, acrosome reaction, and immune surveillance regulation [2, 33, 36]. Seminal plasma extracellular vesicles (Seminal plasma EVs, spEVs) are mainly derived from epididymis and prostate in the male reproductive tract, and a large number of them exist in the seminal plasma. Proteins in the seminal plasma are provided by spEVs [37]. According to the source, there are two main types of spEVs: prostasomes and epididymosomes. A proteomic analysis showed that there are a total of 1474 proteins in seminal plasma-derived exosomes; bioinformatics analysis revealed that these proteins are involved in a variety of biological processes, such as cell growth and maintenance, metabolism, transport, energy pathways, and so on [38]. Human semen exosomes also contain a unique non-coding small RNA library, which may have a potential regulatory function to mediate fertilization by transmitting regulatory signals to the receptors to regulate the female reproductive tract [39].

3.1 EVs derived from the prostate

The first EVs of the reproductive system found in humans are prostasomes, which were observed in human prostatic fluid and seminal plasma in 1978 [40]. Prostasomes are exosomes with a diameter of 30–500 nm, which are released from prostatic epithelial cells and are then transferred to the prostate duct. Prostasomes are characterized by a cholesterol-to-phospholipid ratio of 2:1, of which nearly 50% phospholipid is sphingomyelin, making its bilayer or multilayer lipoprotein plasma membrane very hard. This unique prostate membrane component allows the prostate to fuse with and transfer its contents to other cells [41, 42]. Prostasomes can transport substances such as sphingomyelin, cholesterol, and saturated glycolipids to sperm to reduce their membrane fluidity in order to prevent premature or spontaneous acrosome reaction and premature capacitation [43]. Previous studies have shown that prostasomes may inhibit capacitation and acrosome reaction mainly through cholesterol transfer [44, 45]. Once the appropriate time for acrosome reaction occurs, prostasomes also play important roles in the capacitation and induction of acrosome reactions as mediators of signal transmission to regulate the tyrosine phosphorylation pattern, which is necessary for sperm-oocyte interaction [46].

Sperm motility is affected by intracellular pH and Ca^{2+} concentration. Prostasomes were determined to be associated with Mg^{2+} and Ca^{2+} dependent ATPase activity when they were first identified, suggesting that they are related to sperm energy metabolism [47]. Annexins of prostasome can activate Ca^{2+} channel and increase the level of Ca^{2+} in sperm through carrying CD38 and RyR to sperm to stimulate the production of cyclic adenylate diphosphate, thus affecting sperm motility. After ejaculation, prostasomes interact with sperm and protect them from female reproductive tract acidic environment and regulate sperm motility through PH- dependent manner to maintain the ability of sperm fertilization and prepare for an encounter with oocytes [48, 49]. Prostasomes transfer proteins (such as galactose lectin 3 and CD48) can protect sperm from immune reaction by regulating immune response pathways, such as inhibition of complement pathway, lymphocyte proliferation, and phagocytosis of monocytes and neutrophils in the female reproductive tract [50].

Additionally, prostasomes have antioxidant properties. By interacting with polymorphonuclear neutrophils that produce ROS, prostasomes can reduce the production of ROS, protect sperm and improve sperm survival rate. Prostasomes also carry aminopeptidase N, a protein involved in regulating sperm motility which acts by regulating endogenous opioid peptides (such as enkephalin) [51, 52]. Prostasomes are known to contain a variety of coding and non-coding regulatory RNAs with potential regulatory functions [39]. Currently, there are few studies on nucleic acid cargos of prostasomes and their effects on the male reproductive system.

3.2 EVs derived from the epididymis

Epididymosomes are produced by epididymal epithelial cells through apocrine secretion. They are a kind of exosomes with a diameter of 50–250 nm and characterized by high cholesterol/phospholipid ratio. Epididymosomes were first found in the intraluminal compartment of the epididymis of Chinese hamsters, and then in mice, rats, cattle, sheep, and humans. Epididymosomes can carry and transport numerous proteins with biological functions (including enzymes, adhesion molecules, transport, and signal transduction proteins) and non-coding RNA (such as miRNA). These proteins and non-coding RNA play an important biological role in the process of sperm maturation and fertilization, participating in the acquisition of motility, the acquisition of fertilization ability, and protection against oxidative stress [50, 53–55].

The proteins of epididymosomes including aldo-keto reductase family 1 member B (AKR1B1), phosphatidylethanolamine binding protein 1 (PEBP1), macrophage migration inhibitory factor (MIF), polyol pathway enzymes, glutathione peroxidase 5 (GPX5), plasma membrane Ca^{2+} -ATPase 4a (PMCA4), ubiquitin, and SPAM1/PH-20. AKR1B1 and PEBP1 can jointly regulate the sperm state, keeping sperm at a quiescent state during transportation until ejaculation [56]. MIF and polyol pathway enzymes are involved in sperm maturation and fertilization. GPX5, together with ubiquitin, is transferred to the sperm acrosome region during sperm epididymal transport, which protects sperm from oxidative stress, maintains the integrity of DNA, and prevents premature acrosome reaction. As a calcium efflux pump, PMCA4 carried by epididymosomes can regulate calcium concentration, affect the activation of calcium signaling pathway and maintain the homeostasis of Ca^{2+} in spermatozoa. Sperm delete PMCA4 leads to loss of hyperactivation and capacitation. GPI anchoring proteins include P34h and SPAM1/PH-20. As mentioned earlier, they are sperm binding proteins related to epididymosomes, which locate on the sperm surface during sperm epididymal transport, mediating the docking between epididymosomes and sperm and providing a "highway" for epididymosomes to transport their cargos to participate in the process of fertilization. Similar to their function are the P25b/P26 that facilitate sperm binding to the zona pellucida [33, 57–61].

Epididymosomes are also mediator of cell communication. New surface antigens are obtained on the sperm surface during epididymal transport, which are related to the acquisition of fertilization ability. Epididymosomes transmit Notch signals between epididymal epithelial cells and between epididymis and SPZ to influence sperm motility [62].

Epididymosomes are also rich in non-coding RNA. Epididymosomes contain many kinds of miRNA, such as miR-888, miR-182, miR-24, and miR-15b. MiR-888 of epididymosomes maintains sperm flagella peristalsis and mature sperm structure by regulating SPAG6 and SPAG1. Target gene prediction shows that miR-15b can specifically regulate the expression of IDH3A and control energy metabolism in TCA cycle, while miR-182 and miR-24 can specifically regulate the expression of glycogen synthase kinase 3α (GSK3A), and phosphorylation of GSK3A can affect sperm motility. Functional enrichment analysis shows that these miRNAs play important roles in embryonic development. There are also inflammation-related miRNAs in epididymosomes, such as miR-181a and miR-1224. Studies have shown that miR-181a participates in inflammatory response by regulating B cell differentiation and T cell receptor signals. MiR-146 can inhibit inflammation and innate immune response by down-regulating various pro-inflammatory cytokines, but its specific mechanism needs to be further studied. When inflammation exists, miR-1224 activates the immune response by down-regulating the expression of TNF- α. Epididymosomes also contain miR-29a, which enhances the expression of nuclear autoantigen sperm protein (NASP) and inhibits the proliferation of epididymal epithelial cells [63–68].

In addition, epididymosomes contain a small molecule: tRNA-derived small RNAs (tsRNA). Studies have shown that tsRNAs have the functions of regulating gene transcription, cell proliferation and apoptosis, and stress response [69, 70].

4. Implications of EVs in reproductive pathology

4.1 Polycystic ovarian syndrome

Polycystic ovary syndrome (PCOS) is one of the most common reproductive endocrine diseases in women of childbearing age (global prevalence rate: 4–21%). The clinical features of PCOS are: hyperandrogenemia, polycystic ovary (PCO) morphology, and oligo-ovulation/anovulation [71]. PCOS as a disease of follicular abnormal, there is a close relationship between EVs in follicles and the pathogenesis of PCOS [72].

The amount of EVs (mainly exosomes) in plasma of patients with PCOS are significantly increased and positively correlated with the number of follicles [73]. The expression of non-coding RNAs altered in exosomes of FF in PCOS, including miRNAs, piRNAs, and tRNAs [74]. Further study found that there is differential expression of miRNAs in plasma exosomes of women with PCOS, which was related to the menstrual cycle, antral follicle count (AFC, means the number of antral follicle) and hormone levels [75]. LncRNAs and circRNAs are also differentially expressed in follicular exocrine bodies of patients with PCOS. They may regulate ovarian steroid production, aldosterone synthesis, and secretion, and involved in Jak–STAT signal pathway, hippo signal pathway, and MAPK signal pathway. A high-throughput lncRNAs sequencing study found that there are 1253 upregulated and 613 down-regulated lncRNAs in FF exosomes of patients with PCOS infertility compared with patients with non-PCOS infertility, and nine lncRNAs with significant changes may play an important role in the pathogenesis of PCOS (lncRNA-LINC00173, lncRNA-H19, lncRNA-HDAC6, lncRNA-POP4, lncRNA-PTEN, lncRNAAKT3, lncRNA-DICER1, lncRNA-NF1, and lncRNA-MUM1). High-throughput sequencing of circRNAs also found the expression of 167 circRNAs in FF exosomes of PCOS patients are significantly upregulated and 245 circRNAs are significantly down-regulated compared with the control group, suggesting that these abnormally expressed circRNAs may play some roles in the study of pathophysiological mechanism of PCOS [76–78]. Similarly, proteins in exosomes of FF also changed. A proteomic study of exosomes from FF in patients with PCOS and healthy controls found that exosomes rich in S100-A9 can activate NF-κB signaling pathway in GCs and may play key roles in the progression of PCOS [79].

4.2 Endometriosis

Endometriosis (EMs) is an estrogen-dependent inflammatory disease, characterized by the deposition and growth of endometrial stromal cells (ESCs) outside the uterine cavity, resulting in the appearance of endometrial tissue with growth activity outside the uterine body. Pelvic peritoneum and ovary are the most common sites of ectopic endometrial stromal cells (EuESCs) growth. The main clinical manifestations are pelvic pain, pelvic adhesion, infertility and so on. EMs affects millions of women around the world, and the cause remains to be further determined [80].

The number of EVs in cervical and vaginal samples of rhesus monkeys with EMs decreased, indicating that the synthesis pathway of EVs in EMs has changed [81]. EMs are closely related to the formation of blood vessels. The biologically functional exosomes released by ESCs are transported to other parts through blood countercurrent in the endometrial and peritoneal microenvironment, and mediate intracellular signal transduction to ESCs itself or neighboring cells through the intercellular space in an autocrine or paracrine manner, thus regulating angiogenesis. *In vitro*, exosomes mediate the promotion of angiogenesis by EuESCs in the development of EMs [82]. Previous studies have shown that miRNAs can be extracted from the exosomes of ESCs, and these miRNAs potentially regulate the angiogenesis of ESCs. The expression of miRNA-21 related to angiogenesis in the exosomes of women with EMs is significantly higher than that of women without Ems [83]. Our research group identified the differential expression patterns of exosomal miRNAs in patients with EMs and found that 49 miRNAs expressed differentially in EuESC exosomes compared with normal endometrial stromal cells (NESCs) exosomes. Many exosomal miRNAs may be involved in regulating endometrial receptivity in women with EMs-related infertility through their predicted target genes: homeobox A10 (HOXA10) and leukemia inhibitory factor (LIF) which are essential for normal implantation. Our finding provides a new sight on how EVs participate in the occurrence and development of EMs [84].

4.3 Pregnancy complications

Common pregnancy complications include gestational hypertension, diabetes, and preeclampsia. Serious pregnancy complications may endanger the lives of fetuses and mothers [85]. The concentration and biological activity of EVs changed in a variety of pregnancy complications, such as preeclampsia (PE), gestational diabetes mellitus (GDM), preterm delivery (PTB), intrauterine growth restriction (IUGR), recurrent abortion and unexplained abortion, suggesting that EVs are closely related to pregnancy complications [86, 87].

Maternal obesity is a risk factor for GDM and several other pregnancy complications. Adipose tissue hypertrophy or metabolic stress can change the cargos in EVs (mainly miRNAs), leading to systemic inflammation and insulin resistance (IR) in obese patients with gestational diabetes. These altered EVs may also change the physiological function of placenta and remove the regulation of placental nutrition signal pathway, resulting in obesity-related pregnancy complications [88–90].

There is a class of syncytial nuclear aggregates (SNAs) in placental-derived EVs the level of which increases with the progression of pregnancy and is associated with pregnancy complications such as PE [91]. Numerous placental EVs were detected in the serum of pregnant women with PE. These EVs have pro-inflammatory, anti-angiogenic, and procoagulant activities, which may lead to activation of the blood coagulation system, systemic inflammation, and vascular endothelial dysfunction [92]. In addition, the abundance of syncytin-2 in serum-derived EVs attenuates in women with PE. This may lead to immunosuppressive reduction and pathological inflammation in pregnancy complications [93].

4.4 Male infertility

Infertility has becoming a global health problem, the incidence rate is as high as 15%, of which about 50% of infertility cases are caused by male reproductive disorders [94]. Just as oocytes need normal FF microenvironment to provide nutrition and support, sperm also need SP to provide a safe environment in order to survive and transport in the female reproductive tract. As mentioned earlier, there are a train of semen proteins in SP, most of which are transported by EVs (prostasomes and epididymosomes). Since only EVs proteins from normal sperm can regulate the movement of SPZ and trigger SPZ capacitation, the changes of proteomic characteristics in semen EVs may indicate male reproductive tract dysfunction; at the same time, they can be used as a biomarker of male infertility. In male infertility, the proteins transported by reproduction-related EVs are differentially expressed in azoospermia, asthenospermia, oligozoospermia, teratospermia, or other male infertility compared with normal sperm, and the prostatic proteins related to sperm energy production and sperm activity are under expressed in abnormal sperm, indicating that SP proteome map may be a potential indicator of sperm dysfunction [36].

Azoospermia may be one of the causes of male infertility. There are two types of azoospermia: non-obstructive (NOA) and obstructive azoospermia (OA) caused by seminal tract obstruction [95]. A study found that deficiency of EVs contributes to lower ejection volume, and changes of various nutritional components in semen [96].

At present, a sea of EV cargos has been identified to be differentially expressed in male infertility patients, which can be used as diagnostic markers and treatment of infertility as detailed in the next section.

4.5 Ovarian and cervical cancer

Gynecological malignant tumors such as ovarian cancer (OC) and cervical cancer (CC) are two of the three major tumors of the female reproductive system, the former with the highest mortality rate and the latter with the second incidence and the third fatality rate [97, 98]. The difficulty of early diagnosis, high metastasis rate, and strong drug resistance are still the main obstacles in gynecological cancer diagnosis and treatment. As a member of tumor microenvironment, exosomes not only play important roles in tumor occurrence and development, drug resistance and immunosuppression, but also can be used as a new tumor marker and clinical target molecule in clinical work.

The latest study has proved that there is a difference in the expression of exosomes in the blood of patients with OC and normal people, and the expression in the body fluid of patients with OC is related to the stage of the tumor [99]. A total of 1017 co-expressed proteins were screened from the exosomes secreted by two kinds of OC cells, among which tubulin beta 3 class III (TUBB3), epithelial cell adhesion molecule (EpCAM), claudin 3 (CLDN3), proliferating cell nuclear antigen (PCNA), epidermal growth factor receptor (EGFR), and fatty acid synthase (FASN) were highly enriched in tumor-related signal pathways. Claudin-4 positive exosomes can be used in the diagnosis of OC, with a specificity of 98% and a sensitivity of 51%, while CA125 has a specificity of 98% and a sensitivity of 71%. Although claudin-4 is not a better diagnostic marker than CA125, this study confirms that exosome-related proteins can be used in the diagnosis of OC. These results suggested that the related proteins in exosomes may become markers for the diagnosis of OC [100, 101]. In addition, the level of exosomal proteins extracted from the serum of patients with OC were higher than that of patients with benign ovarian disease and healthy women, and that of patients with advanced OC was higher than that of patients with early stage. The expression of tumor-specific antigen MAGE3/6 and transforming growth factor β1 (TGF- β1) in patients with OC was significantly higher than that in patients with benign ovarian disease and healthy women [102]. CD24 is a marker of poor prognosis in OC and other types of cancer. CD24 positive exosomes were screened from ascites of patients with OC, which is a good marker for early diagnosis of OC [103].

There are a large number of exosomes in cervicovaginal lavage specimens of women with CC, which carry miRNAs playing important roles in CC. It was found that many differentially expressed miRNAs, such as miR-483-5p, miR-1246, miR-1275, microRNA-21, microRNA-146a and miR-222-3p, were up-regulated in vaginal lavage and cell culture of CC, while some miRNAs, such as let-7d-5p, miR-92a-3p, miR-20a-5p, miR-378a-3p, miR-423-3p, miR-7-5p, miR-99-5p, miR-100 5p and miR-320a, were down-regulated. It is suggested that the contents of miRNAs in exosomes may be related to the occurrence of CC and is expected to become a new diagnostic marker of CC [104]. There was a significant difference in the expression of exosome-mediated let-7d-3p and miR-30d-5p between cervical tumors and adjacent normal tissues, suggesting that these two plasma exosomes let-7d-3p and miR-30d-5p can be used as valuable biomarkers for non-invasive screening of CC [105]. The exosomes of cervical squamous cell carcinoma (CSCC) transfers miR-221-3p from cancer cells to vascular endothelial cells and promote angiogenesis by down-regulating thrombospondin-2 (THBS2) [106]. In addition, exosomal miR-221-3p secreted by CC cells promotes the invasion, migration, and angiogenesis of CC microvascular endothelial cells (MVECs) by down-regulating the expression of MAPK10 [107]. The expression of activating transcription factor 1 (ATF1) and RAS genes were significantly upregulated in primary and recurrent CC mouse models, and ATF1 and RAS could also be detected in blood exosomes of mouse models. These results suggest that exosome-mediated ATF1 and RAS may become potential diagnostic markers for CC, which provides a new idea for individual detection and treatment of CC [108].

5. Clinical and therapeutic applications of EVs in reproductive disorders

5.1 EVs as biomarkers

The optimal age to have a baby is a medical social problem that needs to be faced by the whole society. Based on the risk of fetal disease, male and female fertility, and many other factors, it is generally believed that the best childbearing age for men and women is between 25 and 35 years old. With the increase of age, the semen quality of men decreases, the sperm concentration and motility decrease, and the deformity rate increases; the quality of oocytes decreases in women, which limits the success of pregnancy [109]. Therefore, looking for reproductive-related markers that change with age has become a meaningful research direction. It is found that the miRNAs characteristics of follicular EVs vary with the age of women, suggesting that miRNAs of EVs are possible markers and predictors of age-related oocyte quality decline [9]. A study compared the EVs of FF in young women with that in older women and found four significantly different miRNAs: miR-99b, miR-134, and miR-190b were upregulated, and miR-21-5p was down-regulated in older women. These miRNAs regulate genes related to cell apoptosis, p53 signaling, and cytokine-cytokine-receptor interaction, so their changes may affect follicular development and oocyte maturation [110].

There are numerous EVs in SP which can easily detach and collection. A study identified that prostasomal proteins relating to sperm activity and energy production pathways have changed in non-normozoospermic men. Among them, HIST1H2B, KLK2, MIF, MPO, and MSMB are related to liquefaction of semen (in order to break through SP) and sperm-oocyte binding. LDHC, HK1, PNP, APRT, and SLC2A14 are involved in sperm energy production [111]. Others include ELSPBP1/BLVRA, GPX5, SPAM1, P34H, Aldose reductase and sorbitol dehydrogenase, PAP, PSA, TMPRSS2, pTGase, PSCA, KIF5B, ANXA2, and so on [50]. Due to these variations may have a relevant correlation with male infertility, the cargos of EVs can potentially serve as useful biomarkers of male infertility.

In ART, FF can be taken out together with the follicles, so it is an attractive biomarker to detect the quality of oocytes. The expression of miRNAs in EVs is related to fertilization status and embryo quality [112]. It has been found that the overexpression of miR-92a and miR-130b can lead to adverse results of *in vitro* fertilization (IVF), while the differential expression of miR-214, miR-454, and miR-888 is related to high quality embryos [113].

Peripheral blood is the most widely used biological sample in clinical diagnosis. The cargos in EVs in peripheral blood are good indexes for the detection of many diseases [114]. Prostasomes produced by prostate tumor cells may be involved in the spread of prostate cancer. Prostate corpuscles in peripheral plasma and their specific proteins, such as PAP, PSA, TMPRSS2, and PSCA, may be valuable biomarkers of prostate cancer [115]. Since EVs and their content in women's peripheral blood can be detected from early pregnancy, they can be used as biomarkers for the prediction or diagnosis of pregnancy complications, fetal developmental disorders, and preterm birth (PTB). About 11% of infants around the world are born prematurely every year. PTB, defined as delivery before 37 weeks of pregnancy, is the leading cause of neonatal morbidity and mortality. The exosomes in plasma can be used to predict whether pregnant women will develop PTB, so as to prepare for prenatal intervention. Several miRNAs such as hsa-miR-381, hsa-miR-154, hsa-miR-377, and hsa-miR-150-5p in circulating EVs have been predicted to be potential biomarkers of PTB. A longitudinal study analyzed the miRNAs in plasma EVs collected from multiple pregnant women throughout pregnancy and found that the number and expression profile of miRNAs of EVs in maternal plasma of PTB changed significantly compared with the term birth, especially these miRNAs associated with TGF-β, p53, and glucocorticoid receptor signaling. It is worth pointing out that these miRNAs are known to be associated with fetal membrane apoptosis [116–119]. Fetal cell-free DNA (cfDNA) carried by maternal circulating plasma and serum exocrine has been used in non-invasive prenatal diagnosis as a biomarker of pregnancy complications, such as PE and GDM [120].

MVs from vaginal microbes (e.g., Group B Streptococcus) can affect fetal placental tissue and lead to pregnancy complications [121]. Therefore, it may be possible to detect the contents of vaginal MVs to find significantly differential cargos, so as to detect pregnancy complications as soon as possible.

With the deepening of research, studies in mounting numbers have shown that the miRNAs in EVs can be used to diagnose pregnancy-related diseases, including miR-136, miR-155, miR-210, miR-486, miR-494, miR-495, miR-517-5p, miR-520a-5p, miR-525-5p, and miR-548c-5p [122].

Wrong or late diagnosis is one of the main problems of EMs. Therefore, it is of great significance to find biomarkers for early diagnosis. Extracellular nucleases are key enzymes in the process of inflammation, which participate in the occurrence and development of EMs by regulating the levels of extracellular ATP and adenosine.

A study has found that the extracellular nuclease activity in the aspiration fluid of endometrial tumors is significantly higher than that of simple cysts and comes from the exosomes of these fluids, and therefore the exosomes of these lesions may become biomarkers of EMs. The proteins and miRNAs transported by endEVs may be very valuable biomarkers of human endometrial diseases [123].

The expression of DENN domain containing 1A (DENND1A) in urine-derived exosomes of PCOS patients was significantly higher than that of normal controls, suggesting that it can be used as a potential PCOS marker [124].

5.2 Clinical and therapeutic applications

As biological substances are produced by normal human bodies, the in-depth study of EVs in different physiological and disease states is helpful to identify specific proteins related to EVs functional defects, thus promoting the development of new diagnoses and treatment strategies for reproductive dysfunction. EVs have attracted much attention in the field of translational medicine because compared with other commonly used synthetic drug delivery carriers (such as liposomes), bioengineered EVs have not only have inherent targeting ability but also are low immunogenicity, easy to obtain, selective assembly, high modification flexibility, and biological barrier permeability. Considering EVs can deliver functional cargos to target cells, it is hopeful for us to use them as a drug delivery tool. For example, EVs may be able to pass through the tissue barrier (such as endothelial barrier, blood–brain barrier, endothelial barrier, and blood–brain barrier) through endocytosis. At present, the application of exosome therapy has been explored in a variety of reproductive diseases, such as the treatment of female infertility, POI, EMs, IUA, and so on. For instance, exosomes can carry specific miRNAs into receptor cells to target genes. There are also studies to target the treatment of related diseases by putting modified small interference RNA (siRNAs) into exosomes [125, 126].

EuESCs can secrete a large number of exosomes, which contain a variety of miRNAs may not only be closely related to the occurrence, development and complications of EMS, but also have different expression levels from normal endometrial cells. Therefore, exosomal miRNAs have certain application value in the early diagnosis, judgment of disease progression, and prognosis of EMs. In EMs mouse models, the delivery of miR-214 rich exosomes isolated from EuESCs can inhibit fibrosis and regulate the development of EMs [127].

Exosomes derived from endometrial epithelial cells can enhance the adhesion, implantation, and growth ability of embryo *in vivo* [128]. EVs obtained by uterine lavage carries proteins that regulate and predict embryo implantation, and its protein composition, metastatic and invasive properties, and antioxidant function are dynamically regulated throughout the menstrual cycle, so it has the potential to be used as a biomarker of embryonic development, implantation, and successful pregnancy [129].

PCOS and GDM are closely related to obesity. Adipose tissue macrophages (ATM) of obese mice secrete exosomes containing miRNAs, which can lead to impaired glucose tolerance and IR when given to thin mice. In contrast, ATM exosomes obtained from lean mice can improve glucose tolerance and insulin sensitivity when given to obese mice. These miRNAs can be transferred to insulin target cells through paracrine or endocrine regulation, and have a strong effect on cellular insulin action, insulin sensitivity, and global glucose homeostasis [88]. Therefore, the differentially expressed miRNAs of ATM in obese and thin people may be used as biomarkers of PCOS and GDM and used in their treatment.

Premature ovarian failure (POF) is defined as ovarian function lost in women before 40 years of age. The incidence of POF in women is about 1%. The occurrence of POF is closely related to the depletion of ovarian follicles, but the specific mechanism is still not clear [130]. The combination of stem cell therapy and EVs provides a glimmer of hope for the treatment of POF. In one study, exosomes collected from Bone mesenchymal stem cells (BMSCs) were injected into POF mice, and then found that follicles, FF, and corpus luteum increased in mice ovaries, while apoptosis-related genes p53 and caspase3 were down-regulated, indicating that BMSC-derived exosomes can improve the phenotype of POF. Further in *vitro* experiments showed that the expression of miR-664-5p was increased in BMSC-derived exosomes and could reverse the injury of POF granulosa cells, which was regulated by p53 [131]. Another study found that miR-144-5p in BMSC-derived exosomes inhibit POF GCs apoptosis by targeting PTEN and activating the PI3K/AKT pathway in *vitro* [132]. Furthermore, exosomes derived from human amniotic epithelial cells (hAECs), amniotic fluid stem cells (AFSCs), and placenta-derived mesenchymal stem cells (PD-MSCs) can regulate the apoptosis pathway to reduce the apoptosis of ovarian follicles through their different cargos, such as miR-10a, miR-146a, miR-1246, and antioxidant enzymes [133]. Taken together, these studies demonstrated that exosomes can mediate ovarian function, thus pointing out a new direction for the treatment of POF.

6. Conclusions

Gametogenesis, fertilization, implantation and early embryonic development are complex processes that are highly dependent on communication between cells and organs. The transmission of EVs between cells as a newly discovered way of cell communication, is increasingly found to play crucial and multiple roles in the field of reproduction. In the male reproductive system, prostasomes and epididymosomes derived from prostate/epididymis are found in semen, which contribute to maturation, motility, activation, capacitation of SPZ and acrosome reaction (**Figure 2**). In the female reproductive system, the two-way communication between oocytes and

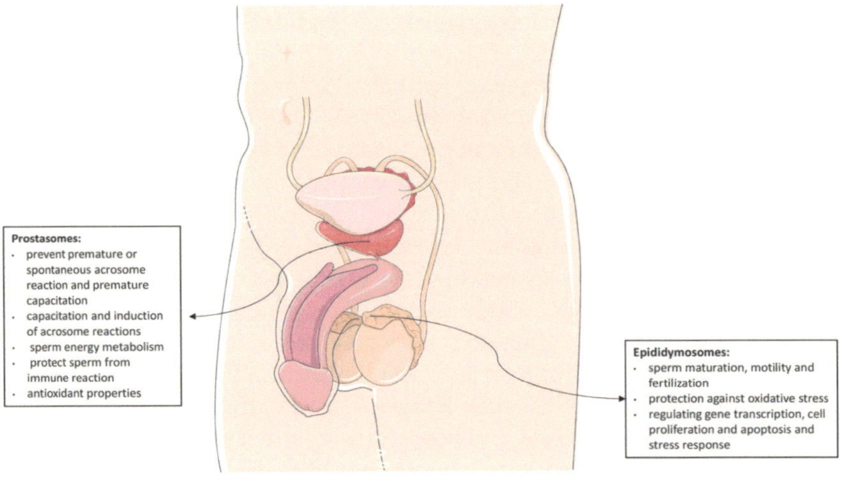

Figure 2.
The functions of EVs in male reproductive system.

somatic cells around them (companion somatic cells) is important for the development of oocytes fertilization and embryogenesis. EVs in ovary, oviduct, endometrium and placenta are the carriers of information during gametogenesis, fertilization and embryo-maternal dialog, helping follicle and oocyte development and maturation. While in fertilization, they can regulate maternal immunity, promote early embryo development, assist implantation, and maintain pregnancy, which are all vital for successful pregnancy. In pathologies, EVs play important roles in female reproductive system disorders, such as PCOS, EMs, pregnancy complications (**Figure 3**).

In the process of EVs synthesis, different RNAs, proteins and other contents are selectively packaged into EVs, therefore the data of genomics, proteomics and

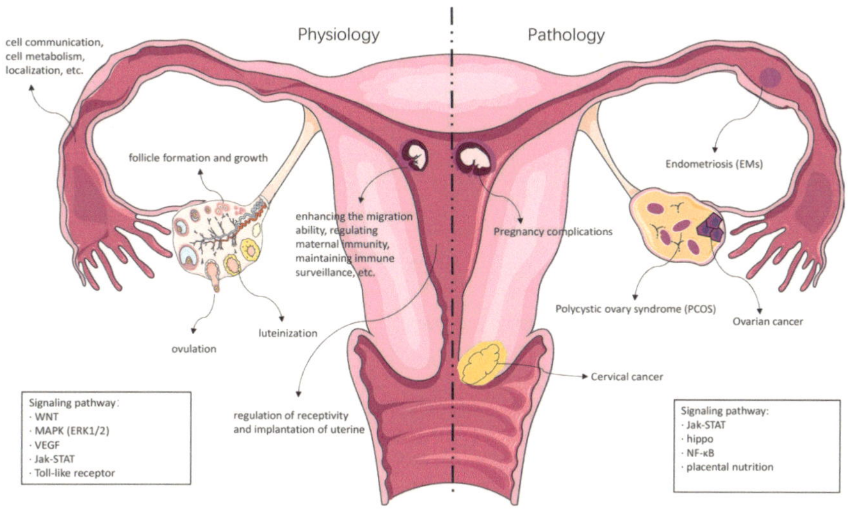

Figure 3.
The functions of EVs in female reproductive physiology and pathology conditions.

Sex	Diseases	EV cargos	References
Female	Polycystic ovary syndrome	LncRNAs: LINC00173, H19, HDAC6, POP4, PTEN, AKT3, DICER1, NF1 and MUM1 Protein: S100-A9	[76–78]
	Endometriosis	MiRNAs: miR-21, miR-615-3p, miR-6873-3p, miR-3195, miR-196a-5p, miR-4483, miR-1273 h-3p, miR-4262, miR-1269a, miR-6859-5p	[83, 84]
	Ovarian and cervical cancer	Proteins: TUBB3, EpCAM, CLDN3, PCNA, EGFR, FASN, MAGE3/6, TGF- β1, ATF1, RAS MiRNAs: miR-483-5p, miR-1246, miR-1275, microRNA-21, microRNA-146a and miR-222-3p, let-7d-5p, miR-92a-3p, miR-20a-5p, miR-378a-3p, miR-423-3p, miR-7-5p, miR-99-5p, miR-100 5p and miR-320a, let-7d-3p, miR-30d-5p, miR-221-3p	[100–102, 104–108]
Male	Male infertility	Proteins: HIST1H2B, KLK2, MIF, MPO, MSMB, LDHC, HK1, PNP, APRT, SLC2A14	[111]

Table 1.
Potential EVs biomarkers in different reproductive diseases.

metabolomics in EVs may be different between normal physiological state and conditions, and the number and composition of EVs may reflect the functional state of its cell origin. In view of the importance of this mode of intercellular communication in the reproductive system, EVs can be used as biomarkers or therapeutic targets for reproductive diseases in the field of reproductive system diseases and assisted reproductive technology (ART) (**Table 1**).

Future research can not only continue to focus on the role of EVs in the normal reproductive physiology, but also focus on the changes of their cargos in a variety of reproductive diseases, so as to find biomarkers for prediction and diagnosis of diseases, and how to transform them into targeted drug therapy and other clinical applications.

Acknowledgements

This work was supported by National Natural Science Foundation of China (No.81871161).

Conflict of interest

The authors declare no conflict of interest.

Author details

Weidong Zhou*, Teng Zhang, Yikai Lian and Wenjing Zhang
The First Affiliated Hospital of Xiamen University, Xiamen, China

*Address all correspondence to: wdzhou@xmu.edu.cn

References

[1] Mittelbrunn M, Sánchez-Madrid F. Intercellular communication: diverse structures for exchange of genetic information. Nature reviews Molecular cell biology. 2012;13(5):328-335. DOI: 10.1038/nrm3335

[2] Simon C, Greening D, Bolumar D, Balaguer N, Salamonsen L, Vilella F. Extracellular Vesicles in Human Reproduction in Health and Disease. Endocrine reviews. 2018;39(3):292-332. DOI: 10.1210/er.2017-00229

[3] Raposo G, Stoorvogel W. Extracellular vesicles: exosomes, microvesicles, and friends. The Journal of cell biology. 2013;200(4):373-383. DOI: 10.1083/jcb.201211138

[4] Pegtel D, Gould S. Exosomes. Annual review of biochemistry. 2019;88:487-514. DOI: 10.1146/annurev-biochem-013118-111902

[5] Cuman C, Menkhorst E, Winship A, Van Sinderen M, Osianlis T, Rombauts L, et al. Fetal-maternal communication: the role of Notch signalling in embryo implantation. Reproduction (Cambridge, England). 2014;147(3):R75-86. DOI: 10.1530/rep-13-0474

[6] Bianchi E, Wright G. Sperm Meets Egg: The Genetics of Mammalian Fertilization. Annual review of genetics. 2016;50:93-111. DOI: 10.1146/annurev-genet-121415-121834

[7] Machtinger R, Laurent L, Baccarelli A. Extracellular vesicles: roles in gamete maturation, fertilization and embryo implantation. Human reproduction update. 2016;22(2):182-193. DOI: 10.1093/humupd/dmv055

[8] Sánchez F, Smitz J. Molecular control of oogenesis. Biochimica et biophysica acta. 2012;1822(12):1896-1912. DOI: 10.1016/j.bbadis.2012.05.013

[9] da Silveira J, Veeramachaneni D, Winger Q, Carnevale E, Bouma G. Cell-secreted vesicles in equine ovarian follicular fluid contain miRNAs and proteins: a possible new form of cell communication within the ovarian follicle. Biology of reproduction. 2012;86(3):71. DOI: 10.1095/biolreprod.111.093252

[10] Bartel D. MicroRNAs: target recognition and regulatory functions. Cell. 2009;136(2):215-233. DOI: 10.1016/j.cell.2009.01.002

[11] Boyer A, Goff A, Boerboom D. WNT signaling in ovarian follicle biology and tumorigenesis. Trends in endocrinology and metabolism: TEM. 2010;21(1):25-32. DOI: 10.1016/j.tem.2009.08.005

[12] Zhang M, Ouyang H, Xia G. The signal pathway of gonadotrophins-induced mammalian oocyte meiotic resumption. Molecular human reproduction. 2009;15(7):399-409. DOI: 10.1093/molehr/gap031

[13] Hung W, Hong X, Christenson L, McGinnis L. Extracellular Vesicles from Bovine Follicular Fluid Support Cumulus Expansion. Biology of reproduction. 2015;93(5):117. DOI: 10.1095/biolreprod.115.132977

[14] Jiang N, Li X. The Complicated Effects of Extracellular Vesicles and Their Cargos on Embryo Implantation. Frontiers in endocrinology. 2021;12:681266. DOI: 10.3389/fendo.2021.681266

[15] Bauersachs S, Mermillod P, Almiñana C. The Oviductal Extracellular Vesicles' RNA Cargo Regulates the Bovine

Embryonic Transcriptome. International journal of molecular sciences. 2020; 21(4). DOI: 10.3390/ijms21041303

[16] Almiñana C, Tsikis G, Labas V, Uzbekov R, da Silveira J, Bauersachs S, et al. Deciphering the oviductal extracellular vesicles content across the estrous cycle: implications for the gametes-oviduct interactions and the environment of the potential embryo. BMC genomics. 2018;19(1):622. DOI: 10.1186/s12864-018-4982-5

[17] Bathala P, Fereshteh Z, Li K, Al-Dossary A, Galileo D, Martin-DeLeon P. Oviductal extracellular vesicles (oviductosomes, OVS) are conserved in humans: murine OVS play a pivotal role in sperm capacitation and fertility. Molecular human reproduction. 2018;24(3):143-157. DOI: 10.1093/molehr/gay003

[18] Almiñana C, Corbin E, Tsikis G, Alcântara-Neto A, Labas V, Reynaud K, et al. Oviduct extracellular vesicles protein content and their role during oviduct-embryo crosstalk. Reproduction (Cambridge, England). 2017;154(3):153-68. DOI: 10.1530/rep-17-0054

[19] Sidrat T, Khan A, Joo M, Wei Y, Lee K, Xu L, et al. Bovine Oviduct Epithelial Cell-Derived Culture Media and Exosomes Improve Mitochondrial Health by Restoring Metabolic Flux during Pre-Implantation Development. International journal of molecular sciences. 2020;21(20). DOI: 10.3390/ijms21207589

[20] Griffiths G, Galileo D, Reese K, Martin-Deleon P. Investigating the role of murine epididymosomes and uterosomes in GPI-linked protein transfer to sperm using SPAM1 as a model. Molecular reproduction and development. 2008;75(11):1627-1636. DOI: 10.1002/mrd.20907

[21] Ng Y, Rome S, Jalabert A, Forterre A, Singh H, Hincks C, et al. Endometrial exosomes/microvesicles in the uterine microenvironment: a new paradigm for embryo-endometrial cross talk at implantation. PloS one. 2013;8(3):e58502. DOI: 10.1371/journal.pone.0058502

[22] Kurian N, Modi D. Extracellular vesicle mediated embryo-endometrial cross talk during implantation and in pregnancy. Journal of assisted reproduction and genetics. 2019;36(2): 189-198. DOI: 10.1007/s10815-018-1343-x

[23] Greening D, Gopal S, Xu R, Simpson R, Chen W. Exosomes and their roles in immune regulation and cancer. Seminars in cell & developmental biology. 2015;40:72-81. DOI: 10.1016/j.semcdb.2015.02.009

[24] Wang H, Cheng H, Shao Q, Dong Z, Xie Q, Zhao L, et al. Leptin-promoted human extravillous trophoblast invasion is MMP14 dependent and requires the cross talk between Notch1 and PI3K/Akt signaling. Biology of reproduction. 2014;90(4):78. DOI: 10.1095/biolreprod.113.114876

[25] Latifi Z, Fattahi A, Ranjbaran A, Nejabati H, Imakawa K. Potential roles of metalloproteinases of endometrium-derived exosomes in embryo-maternal crosstalk during implantation. Journal of cellular physiology. 2018;233(6):4530-4545. DOI: 10.1002/jcp.26259

[26] Greening D, Nguyen H, Elgass K, Simpson R, Salamonsen L. Human Endometrial Exosomes Contain Hormone-Specific Cargo Modulating Trophoblast Adhesive Capacity: Insights into Endometrial-Embryo Interactions. Biology of reproduction. 2016;94(2):38. DOI: 10.1095/biolreprod.115.134890

[27] Kusama K, Nakamura K, Bai R, Nagaoka K, Sakurai T, Imakawa K.

Intrauterine exosomes are required for bovine conceptus implantation. Biochemical and biophysical research communications. 2018;495(1):1370-1375. DOI: 10.1016/j.bbrc.2017.11.176

[28] Vilella F, Moreno-Moya J, Balaguer N, Grasso A, Herrero M, Martínez S, et al. Hsa-miR-30d, secreted by the human endometrium, is taken up by the pre-implantation embryo and might modify its transcriptome. Development (Cambridge, England). 2015;142(18):3210-21. DOI: 10.1242/dev.124289

[29] Tong M, Chamley L. Placental extracellular vesicles and feto-maternal communication. Cold Spring Harbor perspectives in medicine. 2015;5(3): a023028. DOI: 10.1101/cshperspect.a023028

[30] Maas S, Breakefield X, Weaver A. Extracellular Vesicles: Unique Intercellular Delivery Vehicles. Trends in cell biology. 2017;27(3):172-188. DOI: 10.1016/j.tcb.2016.11.003

[31] Nair S, Salomon C. Extracellular vesicles and their immunomodulatory functions in pregnancy. Seminars in immunopathology. 2018;40(5):425-437. DOI: 10.1007/s00281-018-0680-2

[32] Nardi FS, Michelon T, Neumann J, Manvailer L, Wagner B, Horn P, et al. High levels of circulating extracellular vesicles with altered expression and function during pregnancy. Immunobiology. 2016;221(7):753-760. DOI: 10.1016/j.imbio.2016.03.001

[33] Han T, Li W, Li G. Progress in Understanding the Functional Roles of Extracellular Vesicles in Reproduction. Biomedical and environmental sciences: BES. 2020;33(7):518-527. DOI: 10.3967/bes2020.068

[34] Desrochers L, Bordeleau F, Reinhart-King C, Cerione R, Antonyak M. Microvesicles provide a mechanism for intercellular communication by embryonic stem cells during embryo implantation. Nature communications. 2016;7:11958. DOI: 10.1038/ncomms11958

[35] Szekeres-Bartho J, Šućurović S, Mulac-Jeričević B. The Role of Extracellular Vesicles and PIBF in Embryo-Maternal Immune-Interactions. Frontiers in immunology. 2018;9:2890. DOI: 10.3389/fimmu.2018.02890

[36] Candenas L, Chianese R. Exosome Composition and Seminal Plasma Proteome: A Promising Source of Biomarkers of Male Infertility. International journal of molecular sciences. 2020;21(19). DOI: 10.3390/ijms21197022

[37] Samanta L, Parida R, Dias T, Agarwal A. The enigmatic seminal plasma: a proteomics insight from ejaculation to fertilization. Reproductive biology and endocrinology: RB&E. 2018;16(1):41. DOI: 10.1186/s12958-018-0358-6

[38] Yang C, Guo W, Zhang W, Bian J, Yang J, Zhou Q, et al. Comprehensive proteomics analysis of exosomes derived from human seminal plasma. Andrology. 2017;5(5):1007-1015. DOI: 10.1111/andr.12412

[39] Vojtech L, Woo S, Hughes S, Levy C, Ballweber L, Sauteraud R, et al. Exosomes in human semen carry a distinctive repertoire of small non-coding RNAs with potential regulatory functions. Nucleic acids research. 2014;42(11):7290-7304. DOI: 10.1093/nar/gku347

[40] Ronquist G, Brody I, Gottfries A, Stegmayr B. An Mg2+ and Ca2+–stimulated adenosine triphosphatase in

human prostatic fluid--part II. Andrologia. 1978;10(6):427-433. DOI: 10.1111/j.1439-0272.1978.tb03064.x

[41] Arvidson G, Ronquist G, Wikander G, Ojteg A. Human prostasome membranes exhibit very high cholesterol/phospholipid ratios yielding high molecular ordering. Biochimica et biophysica acta. 1989;984(2):167-173. DOI: 10.1016/0005-2736(89)90212-5

[42] Arienti G, Carlini E, Polci A, Cosmi E, Palmerini C. Fatty acid pattern of human prostasome lipid. Archives of biochemistry and biophysics. 1998; 358(2):391-395. DOI: 10.1006/abbi.1998.0876

[43] Burden H, Holmes C, Persad R, Whittington K. Prostasomes--their effects on human male reproduction and fertility. Human reproduction update. 2006;12(3):283-292. DOI: 10.1093/humupd/dmi052

[44] Pons-Rejraji H, Artonne C, Sion B, Brugnon F, Canis M, Janny L, et al. Prostasomes: inhibitors of capacitation and modulators of cellular signalling in human sperm. International journal of andrology. 2011;34:568-580. DOI: 10.1111/j.1365-2605.2010.01116.x

[45] Cross N, Mahasreshti P. Prostasome fraction of human seminal plasma prevents sperm from becoming acrosomally responsive to the agonist progesterone. Archives of andrology. 1997;39(1):39-44. DOI: 10.3109/01485019708987900

[46] Visconti P, Galantino-Homer H, Ning X, Moore G, Valenzuela J, Jorgez C, et al. Cholesterol efflux-mediated signal transduction in mammalian sperm. beta-cyclodextrins initiate transmembrane signaling leading to an increase in protein tyrosine phosphorylation and capacitation. The Journal of biological chemistry. 1999;274(5):3235-3242. DOI: 10.1074/jbc.274.5.3235

[47] Ronquist G, Brody I, Gottfries A, Stegmayr B. An Mg2+ and Ca2+–stimulated adenosine triphosphatase in human prostatic fluid: part I. Andrologia. 1978;10(4):261-272. DOI: 10.1111/j.1439-0272.1978.tb03030.x

[48] Stegmayr B, Ronquist G. Promotive effect on human sperm progressive motility by prostasomes. Urological research. 1982;10(5):253-257. DOI: 10.1007/bf00255932

[49] Saez F, Sullivan R. Prostasomes, post-testicular sperm maturation and fertility. Frontiers in bioscience (Landmark edition). 2016;21:1464-73. DOI: 10.2741/4466

[50] Baskaran S, Panner Selvam M, Agarwal A. Exosomes of male reproduction. Advances in clinical chemistry. 2020;95:149-163. DOI: 10.1016/bs.acc.2019.08.004

[51] Subirán N, Agirregoitia E, Valdivia A, Ochoa C, Casis L, Irazusta J. Expression of enkephalin-degrading enzymes in human semen and implications for sperm motility. Fertility and sterility. 2008;89:1571-1577. DOI: 10.1016/j.fertnstert.2007.06.056

[52] Arienti G, Carlini E, Verdacchi R, Cosmi E, Palmerini C. Prostasome to sperm transfer of CD13/aminopeptidase N (EC 3.4.11.2). Biochimica et biophysica acta. 1997;1336(3):533-538. DOI: 10.1016/s0304-4165(97)00071-8

[53] Yanagimachi R, Kamiguchi Y, Mikamo K, Suzuki F, Yanagimachi H. Maturation of spermatozoa in the epididymis of the Chinese hamster. The American journal of anatomy. 1985; 172(4):317-330. DOI: 10.1002/aja.1001720406

[54] Vernet P, Aitken R, Drevet J. Antioxidant strategies in the epididymis. Molecular and cellular endocrinology. 2004;216:31-39. DOI: 10.1016/j.mce.2003.10.069

[55] Sullivan R, Saez F, Girouard J, Frenette G. Role of exosomes in sperm maturation during the transit along the male reproductive tract. Blood cells, molecules & diseases. 2005;35(1):1-10. DOI: 10.1016/j.bcmd.2005.03.005

[56] Frenette G, Girouard J, D'Amours O, Allard N, Tessier L, Sullivan R. Characterization of two distinct populations of epididymosomes collected in the intraluminal compartment of the bovine cauda epididymis. Biology of reproduction. 2010;83(3):473-480. DOI: 10.1095/biolreprod.109.082438

[57] Frenette G, Thabet M, Sullivan R. Polyol pathway in human epididymis and semen. Journal of andrology. 2006;27(2):233-239. DOI: 10.2164/jandrol.05108

[58] Eickhoff R, Baldauf C, Koyro H, Wennemuth G, Suga Y, Seitz J, et al. Influence of macrophage migration inhibitory factor (MIF) on the zinc content and redox state of protein-bound sulphydryl groups in rat sperm: indications for a new role of MIF in sperm maturation. Molecular human reproduction. 2004;10(8):605-611. DOI: 10.1093/molehr/gah075

[59] Martin-DeLeon P. Epididymal SPAM1 and its impact on sperm function. Molecular and cellular endocrinology. 2006;250:114-121. DOI: 10.1016/j.mce.2005.12.033

[60] Frenette G, Sullivan R. Prostasome-like particles are involved in the transfer of P25b from the bovine epididymal fluid to the sperm surface. Molecular reproduction and development. 2001; 59(1):115-121. DOI: 10.1002/mrd.1013

[61] Al-Dossary A, Strehler E, Martin-Deleon P. Expression and secretion of plasma membrane Ca2+-ATPase 4a (PMCA4a) during murine estrus: association with oviductal exosomes and uptake in sperm. PloS one. 2013;8(11):e80181. DOI: 10.1371/journal.pone.0080181

[62] Murta D, Batista M, Silva E, Trindade A, Henrique D, Duarte A, et al. Notch signaling in the epididymal epithelium regulates sperm motility and is transferred at a distance within epididymosomes. Andrology. 2016;4(2): 314-327. DOI: 10.1111/andr.12144

[63] Rajender S, Meador C, Agarwal A. Small RNA in spermatogenesis and male infertility. Frontiers in bioscience (Scholar edition). 2012;4:1266-74. DOI: 10.2741/s330

[64] Ran M, Chen B, Yin J, Yang A, Jiang M. Advances in miRNA research related to testis development and spermatogenesis. Yi chuan = Hereditas. 2014;36(7):646-54. DOI: 10.3724/sp.J.1005.2014.0646

[65] Belleannée C, Calvo É, Caballero J, Sullivan R. Epididymosomes convey different repertoires of microRNAs throughout the bovine epididymis. Biology of reproduction. 2013;89(2):30. DOI: 10.1095/biolreprod.113.110486

[66] Curry E, Safranski T, Pratt S. Differential expression of porcine sperm microRNAs and their association with sperm morphology and motility. Theriogenology. 2011;76(8):1532-1539. DOI: 10.1016/j.theriogenology.2011.06.025

[67] Li Q, Chau J, Ebert P, Sylvester G, Min H, Liu G, et al. miR-181a is an intrinsic modulator of T cell sensitivity and selection. Cell. 2007;129(1):147-161. DOI: 10.1016/j.cell.2007.03.008

[68] Niu Y, Mo D, Qin L, Wang C, Li A, Zhao X, et al. Lipopolysaccharide-induced miR-1224 negatively regulates tumour necrosis factor-α gene expression by modulating Sp1. Immunology. 2011;133(1):8-20. DOI: 10.1111/j.1365-2567.2010.03374.x

[69] Sharma U, Conine C, Shea J, Boskovic A, Derr A, Bing X, et al. Biogenesis and function of tRNA fragments during sperm maturation and fertilization in mammals. Science (New York, NY). 2016;351(6271):391-6. DOI: 10.1126/science.aad6780

[70] Chen Q, Yan M, Cao Z, Li X, Zhang Y, Shi J, et al. Sperm tsRNAs contribute to intergenerational inheritance of an acquired metabolic disorder. Science (New York, NY). 2016;351(6271):397-400. DOI: 10.1126/science.aad7977

[71] Goodarzi M, Dumesic D, Chazenbalk G, Azziz R. Polycystic ovary syndrome: etiology, pathogenesis and diagnosis. Nature reviews Endocrinology. 2011;7(4):219-231. DOI: 10.1038/nrendo.2010.217

[72] Lian Y, Zhou W. Advances in the research of exosome and exosomal non-coding RNA in the pathogenesis, diagnosis and treatment of polycystic ovary syndrome Acta Pharmaceutica Sinica. 2020;55(10):2256-63. DOI: 10.16438/j.0513-4870.2020-0323

[73] Koiou E, Tziomalos K, Katsikis I, Papadakis E, Kandaraki E, Panidis D. Platelet-derived microparticles in overweight/obese women with the polycystic ovary syndrome. Gynecological endocrinology: the official journal of the International Society of Gynecological Endocrinology. 2013;29(3):250-253. DOI: 10.3109/09513590.2012.743005

[74] Hu J, Tang T, Zeng Z, Wu J, Tan X, Yan J. The expression of small RNAs in exosomes of follicular fluid altered in human polycystic ovarian syndrome. PeerJ. 2020;8:e8640. DOI: 10.7717/peerj.8640

[75] Jiang X, Li J, Zhang B, Hu J, Ma J, Cui L, et al. Differential expression profile of plasma exosomal microRNAs in women with polycystic ovary syndrome. Fertility and sterility. 2021;115(3):782-792. DOI: 10.1016/j.fertnstert.2020.08.019

[76] Wang L, Fan H, Zou Y, Yuan Q, Hu X, Chen X, et al. Aberrant Expression of Long Non-coding RNAs in Exosomes in Follicle Fluid From PCOS Patients. Frontiers in genetics. 2020;11:608178. DOI: 10.3389/fgene.2020.608178

[77] Wang L, Peng X, Lv X, Liu L, Li X, He X, et al. High throughput circRNAs sequencing profile of follicle fluid exosomes of polycystic ovary syndrome patients. Journal of cellular physiology. 2019. DOI: 10.1002/jcp.28201

[78] Huang X, Wu B, Chen M, Hong L, Kong P, Wei Z, et al. Depletion of exosomal circLDLR in follicle fluid derepresses miR-1294 function and inhibits estradiol production via CYP19A1 in polycystic ovary syndrome. Aging. 2020;12(15):15414-35. DOI: 10.18632/aging.103602

[79] Li H, Huang X, Chang X, Yao J, He Q, Shen Z, et al. S100-A9 protein in exosomes derived from follicular fluid promotes inflammation via activation of NF-κB pathway in polycystic ovary syndrome. Journal of cellular and molecular medicine. 2020;24(1):114-125. DOI: 10.1111/jcmm.14642

[80] Klemmt P, Starzinski-Powitz A. Molecular and Cellular Pathogenesis of Endometriosis. Current women's health reviews. 2018;14(2):106-116. DOI: 10.2174/1573404813666170306163448

[81] Muth D, McAlexander M, Ostrenga L, Pate N, Izzi J, Adams R, et al. Potential role of cervicovaginal extracellular particles in diagnosis of endometriosis. BMC veterinary research. 2015;11:187. DOI: 10.1186/s12917-015-0513-7

[82] Sun H, Li D, Yuan M, Li Q, Li N, Wang G. Eutopic stromal cells of endometriosis promote neuroangio-genesis via exosome pathway†. Biology of reproduction. 2019;100(3):649-659. DOI: 10.1093/biolre/ioy212

[83] Harp D, Driss A, Mehrabi S, Chowdhury I, Xu W, Liu D, et al. Exosomes derived from endometriotic stromal cells have enhanced angiogenic effects in vitro. Cell and tissue research. 2016;365(1):187-196. DOI: 10.1007/s00441-016-2358-1

[84] Zhou W, Lian Y, Jiang J, Wang L, Ren L, Li Y, et al. Differential expression of microRNA in exosomes derived from endometrial stromal cells of women with endometriosis-associated infertility. Reproductive biomedicine online. 2020;41(2):170-181. DOI: 10.1016/j.rbmo.2020.04.010

[85] Rai R, Regan L. Recurrent miscarriage. Lancet (London, England). 2006;368(9535):601-11. DOI: 10.1016/s0140-6736(06)69204-0

[86] Salomon C, Scholz-Romero K, Sarker S, Sweeney E, Kobayashi M, Correa P, et al. Gestational Diabetes Mellitus Is Associated With Changes in the Concentration and Bioactivity of Placenta-Derived Exosomes in Maternal Circulation Across Gestation. Diabetes. 2016;65(3):598-609. DOI: 10.2337/db15-0966

[87] Redman C, Tannetta D, Dragovic R, Gardiner C, Southcombe J, Collett G, et al. Review: Does size matter? Placental debris and the pathophysiology of pre-eclampsia. Placenta. 2012:S48-S54. DOI: 10.1016/j.placenta.2011.12.006

[88] Ying W, Riopel M, Bandyopadhyay G, Dong Y, Birmingham A, Seo J, et al. Adipose Tissue Macrophage-Derived Exosomal miRNAs Can Modulate In Vivo and In Vitro Insulin Sensitivity. Cell. 2017;171(2):372-84.e12. DOI: 10.1016/j.cell.2017.08.035

[89] Jayabalan N, Nair S, Nuzhat Z, Rice G, Zuñiga F, Sobrevia L, et al. viaCross Talk between Adipose Tissue and Placenta in Obese and Gestational Diabetes Mellitus Pregnancies Exosomes. Frontiers in endocrinology. 2017;8:239. DOI: 10.3389/fendo.2017.00239

[90] Zhang Y, Yu M, Tian W. Physiological and pathological impact of exosomes of adipose tissue. Cell proliferation. 2016;49(1):3-13. DOI: 10.1111/cpr.12233

[91] Calvert S, Jones C, Sibley C, Aplin J, Heazell A. Analysis of syncytial nuclear aggregates in preeclampsia shows increased sectioning artefacts and decreased inter-villous bridges compared to healthy placentas. Placenta. 2013;34(12):1251-1254. DOI: 10.1016/j.placenta.2013.10.001

[92] Tannetta D, Collett G, Vatish M, Redman C, Sargent I. Syncytiotrophoblast extracellular vesicles - Circulating biopsies reflecting placental health. Placenta. 2017;52:134-138. DOI: 10.1016/j.placenta.2016.11.008

[93] Vargas A, Zhou S, Éthier-Chiasson M, Flipo D, Lafond J, Gilbert C, et al. Syncytin proteins incorporated in placenta exosomes are important for cell uptake and show variation in abundance in serum exosomes from patients with preeclampsia. FASEB journal: official publication of the Federation of American Societies for Experimental Biology. 2014;28(8):3703-3719. DOI: 10.1096/fj.13-239053

[94] Agarwal A, Majzoub A, Parekh N, Henkel R. A Schematic Overview of the Current Status of Male Infertility Practice. The world journal of men's health. 2020;38(3):308-322. DOI: 10.5534/wjmh.190068

[95] Drabovich A, Saraon P, Jarvi K, Diamandis E. Seminal plasma as a diagnostic fluid for male reproductive system disorders. Nature reviews Urology. 2014;11(5):278-288. DOI: 10.1038/nrurol.2014.74

[96] Ballescá J, Balasch J, Calafell J, Alvarez R, Fábregues F, de Osaba M, et al. Serum inhibin B determination is predictive of successful testicular sperm extraction in men with non-obstructive azoospermia. Human reproduction (Oxford, England). 2000;15(8):1734-8. DOI: 10.1093/humrep/15.8.1734

[97] Torre L, Islami F, Siegel R, Ward E, Jemal A. Global Cancer in Women: Burden and Trends. Cancer epidemiology, biomarkers & prevention: a publication of the American Association for Cancer Research, cosponsored by the American Society of Preventive Oncology. 2017;26(4):444-457. DOI: 10.1158/1055-9965.Epi-16-0858

[98] Berman T, Schiller J. Human papillomavirus in cervical cancer and oropharyngeal cancer: One cause, two diseases. Cancer. 2017;123(12):2219-2229. DOI: 10.1002/cncr.30588

[99] Kajdos M, Janas Ł, Kolasa-Zwierzchowska D, Wilczyński J, Stetkiewicz T. Microvesicles as a potential biomarker of neoplastic diseases and their role in development and progression of neoplasm. Przeglad menopauzalny = Menopause review. 2015;14(4):283-91. DOI: 10.5114/pm.2015.56540

[100] Li J, Sherman-Baust C, Tsai-Turton M, Bristow R, Roden R, Morin P. Claudin-containing exosomes in the peripheral circulation of women with ovarian cancer. BMC cancer. 2009;9:244. DOI: 10.1186/1471-2407-9-244

[101] Liang B, Peng P, Chen S, Li L, Zhang M, Cao D, et al. Characterization and proteomic analysis of ovarian cancer-derived exosomes. Journal of proteomics. 2013;80:171-182. DOI: 10.1016/j.jprot.2012.12.029

[102] Szajnik M, Derbis M, Lach M, Patalas P, Michalak M, Drzewiecka H, et al. Exosomes in Plasma of Patients with Ovarian Carcinoma: Potential Biomarkers of Tumor Progression and Response to Therapy. Gynecology & obstetrics (Sunnyvale, Calif). 2013:3. DOI: 10.4172/2161-0932.S4-003

[103] Runz S, Keller S, Rupp C, Stoeck A, Issa Y, Koensgen D, et al. Malignant ascites-derived exosomes of ovarian carcinoma patients contain CD24 and EpCAM. Gynecologic oncology. 2007;107(3):563-571. DOI: 10.1016/j.ygyno.2007.08.064

[104] Liu J, Sun H, Wang X, Yu Q, Li S, Yu X, et al. Increased exosomal microRNA-21 and microRNA-146a levels in the cervicovaginal lavage specimens of patients with cervical cancer. International journal of molecular sciences. 2014;15(1):758-773. DOI: 10.3390/ijms15010758

[105] Zheng M, Hou L, Ma Y, Zhou L, Wang F, Cheng B, et al. Exosomal let-7d-3p and miR-30d-5p as diagnostic biomarkers for non-invasive screening of cervical cancer and its precursors. Molecular cancer. 2019;18(1):76. DOI: 10.1186/s12943-019-0999-x

[106] Wu X, Zhou C, Zhang Y, Yan R, Wei W, Chen X, et al. Cancer-derived exosomal miR-221-3p promotes angiogenesis by targeting THBS2 in

cervical squamous cell carcinoma. Angiogenesis. 2019;22(3):397-410. DOI: 10.1007/s10456-019-09665-1

[107] Zhang L, Li H, Yuan M, Li M, Zhang S. Cervical Cancer Cells-Secreted Exosomal microRNA-221-3p Promotes Invasion, Migration and Angiogenesis of Microvascular Endothelial Cells in Cervical Cancer by Down-Regulating MAPK10 Expression. Cancer management and research. 2019;11: 10307-10319. DOI: 10.2147/cmar.S221527

[108] Shi Y, Wang W, Yang B, Tian H. ATF1 and RAS in exosomes are potential clinical diagnostic markers for cervical cancer. Cell biochemistry and function. 2017;35(7):477-483. DOI: 10.1002/cbf.3307

[109] Bellieni C. The Best Age for Pregnancy and Undue Pressures. Journal of family & reproductive health. 2016;10(3):104-107.

[110] Diez-Fraile A, Lammens T, Tilleman K, Witkowski W, Verhasselt B, De Sutter P, et al. Age-associated differential microRNA levels in human follicular fluid reveal pathways potentially determining fertility and success of in vitro fertilization. Human fertility (Cambridge, England). 2014;17(2):90-8. DOI: 10.3109/14647273.2014.897006

[111] García-Rodríguez A, de la Casa M, Peinado H, Gosálvez J, Roy R. Human prostasomes from normozoospermic and non-normozoospermic men show a differential protein expression pattern. Andrology. 2018;6(4):585-596. DOI: 10.1111/andr.12496

[112] Revelli A, Delle Piane L, Casano S, Molinari E, Massobrio M, Rinaudo P. Follicular fluid content and oocyte quality: from single biochemical markers to metabolomics. Reproductive biology and endocrinology: RB&E. 2009;7:40. DOI: 10.1186/1477-7827-7-40

[113] Martinez R, Liang L, Racowsky C, Dioni L, Mansur A, Adir M, et al. Extracellular microRNAs profile in human follicular fluid and IVF outcomes. Scientific reports. 2018;8(1):17036. DOI: 10.1038/s41598-018-35379-3

[114] Revenfeld A, Bæk R, Nielsen M, Stensballe A, Varming K, Jørgensen M. Diagnostic and prognostic potential of extracellular vesicles in peripheral blood. Clinical therapeutics. 2014;36(6):830-846. DOI: 10.1016/j.clinthera.2014.05.008

[115] Aalberts M, Stout T, Stoorvogel W. Prostasomes: extracellular vesicles from the prostate. Reproduction (Cambridge, England). 2014;147(1):R1-14. DOI: 10.1530/rep-13-0358

[116] Blencowe H, Cousens S, Oestergaard M, Chou D, Moller A, Narwal R, et al. National, regional, and worldwide estimates of preterm birth rates in the year 2010 with time trends since 1990 for selected countries: a systematic analysis and implications. Lancet (London, England). 2012; 379(9832):2162-72. DOI: 10.1016/s0140-6736(12)60820-4

[117] Goldenberg R, Culhane J, Iams J, Romero R. Epidemiology and causes of preterm birth. Lancet (London, England). 2008;371(9606):75-84. DOI: 10.1016/s0140-6736(08)60074-4

[118] Menon R, Debnath C, Lai A, Guanzon D, Bhatnagar S, Kshetrapal P, et al. Circulating Exosomal miRNA Profile During Term and Preterm Birth Pregnancies: A Longitudinal Study. Endocrinology. 2019;160(2):249-275. DOI: 10.1210/en.2018-00836

[119] Hashimoto A, Sugiura K, Hoshino A. Impact of exosome-mediated feto-maternal interactions on pregnancy maintenance and development of obstetric complications. Journal of

biochemistry. 2021;169(2):163-171. DOI: 10.1093/jb/mvaa137

[120] Konečná B, Tóthová Ľ, Repiská G. Exosomes-Associated DNA-New Marker in Pregnancy Complications? International journal of molecular sciences. 2019;20(12). DOI: 10.3390/ijms20122890

[121] Surve M, Anil A, Kamath K, Bhutda S, Sthanam L, Pradhan A, et al. Membrane Vesicles of Group B Streptococcus Disrupt Feto-Maternal Barrier Leading to Preterm Birth. PLoS pathogens. 2016;12(9):e1005816. DOI: 10.1371/journal.ppat.1005816

[122] Zhang J, Li H, Fan B, Xu W, Zhang X. Extracellular vesicles in normal pregnancy and pregnancy-related diseases. Journal of cellular and molecular medicine. 2020;24(8):4377-4388. DOI: 10.1111/jcmm.15144

[123] Texidó L, Romero C, Vidal A, García-Valero J, Fernández Montoli M, Baixeras N, et al. Ecto-nuclcotidascs activities in the contents of ovarian endometriomas: potential biomarkers of endometriosis. Mediators of inflammation. 2014;2014:120673. DOI: 10.1155/2014/120673

[124] McAllister J, Modi B, Miller B, Biegler J, Bruggeman R, Legro R, et al. Overexpression of a DENND1A isoform produces a polycystic ovary syndrome theca phenotype. Proceedings of the National Academy of Sciences of the United States of America. 2014;111(15): E1519-E1527. DOI: 10.1073/pnas.1400574111

[125] Sun X, Ma X, Yang X, Zhang X. Exosomes and Female Infertility. Current drug metabolism. 2019;20(10):773-780. DOI: 10.2174/138920022066619 1015155910

[126] Weng Z, Zhang B, Wu C, Yu F, Han B, Li B, et al. Therapeutic roles of mesenchymal stem cell-derived extracellular vesicles in cancer. Journal of hematology & oncology. 2021;14(1):136. DOI: 10.1186/s13045-021-01141-y

[127] Wu D, Lu P, Mi X, Miao J. Exosomal miR-214 from endometrial stromal cells inhibits endometriosis fibrosis. Molecular human reproduction. 2018;24(7):357-365. DOI: 10.1093/molehr/gay019

[128] Gurung S, Greening D, Catt S, Salamonsen L, Evans J. Exosomes and soluble secretome from hormone-treated endometrial epithelial cells direct embryo implantation. Molecular human reproduction. 2020;26(7):510-520. DOI: 10.1093/molehr/gaaa034

[129] Rai A, Poh Q, Fatmous M, Fang H, Gurung S, Vollenhoven B, et al. Proteomic profiling of human uterine extracellular vesicles reveal dynamic regulation of key players of embryo implantation and fertility during menstrual cycle. Proteomics. 2021;21:e2000211. DOI: 10.1002/pmic.202000211

[130] Jankowska K. Premature ovarian failure. Przeglad menopauzalny = Menopause review. 2017;16(2):51-6. DOI: 10.5114/pm.2017.68592

[131] Sun B, Ma Y, Wang F, Hu L, Sun Y. miR-644-5p carried by bone mesenchymal stem cell-derived exosomes targets regulation of p53 to inhibit ovarian granulosa cell apoptosis. Stem cell research & therapy. 2019;10(1):360. DOI: 10.1186/s13287-019-1442-3

[132] Yang M, Lin L, Sha C, Li T, Zhao D, Wei H, et al. Bone marrow mesenchymal stem cell-derived exosomal miR-144-5p improves rat ovarian function after chemotherapy-induced ovarian failure by targeting PTEN. Laboratory

investigation; a journal of technical methods and pathology. 2020;100(3): 342-52. DOI: 10.1038/s41374-019-0321-y

[133] Esfandyari S, Elkafas H, Chugh R, Park H, Navarro A, Al-Hendy A. Exosomes as Biomarkers for Female Reproductive Diseases Diagnosis and Therapy. International journal of molecular sciences. 2021;22(4). DOI: 10.3390/ijms22042165

Chapter 8

Exosomes and HIV-1 Association in AIDS-Defining Patients

Sushanta Kumar Barik, Sanghamitra Pati, Keshar Kunja Mohanty, Sashi Bhusan Mohapatra, Srikanta Jena and Srikanth Prasad Tripathy

Abstract

Exosomes are membranous nanovesicles of endocytic origin that help to facilitate cell-to-cell communication by transporting cellular cargo locally or systemically to a recipient cell. These are subsequently fused and internalised by recipient cells. Exosomes are secreted from all cell types in HIV-1 infected patients. Recent studies reveal that exosomes from various sources modulate the pathophysiology of HIV-1, and conversely, exosomes are also targeted by HIV-1 factors. Semen or plasma exosomes could suppress/inhibit HIV-1 replication in humans and rodent models. Exosomal cargo components could be used as a biomarker in HIV-1patients and AIDS-defining patients. Exosome in semen and plasma is a useful tool for the diagnosis of HIV-1 and an alternative therapeutic tool for antiretroviral therapy.

Keywords: exosome, HIV-1, AIDS, semen, plasma

1. Introduction

Extracellular vesicles are membrane vesicles. The extracellular vesicles are secreted by most cell types of any living organism from archaea, bacteria, and eukarya. Extracellular vesicles are isolated from biological fluids such as blood, semen, cerebrospinal fluid, and saliva [1, 2]. Exosomes are formed by the fusion of multivesicular bodies (MVBs) with the cell membrane and then released into the extracellular space [3]. Exosomes are extracellular vesicles with a diameter of 30–100 nm. Exosomes are used in therapeutic applications like a delivery system. The exosomes and retroviral vectors are efficient transporters of biotransformation. The exosomes are used in immunotherapies and biotherapies. Exosomes and lentiviruses are excellent vehicles to carry encapsulated cargo with reduced immunogenicity with multiple cell types and tissue barriers, including the blood-brain barrier. The inherent properties of exosomes are to enclose nucleic acids, proteins, lipids, and chemical agents. Exosomes have been considered to carry the coding, and non-coding RNA (miRNA and siRNA) along with the lentiviral vectors which give a stable integration into target cells [4–9]. Some findings proposed that the human immune deficiency virus (HIV) particles like exosomes contribute to HIV pathogenesis, but the recent findings reported that

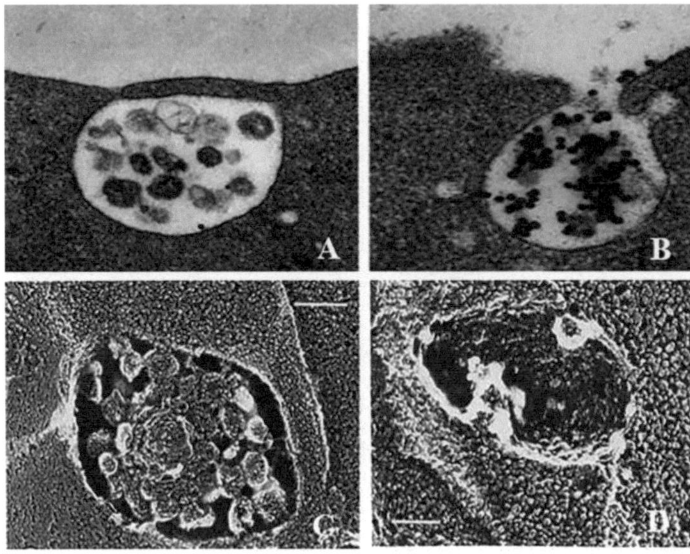

Figure 1.
A: View of an Multivesicular endosome (MVE) sparsely labelled with AuTf after a 20 min incubation at 37°C. note the apparent fusion of the MVE and the plasma membrane. This may represent incipient MVE exocytosis. Bar, 100 nm. × 107,000. Fig. B: View of MVE exocytosis in an unfixed reticulocyte. This cell was incubated for 30 min with AuTf, subjected to a 20 min chase with unconjugated transferrin, and then quick-frozen without prior fixation and freeze-substituted. Bar, 200 rim. × 61,000. Fig C: Freeze-fracture, deep-etch view of a large MVE filled with numerous inclusions. The remaining free lumen possesses a markedly lower protein concentration than the surrounding cytoplasm and is deeply etched. Bar, 200 nm. × 63,000. Fig. D: View of an MVE lightly labelled by AuTf after a 20 min incubation at 37°C. AuTf (white dots) is bound to the walls of the MVE and is clustered heavily around the inclusions, making the outline of the individual gold beads difficult to discern. Bar, 100 nm. × 127,000. All A, B, C, D figures are adopted from Harding et al. [11], figure source has permitted from the author).

the exosomes exert an effect on the replication of HIV [10]. The potential research on exosomes will be helpful in the vaccine or therapeutic design to treat HIV patients along with antiretroviral therapy. The exosome adopted picture is given in **Figure 1**.

2. The exosomes association with HIV-1 patient

The circulated exosomes in plasma are small and nanometre in size. Due to the small size, the exosomes are easily separated from human plasma through low density, differential ultracentrifugation, or ultrafiltration [12]. The exosomes are isolated from the semen for HIV-1 infection assays [13]. The exosomes in the host play an important role in HIV-1 infection, releasing intracellular and extracellular material and establishing cell-cell communication. The exosome study in HIV research is a hotspot because it has great significance in predicting acquired immunodeficiency syndrome (AIDS). The microparticles containing the chemokine receptor CCR5, the principal coreceptor for macrophage-tropic human immuno-deficiency virus-1 was released from the surface of the $CCR5^{+}$Chinese hamster ovary cells and peripheral blood mononuclear cells. It proves that intercellular communication is associated with the cellular membranes of the HIV-1 infection [14]. Exosomes utilise the primary and secondary host cells. The human T cell immunoglobin mucin (TIM) proteins are used by retroviruses, such as HIV, for viral entry. The TIM1 proteins promote the infection of multiple viruses, such as HIV, through virion-associated phosphatidylserine [15]. As AIDS is incurable by antiretroviral therapy, a hypothesis is explained that the 'Trojan exosomes' can be a novel approach to vaccine development [16].

HIV-1 patients in antiretroviral therapy may develop the risk of non-AIDS-defining-cancers (NDCs). The transactivation response RNA containing exosomes from the HIV-1 infected T cells promotes the growth and progression of particular NADCs. In the host, the HIV-infected T cell exosome enters through the epidermal growth factor receptor by stimulating ERK1/2 phosphorylation via the EGF/TLR3 dependent manner leading to NADCs [17]. HIV-1 infected patients are more likely aggressive to develop cervical cancer. The HIV-1-associated exosomes promote the growth and progression of cervical cancer in HIV-1 patients [18]. A study reported the presence of the transactivation response element (TAR) miRNA encoded by HIV-1 infected cells and patient sera. This TAR miRNA was isolated from the exosomes derived from the HIV-1 infected cells. From the study, it was observed that the HIV-1 RNA molecules within the exosome helped in the intercellular viral spread in infected host cells [19]. This TAR RNA stimulated the production of proinflammatory cytokines from exosomes derived from HIV-1 infected cells [20]. The stable epigenetic repression of the HIV-1 expression *in vivo* was observed by the exosome-mediated systematic delivery of therapeutic cargo [21]. An engineered exosome Tat protein was used as a new class of biologic product to activate the latently HIV-1 infected primary $CD4^{+}$ lymphocytes [22]. A research study found that the *Mycobacterium tuberculosis* drug-resistant strains secreted from exosomes by macrophages reactivate HIV-1 induced through oxidative stress. Proteomic information on several host factors, such as galectins, HSP90, HIF-1α in the *Mtb* specific reactivation, promotes the HIV-1 reactivation in the host. This study of this redox and bioenergetics basis of HIV-1 TB coinfection would be an effective therapeutic strategy [23].

Exosomes are nano-sized membrane vesicles released by fusion of the multivesicular body or the organelle of the endocytic pathway origin with the plasma membrane of the cells. The exosome biogenesis is an endosomal sorting complex required for transport (ESCRT) dependent or ESCRT independent. The endosomal pathways are not entirely separated but the different subpopulations of exosomes work synergistically in different cell types of machinery. The cell type or cellular machinery is an important factor that controls the secretion of exosomes [24]. The formation of the exosome is behind with multiple mechanisms. In the early endosome formation, the best mechanism involves the recruitment of the endosomal sorting complex required for transport (ESCRT) machinery to the ubiquitinated proteins. The ESCRT complex is made up of four protein complexes (ESCRT-0, -I, -II, and -III) along with accessory proteins (Alix, VPS4, and VTA-1). These protein complexes bind to the exosome cargos through the intraluminal vesicles (ILVs). One of the ESCRT-III complexes induces the inward budding and fission of the vesicles to form the microvesicular bodies [25–28]. The exosome cargos formation depends on the synthesis of ceramide as a mechanism to induce the vesicle curvature and budding [29]. The exosome formation depends on the synthesis of the tetraspanin-mediated organisation of specific proteins, such as the amyloidogenic protein and pre-melanosome protein [30, 31]. After the formation of exosomes, the exosomes played an important role in cellular functions such as cellular migration and invasion, immunity, normal development and adult physiology, fertilisation and mating behaviour, nervous system [24]. Exosomes and HIV-1 particles share important features in biogenesis, biophysical, molecular, and cellular uptake mechanisms. Both HIV-1 and exosomes are surrounded by a phospholipid bilayer in a size of 100–200 nm but their separation is a technical challenge by centrifugations. Exosomes are membrane-derived vesicles that harbour genomic, proteomic, and lipid cargos and also participate in the progression of HIV-1 pathogenesis. Exosomes derived from the HIV-1 infected dendritic cells were more infective than the cell-free HIV-1 or exosome-derived T cells. During HIV-1 infection, the exosomes

Categories of protein	Names
Tetraspanins	CD9, CD63, CD81, CD82, CD37, CD53
Heat shock proteins (HSP)	HSP90, HSP70, HSP27, HSP60
Cell adhesion	Integrins, Lactadherin, Intercellular Adhesion Molecule
Antigen presentation	Human leukocyte antigen I and II /peptide complexes
Multivesicular body Biogenesis	Tsg101, Alix, Vps, Rab proteins
Membrane transport	Lysosomal-associated membrane protein 1/2, CD13, PG regulatory-like protein
Signalling proteins	GTPase HRas, Ras-related protein, furloss, extracellular signal-regulated kinase, Src homology 2 domain phosphatase, GDP dissociation inhibitor, Syntenin-1, 14-3-3 Proteins, Transforming protein RhoA
Cytoskeleton components	Actins, Cofilin-1, Moesin, Myosin, Tubulins, Erzin, Radixin, Vimentin
Transcription and protein synthesis	Histone1, 2, 3, Ribosomal proteins, Ubiquitin, major vault protein, Complement factor 3
Metabolic enzymes	Fatty acid synthase Glyceraldehyde-3-phosphate dehydrogenase Phosphoglycerate kinase 1 Phosphoglycerate mutase 1 Pyruvate kinase isozymes M1/M2 ATP citrate lyase ATPase Glucose-6-phosphate isomerase Peroxiredoxin 1 Aspartate aminotransferase Aldehyde reductase
Trafficking and membrane fusion	Ras-related protein 5, 7 Annexins I, II, IV, V, VI Synaptosomal-associated protein Dynamin, Syntaxin-3
Antiapoptosis	Alix, Thioredoxin, Peroxidase
Growth factors and cytokine	Tumour necrosis factor-α, TNF receptors, Transforming growth factor-β
Death receptors	FasL, TNF-related apoptosis-inducing ligand
Iron transport	Transferrin receptor
Lipids	*Lipid related enzymes*
LTA4, LTB4, LTC4, PGE2, 15d-PGJ2, PGE2, PA, Ceramides, Cholesterol, Bis(monoacylglycero)phosphate, Phosphatidyl serine, Spingomyelin	LTA4 hydrolase, LTC4 synthase, COX-1, COX-2, PGE synthase, PLD2, DGK, nSMase2

Table 1.
Types of proteins and lipids in exosomes: The author has permitted the source [42].

from the dendritic cells contain fibronectin and galectin-3. In the T-cells exposed to exosomes cargo derived from HIV-1 infected dendritic cells, the gene expression of the pro-inflammatory cytokines IFN-γ, TNF-α, IL-1β, and RANTES and activation of p38/Stat pathways was observed [32]. A class of proteins called 'Tetraspanins' provides a powerful approach to distinguish exosomes and HIV-1 infected cells during the progression of the disease [33]. The Epstein–Barr virus (EBV) that causes tumours in humans induces the transfer of viral oncoprotein, LMP1, and a virus-encoded miRNA through exosome [34]. Exosomes are involved in viral transmission and immune sensing in the host. Hepatitis-A and hepatitis-C viruses use exosomes for viral transmission and antibody-mediated immune responses in the host [35]. The cytomegalovirus, the herpes simplex virus, human papillomavirus, the respiratory syncytial virus was mediated exosome cargo of various molecules, such as dendritic cell-specific ICAM-3 grabbing nonintegrin (DC-SIGN) and MHC-I and II [36].

The architecture of the exosome is very complex and contains proteins, nucleic acids, and lipids. Various databases are available for information on exosomal proteins, mRNA, MicroRNAs and lipids [37, 38]. The integrated database provides high-throughput datasets on vesicular components of prokaryotes, mammalian and nonmammalian eukaryotes extracellular vesicles [38]. An exosome contains several tetraspanins such as CD9, CD63, CD81, CD82 takes part in cell penetration, invasion and fusion events, heat shock proteins HSP70 and HSP90 involved in stress response in antigen binding and presentation, Alix, TSG101 proteins are involved in exosome release, annexins and Rab proteins are involved in membrane transport and fusion. The exosome biogenesis proteins Alix, flotillin, and TSG101 participate in exosome biogenesis. The proteins, such as TSG101, HSP70, CD81, and CD63, are highly enriched in exosomes and considered as the exosome marker proteins. Exosomes contain all types of RNA such as miRNA, ribosomal RNA, long non-coding RNA, piwi interacting RNA, transfer RNA, small nuclear and nucleolar RNA, etc. [39, 40]. Besides proteins and nucleic acids, the exosomes contain the lipids, such as phosphatidylserine, phosphatidic acid, arachidonic acid, cholesterol, and sphingomyelin. The human plasma and serum exosomes contain high-density lipoproteins (HDL), low/very low-density lipoproteins (LDL/VLDL), Apo AI, Apo-B100, etc. [41]. The various proteins and lipid components of the exosome are given in **Table 1**.

Exosomes contain receptors for virus entry. SARS-CoV-2 spreaders are contributed by exosomes as they transfer the receptors CD9 and ACE2 to recipient cells susceptible to the SARS-CoV-2. During the SARS-CoV-2 infection, the SARS-CoV-2 virus is directed into the exosome pathways and the components are packaged into exosomes for secretion [43].

3. Semen exosomes in HIV-1 patient

HIV-1 could transmit via the semen through sexual intercourse and is the leading cause of HIV-1 transmission worldwide during vaginal and rectal intercourse by human beings. The major route of transmission is the vaginal secretion during the sexual intercourse to the person who is involved in this period [44]. Exosomes play an important role in intracellular communication. Semen exosomes are epididymosomes and prostasomes. Epididymosomes and prostasomes are exosomes involved in the acquisition of fertilising ability, modulation of motility, maturation of spermatozoa and could be a possible biomarker in the early diagnosis of male infertility [45].

The identified proteins in the blood exosome and semen exosome are useful as a biomarker applicable to different fields of medicine, especially in reproduction and infectious diseases. The proteomic profiling of semen exosomes and blood exosomes from HIV-infected and uninfected individuals was compared. The compositional and functional variabilities were observed. Semen exosomes were enriched in clusterin protein compared with the blood exosome in HIV-infected and uninfected patients [46]. The antiretroviral activity of the human semen exosome was reported. The human semen exosome inhibits the HIV-1 replication, and it could be useful as a therapeutic purpose for HIV-1. Semen exosomes restrict HIV-1 transmission by vaginal cells. This study of the semen exosomes helps the development of the therapeutic diagnosis during the transmission of HIV-1 [13]. Semen and vaginal fluid exosomes of healthy individuals could inhibit the HIV-1 infection and block the transfer of HIV-1 into vaginal cells [47]. *In vitro* and *in vivo* study of HIV-1 infected, uninfected patients, ART naïve, ART patients' semen exosome were isolated and found the inhibition of replication of HIV-1 [48]. The semen exosomes in HIV-1 patients could be an option for the development of novel therapeutic biomarkers in the field of infertility and the diagnosis of AIDS.

4. Plasma exosomes in HIV-1 patient

Exosomes play an important role in pathogenesis in HIV-1 patients. The exosomes in HIV-1 patients and their relationship in immune and oxidative stress response have been documented [49]. Exosome plasma proteomic profiles were identified and characterised in HIV patients of alcohol drinkers and cigarette smokers. The exosome proteins, such as hemopexin and properdin, could be used as a potential biomarker for physiological effects that may arise in HIV-infected individuals tobacco and alcohol abusers [50]. Exosome-associated most cytokines were observed in the HIV-1 infected patients. The exosomes purified from HIV-positive patients induced the CD38 expression in naive and central memory $CD4^+$ and CD^+ T cells leads to inflammation and viral propagation [12]. The expression levels of miR21 were lower in the plasma-derived exosome of HIV-1 elite controller with decreasing CD4 T cell count [51].

5. Exosome and HIV-associated neurological disorders (HANDS)

In the HIV-1-associated neurocognitive disorders, the HIV-exosomes accelerate the dysfunction of primary human brain microvascular endothelial cells (HBMVECs) by inducing mitochondrial hyperfusion. HIV-1 exosome increases the mitochondrial hyperfusion due to loss of phosphorylated dynamin-related protein1(p-DRP1). HIV-exosomes dysregulate the mitochondrial function could adversely change the effect of the brain microvascular endothelium [52]. The important role of exosomes in HAND has been reported to lead to a disease like Alzheimer's. The HIV patients on ART are the main cause of the deposition of Beta-amyloid (Aβ) leads to dementia. The neuroinflammation is induced by HIV proteins, such as Tat, gp120, and Nef. Exosomes serve as a link between HIV and Alzheimer's disease by packaging and transporting the toxic proteins [53]. The exosome played an important role in signalling in the central nervous system and acts as a potential vehicle to deliver various therapeutics to treat HIV neuroinflammation [54]. The exosomes played an important role in transporting cargo in many neurological disorders, such as mental disorders, brain injury, abnormal neuronal development, neurodegenerative diseases, epilepsy, stroke, and brain cancer [55].

6. Exosomes used as therapy

Exosomes in semen and plasma could be used as a biomarker for the diagnosis of HIV-1 patients. Exosomes could be used as an immunological and oxidative stress marker in HIV patients. Exosomes either promote or inhibit HIV-1 pathogenesis in the human body. The exosomes could be used as a diagnostic biomarker, therapeutic tool for successful implementation, and clinical application in HIV-1 patients [56]. Plasma neuronal exosomes are used as biomarkers in HIV-associated neurological disorders and Alzheimer's disease [57]. In a recent study, the researchers used the exosome-based strategy to block HIV in mice in a block and lock approach [58].

7. Conclusion

The exosome formation and secretion after a retrovirus infection is a very complex process. Further study of the virally complex microvesicles will clarify their role in infection. The main function of the exosome in semen and plasma is immune activation, immune suppression, transfer of functional miRNA, intracellular and intercellular communication. The exosome study will manage the broad biological and medical implications towards the pathogenesis of HIV-1 infected individuals. The current understanding of the exosome biology and function with regards to the HIV-1 infection is in early-stage now. The role of exosomes in the host-HIV-1 infections can open a new era of understanding of disease mechanisms, therapeutic interventions, and future diagnostic purposes in the field of medical science research.

Acknowledgements

Indian Council of Medical Research (ICMR), Govt. of India.

Conflict of interest

No conflict of interest by the authors.

Key points

1. The exosomes are nanovesicles derived from various cell types under the physiological or pathological conditions of the body of human beings.

2. Basically, exosomes carry the host and pathogen-derived genomic, proteomic, lipidomic materials. The exosomes exchange information between the cells during the progression of AIDS.

3. Exosomes in the HIV-1 patients have the dual activity that could inhibit HIV-1 infection or progress the HIV-1 infection within the host.

4. Exosomes may activate the resting $CD4^{+}$T lymphocytes and, by activating these cells, lead to chronic infection.

5. The study of biological information in an exosome could be used as a marker in HIV-1 patients and could be helpful in diagnosis and treatment.

6. Semen exosomes are rich in nucleic acids and protein binding factors than blood exosomes.

Author details

Sushanta Kumar Barik[1*], Sanghamitra Pati[2], Keshar Kunja Mohanty[1*], Sashi Bhusan Mohapatra[3], Srikanta Jena[4] and Srikanth Prasad Tripathy[5]

1 ICMR-National JALMA Institute for Leprosy and Other Mycobacterial Diseases, Agra, Uttar Pradesh, India

2 ICMR-Regional Medical Research Centre, Bhubaneswar, Odisha, India

3 Department of Zoology, Banki (Auto) College, Banki, Odisha, India

4 Department of Zoology, Ravenshaw University, Cuttack, Odisha, India

5 Dr. D.Y. Patil Medical College, Hospital and Research Centre, Pune, Maharashtra, India

*Address all correspondence to: sushantakumarbarik82@gmail.com; keshar63@yahoo.com

References

[1] Cocucci E, Meldolesi J. Ectosomes and exosomes: Shedding the confusion between extracellular vesicles. Trends in Cell Biology. 2015;**25**(6):364-372

[2] Raposo G, Stoorvogel W. Extracellular vesicles: Exosomes, microvesicles, and friends. Journal of Cell Biology. 2013;**200**(4):373-383

[3] Colombo M, Raposo G, Théry C. Biogenesis, secretion, and intercellular interactions of exosomes and other extracellular vesicles. Annual Review of Cell and Developmental Biology. 2014;**30**:255-289

[4] Breckpot K, Aerts JL, Thielemans K. Lentiviral vectors for cancer immunotherapy: Transforming infectious particles into therapeutics. Gene Therapy. 2007;**14**(11):847-862

[5] Ferguson SW, Nguyen J. Exosomes as therapeutics: The implications of molecular composition and exosomal heterogeneity. Journal of Controlled Release. 2016;**228**:179-190

[6] Conlan RS, Pisano S, Oliveira MI, Ferrari M, Pinto IM. Exosomes as reconfigurable therapeutic systems. Trends in Molecular Medicine. 2017;**23**(7):636-650

[7] Funke S, Maisner A, Mühlebach MD, Koehl U, Grez M, Cattaneo R, et al. Targeted cell entry of lentiviral vectors. Molecular Therapy. 2008;**16**(8):1427-1436

[8] Lee Y, El Andaloussi S, Wood MJ. Exosomes and microvesicles: Extracellular vesicles for genetic information transfer and gene therapy. Human Molecular Genetics. 2012;**21**(R1):R125-R134

[9] Escors D, Breckpot K. Lentiviral vectors in gene therapy: Their current status and future potential. Archivum Immunologiae et Therapiae Experimentalis. 2010;**58**(2):107-119

[10] Patters BJ, Kumar S. The role of exosomal transport of viral agents in persistent HIV pathogenesis. Retrovirology. 2018;**15**(1):1-13

[11] Harding C, Heuser J, Stahl P. Receptor-mediated endocytosis of transferrin and recycling of the transferrin receptor in rat reticulocytes. Journal of Cell Biology. 1983;**97**(2):329-339

[12] Konadu KA, Chu J, Huang MB, Amancha PK, Armstrong W, Powell MD, et al. Association of cytokines with exosomes in the plasma of HIV-1–seropositive individuals. The Journal of Infectious Diseases. 2015;**211**(11):1712-1716

[13] Madison MN, Welch JL, Okeoma CM. Isolation of exosomes from semen for in vitro uptake and HIV-1 infection assays. Bio-Protocol. 5 Apr 2017;7(7):e2216. DOI: 10.21769/BioProtoc.2216.

[14] Mack M, Kleinschmidt A, Brühl H, Klier C, Nelson PJ, Cihak J, et al. Transfer of the chemokine receptor CCR5 between cells by membrane-derived microparticles: A mechanism for cellular human immunodeficiency virus 1 infection. Nature Medicine. 2000;**6**(7):769-775

[15] Jemielity S, Wang JJ, Chan YK, Ahmed AA, Li W, Monahan S, et al. TIM-family proteins promote infection of multiple enveloped viruses through virion-associated phosphatidylserine. PLoS Pathogens. 2013;**9**(3):e1003232

[16] Hildreth JE. HIV as Trojan exosome: Immunological paradox explained? Frontiers in Immunology. 2017;**8**:1715

[17] Chen L, Feng Z, Yue H, Bazdar D, Mbonye U, Zender C, et al. Exosomes derived from HIV-1-infected cells promote growth and progression of cancer via HIV TAR RNA. Nature Communications. 2018;**9**(1):1-12

[18] Li H, Chi X, Li R, Ouyang J, Chen Y. HIV-1-infected cell-derived exosomes promote the growth and progression of cervical cancer. International Journal of Biological Sciences. 2019;**15**(11):2438

[19] Narayanan A, Iordanskiy S, Das R, Van Duyne R, Santos S, Jaworski E, et al. Exosomes derived from HIV-1-infected cells contain trans-activation response element RNA. Journal of Biological Chemistry. 2013;**288**(27):20014-20033

[20] Sampey GC, Saifuddin M, Schwab A, Barclay R, Punya S, Chung MC, et al. Exosomes from HIV-1-infected cells stimulate production of pro-inflammatory cytokines through trans-activating response (TAR) RNA. Journal of Biological Chemistry. 2016;**291**(3): 1251-1266

[21] Shrivastava S, Ray RM, Holguin L, Echavarria L, Grepo N, Scott TA, et al. Exosome-mediated stable epigenetic repression of HIV-1. Nature Communications. 2021;**12**(1):1-14

[22] Tang X, Lu H, Dooner M, Chapman S, Quesenberry PJ, Ramratnam B. Exosomal Tat protein activates latent HIV-1 in primary, resting CD4+ T lymphocytes. JCI Insight. 2018;**3**(7):e95676. Published online 2018 Apr 5. DOI: 10.1172/jci.insight.95676

[23] Tyagi P, Pal VK, Agrawal R, Singh S, Srinivasan S, Singh A. Mycobacterium tuberculosis reactivates HIV-1 via exosome mediated resetting of cellular redox potential and bioenergetics. MBio. 2020;**11**(2):e03293-e03219

[24] Maas SL, Breakefield XO, Weaver AM. Extracellular vesicles: Unique intercellular delivery vehicles. Trends in Cell Biology. 2017;**27**(3):172-188

[25] Christ L, Raiborg C, Wenzel EM, Campsteijn C, Stenmark H. Cellular functions and molecular mechanisms of the ESCRT membrane-scission machinery. Trends in Biochemical Sciences. 2017;**42**(1):42-56

[26] Chiaruttini N, Redondo-Morata L, Colom A, Humbert F, Lenz M, Scheuring S, et al. Relaxation of loaded ESCRT-III spiral springs drives membrane deformation. Cell. 2015;**163**(4):866-879

[27] McCullough J, Clippinger AK, Talledge N, Skowyra ML, Saunders MG, Naismith TV, et al. Structure and membrane remodeling activity of ESCRT-III helical polymers. Science. 2015;**350**(6267):1548-1551

[28] Lee IH, Kai H, Carlson LA, Groves JT, Hurley JH. Negative membrane curvature catalyzes nucleation of endosomal sorting complex required for transport (ESCRT)-III assembly. Proceedings of the National Academy of Sciences. 2015;**112**(52): 15892-15897

[29] Trajkovic K, Hsu C, Chiantia S, Rajendran L, Wenzel D, Wieland F, et al. Ceramide triggers budding of exosome vesicles into multivesicular endosomes. Science. 2008;**319**(5867):1244-1247

[30] Theos AC, Truschel ST, Tenza D, Hurbain I, Harper DC, Berson JF, et al. A lumenal domain-dependent pathway for sorting to intralumenal vesicles of multivesicular endosomes

involved in organelle morphogenesis. Developmental Cell. 2006;**10**(3): 343-354

[31] Van Niel G, Charrin S, Simoes S, Romao M, Rochin L, Saftig P, et al. The tetraspanin CD63 regulates ESCRT-independent and-dependent endosomal sorting during melanogenesis. Developmental Cell. 2011;**21**(4):708-721

[32] Kulkarni R, Prasad A. Exosomes derived from HIV-1 infected DCs mediate viral trans-infection via fibronectin and galectin-3. Scientific Reports. 2017;**7**(1):1-14

[33] Dias MV, Costa CS, DaSilva LL. The ambiguous roles of extracellular vesicles in HIV replication and pathogenesis. Frontiers in Microbiology. 2018;**9**:2411

[34] Meckes DG, Shair KH, Marquitz AR, Kung CP, Edwards RH, Raab-Traub N. Human tumor virus utilizes exosomes for intercellular communication. Proceedings of the National Academy of Sciences. 2010;**107**(47):20370-20375

[35] Longatti A. The dual role of exosomes in hepatitis A and C virus transmission and viral immune activation. Viruses. 2015;**7**(12): 6707-6715

[36] Wang J, Yao Y, Chen X, Wu J, Gu T, Tang X. Host derived exosomes-pathogens interactions: Potential functions of exosomes in pathogen infection. Biomedicine and Pharmacotherapy. 2018;**108**:1451-1459

[37] Mathivanan S, Fahner CJ, Reid GE, Simpson RJ. ExoCarta 2012: Database of exosomal proteins, RNA and lipids. Nucleic Acids Research. 2012;**40**(D1):D1241-D1244

[38] Kim DK, Kang B, Kim OY, Choi DS, Lee J, Kim SR, et al. EVpedia: An integrated database of high-throughput data for systemic analyses of extracellular vesicles. Journal of Extracellular Vesicles. 2013;**2**(1):20384

[39] Vlassov AV, Magdaleno S, Setterquist R, Conrad R. Exosomes: Current knowledge of their composition, biological functions, and diagnostic and therapeutic potentials. Biochimica et Biophysica Acta (BBA)-General Subjects. 2012;**1820**(7):940-948

[40] Huang X, Yuan T, Tschannen M, Sun Z, Jacob H, Du M, et al. Characterization of human plasma-derived exosomal RNAs by deep sequencing. BMC Genomics. 2013; **14**(1):1-14

[41] Sun Y, Saito K, Saito Y. Lipid profile characterization and lipoprotein comparison of extracellular vesicles from human plasma and serum. Metabolites. 2019;**9**(11):259

[42] Zhang Y, Liu Y, Liu H, Tang WH. Exosomes: Biogenesis, biologic function and clinical potential. Cell & Bioscience. 2019;**9**(1):1-18

[43] Hassanpour M, Rezaie J, Nouri M, Panahi Y. The role of extracellular vesicles in COVID-19 virus infection. Infection, Genetics and Evolution. 2020;**85**:104422

[44] Barreto-de-Souza V, Arakelyan A, Margolis L, Vanpouille C. HIV-1 Vaginal Transmission: Cell-Free or Cell-Associated Virus? American Journal of Reproductive Immunology. 2014;**71**(6):589-599

[45] Baskaran S, Selvam MKP, Agarwal A. Exosomes of male reproduction. Advances in Clinical Chemistry. 2020;**95**:149-163

[46] Kaddour H, Lyu Y, Welch JL, Paromov V, Mandape SN, Sakhare SS,

et al. Proteomics profiling of autologous blood and semen exosomes from HIV-infected and uninfected individuals reveals compositional and functional variabilities. Molecular and Cellular Proteomics. 2020;**19**(1):78-100

[47] Ouattara LA, Anderson SM, Doncel GF. Seminal exosomes and HIV-1 transmission. Andrologia. 2018;**50**(11):e13220

[48] Welch JL, Kaddour H, Winchester L, Fletcher CV, Stapleton JT, Okeoma CM. Semen extracellular vesicles from HIV-1 infected individuals inhibit HIV-1 replication in vitro, and extracellular vesicles carry antiretroviral drugs in vivo. Journal of Acquired Immune Deficiency Syndromes. 2020;**83**(1):90

[49] Chettimada S, Lorenz DR, Misra V, Dillon ST, Reeves RK, Manickam C, et al. Exosome markers associated with immune activation and oxidative stress in HIV patients on antiretroviral therapy. Scientific Reports. 2018;**8**(1):1-16

[50] Kodidela S, Wang Y, Patters BJ, Gong Y, Sinha N, Ranjit S, et al. Proteomic profiling of exosomes derived from plasma of HIV-infected alcohol drinkers and cigarette smokers. Journal of Neuroimmune Pharmacology. 2020;**15**(3):501-519

[51] Ruiz-de-León MJ, Jiménez-Sousa MA, Moreno S, García M, Gutiérrez-Rivas M, León A, et al. Lower expression of plasma-derived exosome miR-21 levels in HIV-1 elite controllers with decreasing CD4 T cell count. Journal of Microbiology, Immunology and Infection. 2019;**52**(4):667-671

[52] Chandra PK, Rutkai I, Kim H, Braun SE, Abdel-Mageed AB, Mondal D, et al. Latent HIV-exosomes induce mitochondrial hyperfusion due to loss of phosphorylated dynamin-related protein 1 in brain endothelium. Molecular Neurobiology. 2021;**58**(6):2974-2989

[53] Kodidela S, Gerth K, Haque S, Gong Y, Ismael S, Singh A, et al. Extracellular vesicles: A possible link between hiv and alzheimer's disease-like pathology in hiv subjects? Cell. 2019;**8**(9):968

[54] Mahajan SD, Ordain NS, Kutscher H, Karki S, Reynolds JL. HIV neuroinflammation: The role of exosomes in cell signalling, prognostic and diagnostic biomarkers and drug delivery. Frontiers in Cell and Developmental Biology. 2021;**9**:504

[55] Xiao L, Hareendran S, Loh YP. Function of exosomes in neurological disorders and brain tumors. Extracellular Vesicles and Circulating Nucleic Acids. 2021;**2**(1):55-79

[56] Pang SW, Teow SY. Emerging therapeutic roles of exosomes in HIV-1 infection. In: Exosomes. Cambridge, MA: Academic Press; 2020. pp. 147-178

[57] Pulliam L, Sun B, Mustapic M, Chawla S, Kapogiannis D. Plasma neuronal exosomes serve as biomarkers of cognitive impairment in HIV infection and Alzheimer's disease. Journal of Neurovirology. 2019;**25**(5):702-709

[58] NIH Researcher. Researchers Use Exosome-Based Strategy to Block HIV in Mice. Bethesda, MD: NIH Researcher; September 21, 2021. Press Release Office of Science Policy, Planning, and Communications Science Writing, Press, and Dissemination Branch 6001 Executive Boulevard Bethesda, MD 20892

Chapter 9

The Role of Extracellular Vesicles in Immunomodulation and Pathogenesis of *Leishmania* and Other Protozoan Infections

Zeynep Islek, Batuhan Turhan Bozkurt, Mehmet Hikmet Ucisik and Fikrettin Sahin

Abstract

Extracellular vesicles (EVs) have lately emerged as crucial mediators in parasite infections. Recent research suggests that protozoan parasites, including *Leishmania*, employ EVs as transport vehicles to deliver biologically active effector molecules such as parasitic virulence factors to modulate the host immune system and their micro-environment. The immunomodulatory effects of EVs play an essential role in the formation and progression of parasitic diseases. The immunomodulatory strategies applied by EVs of protozoan origin have similarities to the development and progression of other infections or diseases such as cancer. In this chapter, we will provide recent insights into the role of EVs in host-pathogen interactions, intercellular-communication, immunomodulation and pathogenesis of *Leishmania* and other protozoan parasites, including *Plasmodium spp., Toxoplasma spp*. and *Trypanosoma spp.* In addition, biologically inspired by the immunomodulation strategies of protozoan parasites, new immunotherapeutic models are being currently investigated to implement EVs more intensively in both therapy and diagnostics. Therefore, besides highlighting the role of EVs in protozoan infections, this chapter sheds light briefly on new immunotherapeutic approaches utilizing the strategies of protozoan EVs in medicine.

Keywords: extracellular vesicles, immunomodulation, pathogenesis, protozoan, *Leishmania*, infectious disease

1. Introduction

Cellular communication is essential for all life forms to observe, comprehend and affect their surroundings [1–6]. One pathway that cells employ for the transfer of information is the use of extracellular vesicles (EVs) – lipid-bilayered secreted vesicles that carry lipids, nucleic acids and proteins that can cause physiological changes in other cells. The use of EVs for cellular communications is a highly conserved process of life. The EV secretion was observed in all types of cells and organisms studied up to date, including plants [7–9], prokaryotes [3, 10, 11] and protozoans [12–21]. Moreover, evidence suggests that EVs can affect cells of different species, even across different kingdoms [10, 11, 13, 16, 20]. Cross-kingdom EV interactions were shown to take part in the pathogenesis of some parasitic diseases such as those caused by protozoan parasites [22, 23].

Protozoan parasites, also known as first animals, are single-celled organisms that display diversity among unicellular eukaryotic organisms with a complex life cycle on the host system [20]. They have developed many strategies not only to provide their survival and reproduction, but also to enable the invasion into the hosts by means of immune strategies including change in host antigens, development of self-tolerance, immune inactivation, immunosuppression and intervention of molecule-mimetic mechanisms between parasites and host antigens [16, 24, 25]. Recent studies propose that the parasites actually utilize the extracellular vesicles as one infection strategy [18, 20, 21, 26–31], where the questions are arisen on how EVs modulate the host immune system and ultimately cause the infection. Based on the cell of origin, the release mechanisms of EVs from different protozoan parasites, including Apicomplexa and Kinetoplastids such as *Leishmania* species (spp.) [22, 23, 26, 32–35], *Plasmodium spp.* [31, 36–41], *Toxoplasma spp.* [36, 42, 43] and *Trypanosoma spp.* [44–49] were described, where the parasitic infections were studied in detail for leishmaniasis, malaria, toxoplasmosis and Chagas disease independently.

Among the many species and subspecies of protozoa, *Leishmania* are digenetic intracellular protozoan parasite that cause leishmaniasis through the localization either in mononuclear phagocytes of vertebrates as amastigote form or in the sandfly vector as promastigote form. There are three main forms of leishmaniasis, including a localized form- cutaneous leishmaniasis (CL) or mucocutaneous leishmaniasis (MCL), and a life-threatening form – visceral leishmaniasis (VL) (also known as "Kala-azar") [50].

The EVs released from parasites or infected cells play a significant role in host-pathogen communications and thus contribute to pathogenesis [12, 13, 15, 16, 18–21, 51]. Studies indicated that *Leishmania* exosomes can modulate the host immune system through monocyte cytokine production occurring in response to *Leishmania* infection, which in return further exacerbates *Leishmania* infection [14, 21–23, 26, 32–35, 52–54]. Likewise, Evs' role in the occurrence of infection was also confirmed later for more protozoan family members such as *Plasmodium spp.* [31, 36–41], *Toxoplasma spp.* [36, 42, 43] and *Trypanosoma spp.* [44–49], which further directed the attention of researchers on protozoan EVs and their mechanism of action.

This chapter largely focuses on the role of EVs in *Leishmania*-host interaction, immunomodulation of the host immune system by *Leishmania* EVs, manipulation of the cellular microenvironment in favor of *Leishmania* species. In addition, the role of EVs in the pathogenesis of other protozoan parasites including *Plasmodium spp.*, *Toxoplasma spp.* and *Trypanosoma spp.* are discussed and compared at the biological level to get a better insight on strategies in immunomodulation mechanisms. At the end of the chapter, novel and potential immunotherapeutic approaches utilizing the strategies of protozoan EVs are briefly discussed.

2. Extracellular vesicles (EVs)

Extracellular vesicles are nano-sized messengers secreted by all cell types. They consist of a lipid bilayer membrane, proteins, nucleic acids and other biomolecules, which together make up the "message" to be conveyed to other cells. The composition of molecules that control the message differs in different cell types, and under different physiological conditions.

EVs' size ranges between 20 and 1000 nm in diameter, and they can be produced through a variety of different biogenesis pathways, with different physical and structural properties. Budding from the cellular membrane generally forms larger vesicles called microvesicles – however, this biogenesis pathway may also form vesicles that are smaller than 200 nm. Small extracellular vesicles can also be formed through the invagination of the cellular membrane into endosomes, collected and secreted together in multivesicular bodies (MVBs), or so-named exosomes [55]. However, it should be noted that most of the EV isolation methods used today cannot separate exosomes from small EVs formed through membrane budding, resulting in mixed populations of EVs in the working medium. The full extent of the biogenesis pathways remains to be unknown to researchers, and this is even more apparent in non-mammalian EVs [12]. However, evidence indicates that parasites secrete EVs through both the membrane budding and the multivesicular body pathways, mimicking the previously studied EV secretion pathways of mammalian cells [45].

3. Immunomodulation and pathogenesis by EVs from *Leishmania* species and other protozoan parasites

While the study of EVs in eukaryotes other than mammalians has been gaining momentum, the methods used in these studies were developed with mammalian EVs in mind. The International Society for EVs has listed the minimal requirements for categorizing a particle as an extracellular vesicle as reporting the size distribution of the population at a single-vesicle resolution, and detecting the presence of transmembrane and cytosolic proteins in the sample while testing for a non-vesicle related protein as negative control [6, 12]. While the physical characteristics of non-mammalian EVs do not differ greatly from their mammalian counterparts, the literature lacks the necessary amount of data to decide on protein biomarkers for most non-mammalian samples. These experimental results are also required for the characterization of *Leishmania* EVs and other protozoan parasites, including *Plasmodium spp., Toxoplasma spp.* and *Trypanosoma spp.*

3.1 *Leishmania species (spp.)*

Leishmania spp. are protozoan parasites belonging to the Trypanosomatidae family in the Kinetoplastidae order, belonging to the characteristics of a kinetoplast. They are obligated intracellular parasites that primarily infect macrophages in the mammalian through the transmission of the bite of an infected sand fly and cause leishmaniasis. Moreover, they are digenetic organisms that survive and replicate either as the promastigote, i.e., the extracellular form existing in the insect midgut or as the amastigote, i.e. intracellular form lodged within phagolysosome-like vacuoles inside the macrophages [50, 56].

The promastigote form of parasites inoculate in the dermis by the bite of a sandfly (*Lutzmoyia spp., Phlebotomus spp.*) are thought to infect macrophages and/or dendritic cells (DCs) of the skin where they transform into amastigotes and might protect their host cell from apoptosis [25]. Studies have shown that exosomes released from *Leishmania spp.* promastigote and amastigotes play a crucial role in host-pathogen interactions and intercellular communication, leading to the development of infection (pathogenesis) and immunomodulation [14, 21–23, 26, 32–35, 52–54].

3.1.1 Leishmaniasis

Leishmaniasis is a neglected tropical disease caused by vector-borne parasites of the genus *Leishmania.* There are over 20 species of *Leishmania* that cause life-threatening disorders widely distributed in 98 tropical and subtropical regions including Asia, South America, Northern Africa, Southern Europe and the Middle East. According to the recent WHO report, more than 350,000 people are estimated at risk and 1.3 million new cases of leishmaniasis occur every year [50].

Leishmaniasis can be grouped into three main clinical forms: cutaneous leishmaniasis (CL), visceral leishmaniasis (VL), also known as "Kala-azar", and mucocutaneous leishmaniasis (MCL), depending on which species is involved in the infection [50]. CL is a benign but often disfiguring condition that is caused by the multiplication of *Leishmania* in the phagocytes of the skin and has a tendency toward spontaneous resolution. The coexistence of these clinical forms in the same patient is rare. MCL is a metastatic form of localized CL infections occurring during the first episode of CL within 5 years. Lymphatic or hematogenous dissemination of the amastigotes from the skin to the naso-oropharyngeal mucosa results in the destruction of the nose and mouth to the pharynx and larynx. Untreated infections can result from severe disfiguration or even death. VL is a severe condition that results from the dissemination of *Leishmania* in the phagocytes, mainly macrophages, and is fatal in almost all cases if left untreated. VL is characterized by irregular bouts of fever, substantial weight loss, swelling of the spleen and liver and serious anemia [50].

The outcomes of the infection are highly dependent on both host and pathogen factors involved in a molecular battle where the fittest survive and continue. In this context, it is well established that macrophages play an important role in defense against various parasites by regulating their invasion and progression within the potential host. However, like other pathogens, most *Leishmania* species have developed effective strategies to circumvent the innate immune response in the early moments of infection, provided by rapidly blocking the induction and regulation of major host cell functions including nitric oxide (NO) production, tumor necrosis factor-alpha (TNF-α), interleukin-12 (IL-12), radical oxygen species (ROS) [57–60].

Recent studies have investigated that EVs released from *Leishmania* can involve in the pathogenesis by delivering the virulence factors – GP63, Elongation Factor 1-alpha (EF-1α) and others – to mammalian host cells, modulating their microenvironment and inferring on host signaling pathways [26, 34, 61, 62].

3.1.2 Secretion of EVs containing Leishmania proteins

EVs carry biological messages in the form of the lipids, proteins and nucleic acids they are composed of. Both the cargo enclosed within the EV and the structural molecules of the EV itself can initiate cellular responses. The lipids and membrane proteins of EVs are capable of interacting with the surface receptors of a recipient cell, allowing the EV to initiate cell-to-cell contact-dependent responses by acting as a surrogate to their cell-of-origin. Cells tailor the cargo of their EVs for them to initiate the desired response on recipient cells [55].

Protein interactions are one of the primary ways for EVs to affect target cells. Hence, the proteomic analysis of protozoan EVs becomes crucial in determining

Evs' biological functions. Proteomic analysis indicate that parasite EVs are enriched in proteases [33, 45, 63–65], stress response proteins [45, 64, 66] and transcription factors [45, 67].

One of the most common types of proteins found in parasite EVs are proteases. Proteases are a large family of hydrolytic enzymes that take part in a large majority of biological processes. Through the breakdown of specific peptides, proteases allow the activation and removal of various proteins, regulating biological reactions associated with them [68]. Proteases are considered as one of the virulence factors of parasites increasing the infectivity by inactivating the complement system and cleaving transcription factors that aid macrophage activation. *Leishmania* parasites and other trypanosomatids employ *Leishmania* virulence factors, such as metalloprotease GP63 and other immunosuppressive proteins, as well as the ER/Golgi-mediated secretion pathway to exit the host cell post-transfection [21]. An example of this process was shown with *L. mexicana*, where cysteine proteases were sorted into lysosomes and subsequently released via the flagellar pocket when they reached the Golgi apparatus [21, 29].

Initial clues for the existence of EV-mediated non-conventional protein secretion in parasites came from a study of the *Leishmania* parasites, where hydrophilic acylated surface protein B (HSAPB) was found to be present on the parasites' membrane despite not having a signal peptide, transmembrane domain or GPI-anchor site [21]. A study by Denny et al. discovered a novel sequence of 18 amino acids that act as a "special" signal peptide, which allows the transfer of the protein to the cellular membrane [21]. The study also showed that the transfer of HSAPB continued even after the transfection of mammalian cells, with the protein being observed on the cell surface. This non-conventional secretion pathway of proteins is a characteristic feature of EVs and is crucial for the ability of parasite EVs in manipulating the hosts' microenvironment.

The evidence of *Leishmania* exosome secretion was demonstrated in the study of *L. mexicana* exoproteome associated with proteases [69]; however, the first report on the certain secretion of *Leishmania* exosomes was issued by Silverman et al. [54]. Also, proteomic analysis of parasite EVs reveals that different types of proteases are among the most abundant type of proteins in their proteome [62, 64, 65]. The enrichment of proteases in EVs occurs during the entire lifecycle of the parasites during the avirulent procyclic and virulent metacyclic phases [62]. However, metacyclic parasite EVs were shown to contain a higher concentration of proteases than EVs of avirulent procyclic parasites, suggesting a link between proteases and infectivity (34). Another study showed that *Leishmania* species can also hijack host proteases through plasminogen binding proteins that bind plasmin-precursor plasminogen to the parasite cell membrane. One such plasminogen binding protein, discovered in *Leishmania mexicana* EVs, is enolase, a highly conserved EV protein that may allow immune avoidance and parasite dissemination [63].

On the other hand, the EVs of different parasites have similar physical and biochemical properties with each other as well as with EVs of mammalian origin [54]. TEM micrographs captured the secretion of *Leishmania* exosomes through the fusion of MVBs with the parasite membrane [53] and orthologues to key proteins commonly associated with EV formation, such as Rab GTPases, Alix, and ESCRT proteins were found in the proteome of *Leishmania* EVs.

Another category of proteins commonly found in parasite EVs are stress-response proteins. Parasites face various stress conditions in both their insect and vertebrate hosts, and the proteomic profile of the parasite reflects that suitably. Oxidoreductase proteins may protect the parasite from the free radicals of the immune system [45], while chaperone proteins such as the ER chaperone glucose-regulated protein (GRP), heat shock protein 70 (HSP70) are commonly reported as upregulated in parasite EVs [45, 66]. Their presence in the EVs may be due to the elevated expression of these proteins in the parasite itself, instead of an EV-specific sorting mechanism.

Transcription and translation factors detected in parasite EVs may also have roles in parasite infectivity and resilience against stress factors [45, 67]. While it is not clear whether or not if these factors are specifically packaged into EVs for a function, or present due to their abundance in the cytoplasm, studies note that proteins such as EF 1 or 2 were shown to be pro-infective in the parasite itself [70].

A recent study indicated that *Leishmania donovani* infection led to a quantitative and qualitative change in the protein profile of EVs released by the infected macrophages, confirmed by mass spectrometry and western blot analysis. Through the protein analysis, 59 parasite-derived proteins in EVs were found, which promote angiogenesis by inducing endothelial cells to release angiogenesis-promoting mediators [32].

EVs' role in exposed drug resilience of particular strains was also investigated. *L. infantum* strains resistant to various *Leishmania* drugs were found to secrete EVs with different physical and proteomic profiles and secreted more EVs than wild-type parasites [67]. Different histone and ribosomal proteins were found to be enriched in the EVs of drug-resistant strains, which might be a non-specific adaptation of the parasite to increase its fitness in general. This knowledge may be used to diagnose whether or not a patient is infected with a drug resilient strain of the parasite, and could potentially allow identification and prediction of the drug-resistance mechanism of the strain before starting the therapy [45, 67].

3.1.3 The evidence of the EVs released from Leishmania spp.

Leishmania parasites secrete EVs both *in vitro* and *in vivo* in the sandfly midgut [53] and these EVs display immunomodulatory and signal-triggering events on the host system, associating with the parasite virulence factors. Studies with mice and immune cells showed that EVs released from *Leishmania spp.* and infected cells may affect and contribute to the clinical form and severity of the disease regarding the multitude of factors [21].

Originally, the presence of exosomes-like vesicles secreted from *Leishmania* parasites was suggested in the supernatant of infected macrophage cultures by proteomic analysis of the secretome of *Leishmania donovani* [64]. Silverman and colleagues proposed that *L. donovani* utilizes the alternative non-classical secretion pathways and targeting mechanism rather than the classical secretion signal to direct the secreted protein export [64]. Based on this study, exosomes from *Leishmania* parasites are involved in the delivery of proteins into host target cells [54, 64].

On the other hand, the first report on the release of the exosomes from the protozoan pathogens and their use as a vehicle for protein secretion and uptake by macrophages was established by Silverman et al. [30]. This study demonstrated that L. *donovani* and *L. major* can release exosomes that were detected in cytosol of the infected macrophages and selectively induced secretion of IL-8 from macrophages [30]. Furthermore, exosome release was significantly detected in the culture supernatant of *L. donovani, L. mexicana* and *L. major spp.*, under high temperature (37°C) and low pH in which condition required for promastigote differentiation into amastigotes. In another study, using *Leishmania* expressing green fluorescent protein

(GFP), they found a release of *Leishmania* GFP+ vesicles into infected cells and an uptake fluorescence vesicles by non-infected cells, with the collection of GFP and parasite proteins in structures consistent with MVBs within the cytosol of infected macrophages [30].

In addition to studies on EVs from *Leishmania* within mammalian hosts, the secretion of EVs from *Leishmania* residing within the sandfly midgut was also demonstrated by Atayde et al. [53]. Moreover, the detailed characterization of EVs isolated from infected sandfly midguts was investigated. *Leishmania* EVs isolated from infected sandfly midguts were also compared with previously described *in vitro*-isolated *Leishmania* EVs.

3.1.4 Host manipulation and immunomodulation by EVs from Leishmania spp.

Leishmania inhibits normal macrophage functions and also interferes with the innate and acquired (both cell-mediated and humoral) immunity [60]. The uptake of promastigotes by the host-immune cells involves several different strategies that allow the parasite's protective mechanism to evade their immune systems [71]. To survive and evade the host defense mechanism, transmission begins with the differentiation of the intracellular amastigote form of *Leishmania* that replicates within macrophages in the vertebrate hosts to the extracellular promastigote form in the sandfly vector [60, 72].

Briefly, the life cycle of *Leishmania* begins with an infection of the female sandflies after ingesting blood meal in *Leishmania*-infected vertebrate hosts, as illustrated in **Figure 1**. In the sandfly vector, within the midgut, ingested amastigotes proliferate and then migrate to the foregut to differentiate into metacyclic promastigotes presented on the salivary glands of the sandfly vector. Once delivered to a vertebrate host

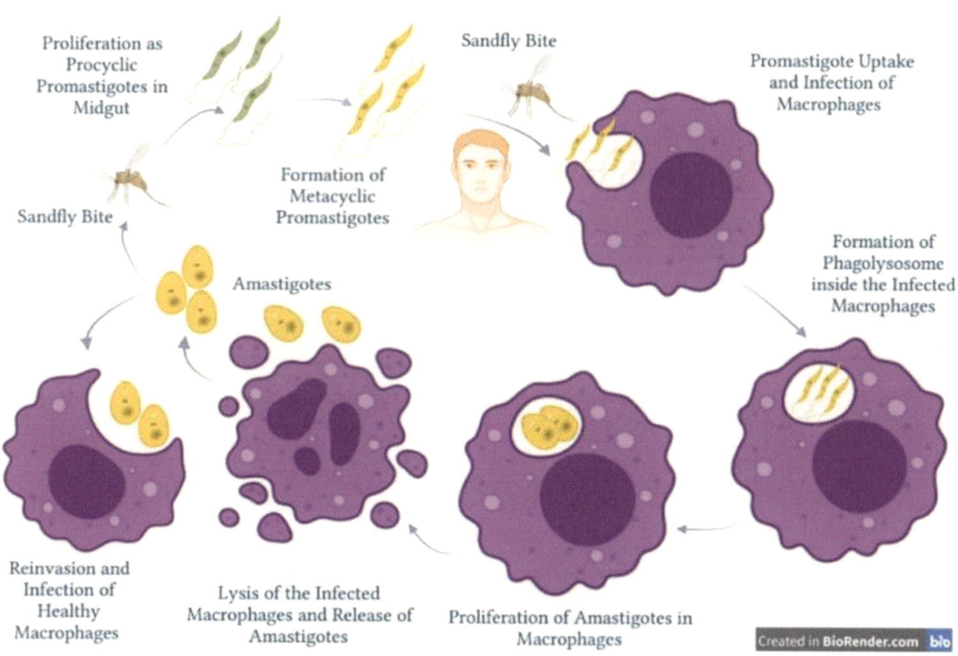

Figure 1.
The lifecycle of Leishmania parasites. Biorender software was used to create this figure under an academic license.

by the bite of an infected sandfly, promastigotes attach to phagocytic cells, macrophages, and are readily engulfed. Parasite-containing parasitophore vacuoles fuse with lysosomes forming a "phagolysosomes" in which promastigotes differentiate into the vertebrate stage, a flagellate form of amastigote [60, 73] (**Figure 1**). When a sandfly ingests a blood meal from an infected host, amastigotes differentiate back into promastigotes and become metacyclic. The metacyclic promastigotes that inoculate in the dermis by the bite of a sandfly (*Lutzmoyia spp*., *Phlebotomus spp*.) are thought to infect macrophages and/or DCs of the skin, where they transform into amastigotes into macrophages and might protect their host cell from apoptosis [74].

Once *Leishmania* metacyclic promastigotes (infective form) with sandfly saliva components are delivered into the mammalian hosts by an infected sandfly, promastigotes have to evade the complement-mediated cell-lysis before being eliminated by phagocytosis and must survive the impact of the innate immune system (**Figure 2**). For phagocytosis, macrophages are the main immune population involved in the elimination and clearance of the parasites. Although macrophages are the main host cell for *Leishmania* parasites, monocytes, DCs and neutrophils can be infected and contribute differentially to the immune response and the outcome of the infection [75] (**Figure 2**). As the first cell to be recruited to the infection site, neutrophils have delivered promastigotes to the macrophages through facilitating a silence entry, proposed as "Trojan Horse" [76] (**Figure 2**). Neutrophils infiltration and recruitment are contributed by various factors such as the leishmania chemotactic factor inducing IL-8 secretion by human neutrophils or interleukin-17 (IL-17), a hallmark of

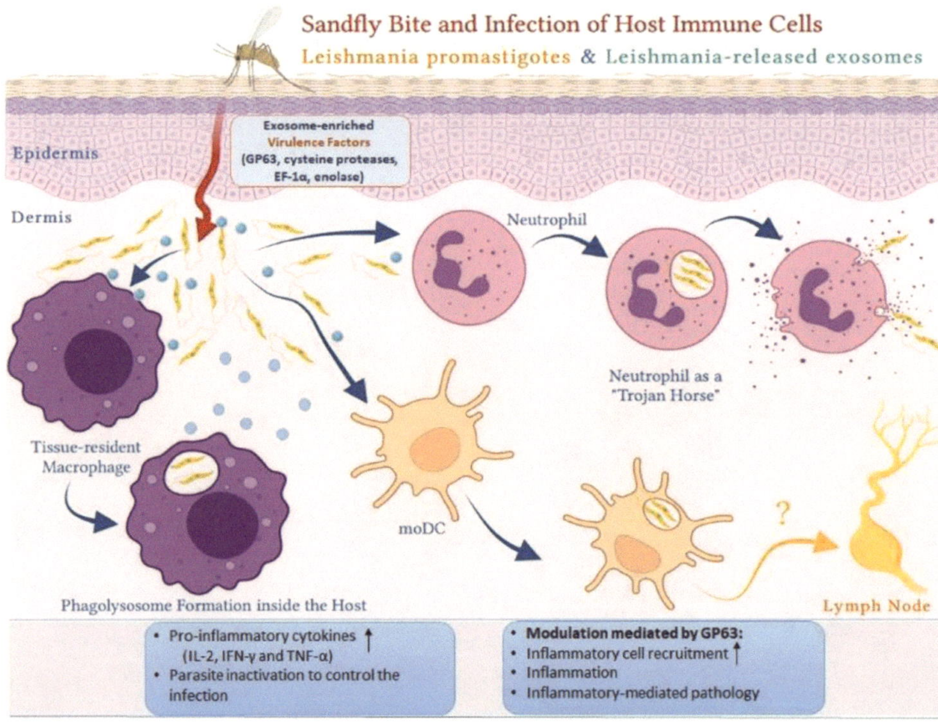

Figure 2.
The interaction of innate immune cells during Leishmania infection. Biorender software was used to create this figure under an academic license.

T helper 17 (Th17) inflammation in later phases of mucocutenous infection [77, 78]. Although parasites can readily be found in neutrophils, it is within mononuclear phagocytes that there is the best evidence for their replication and long-term survival. In a previous study, two-photon intravital imaging of mouse skin following needle injection of *L. major* has revealed that promastigotes were taken up by resident DCs like Langerhans within the first 4 h of infection and stimulating the activation of cytotoxic CD8-T cells [79]. DCs play a critical role in development of the immune response and coordinating an effector T helper 1 (Th1) adaptive immunity over the secretion of cytokines. Pro-inflammatory cytokines such as interleukin-2 (IL-2), interferon-gamma (IFN-γ) and TNF-α can activate the anti-parasitic mechanisms of the macrophages, leading to parasite inactivation and secretion of the cytokines such as IL-4, IL-5 and IL-13 to control the infection [71] (**Figure 2**). On the other hand, as the numbers of DCs and resident macrophages in the skin are too limited to sustain parasite multiplication, the progression of infection requires the recruitment of monocytes (**Figure 2**). DCs can become monocyte-derived DCs (moDCs) that express the major histocompatibility complex class II (MHC class II) molecules, which are critical for the secretion of IL-12 leading to the activation of a host-protective Th1- type response [80].

Several studies indicated that *Leishmania* exosomes can modulate monocyte cytokine production in response to *Leishmania* infection by influencing the innate and adaptive immune systems [22, 26, 30, 52, 54, 61] (**Figure 2**). Silverman and colleagues found that *L. donovani* exosomes could be predominantly immunosuppressive regarding cytokine responses on IFN-γ inhibition and IL-10 production by human moDCs [54]. In addition, exosomes released from heat shock protein 100 (HSP100) null *Leishmania donovani* in contrast to wild type *L. donovani* exosomes, are highly proinflammatory on immune cells, enabling the differentiation of naive CD4 lymphocytes into Th1 cells [54]. Similarly, pretreatment of mice with *L. donovani*- and *L. major*-released exosomes led to exacerbated infection and pathogenesis *in vivo*, related with IL-10 production and impaired generation of inflammatory Th2 cell response for parasite elimination and clearance [54].

In addition, studies on *Leishmania* EVs showed that EVs can involve in the pathogenesis by modulating the microenvironment of the mammalian hosts which is at a high temperature and a low pH than the midgut of the sandfly, and thus causing the disease [30, 61, 69]. Regarding the effect of the host microenvironment on *Leishmania* EVs, three independent studies have reported on temperature-dependent vesicle release from *Leishmania spp*. with different perspectives [30, 69, 81]. Accordingly, the release of *L. donovani* EVs was increased 3-fold by heat shocked-stationary phase promastigotes at a temperature mimicking the human body (37°C) [30]. In another study, increased temperature triggered the secretion of vesicles with the exposure of 4 h heat shocks [69]. However, contrary to temperature-induced vesicle release, Barbosa and colleagues indicated that the temperature shift (ambient temperatures of 25–26°C and 37°C) reduced the secretion of EVs from promastigotes and increasing temperature decreased parasite viability and morphology, hence affecting the release of EVs [81].

Up-regulation of EV secretion induced by infection-like temperatures suggested that these vesicles are released into the extracellular environment, before the invasion of a host such as macrophage, neutrophil, or DC occurs. These EVs may be secreted from either inoculated metacyclic promastigotes within the sand-fly salivary gland, free amastigotes in the mammalian hosts, or both [26, 32, 53, 64]. A study of Atayde et al. [53] demonstrated that *in vivo* secreted *Leishmania* EVs in the sand fly midgut were egested by the sand fly during the bite, and these vesicles may have a role in the establishment and pathology of the CL [53]. Co-injection of mice footpads with metacyclic *L. major* promastigotes plus midgut-isolated or *in vitro*-isolated *L. major* EVs led to a significant increase in footpad swelling, and produce exacerbated lesions up to 6 weeks post-infection through over induction of inflammatory cytokines, in particular IL-17a (which is related to neutrophil infiltration) [53, 78]. On the other hand, a recent study indicates that *L. donovani* infection may promote angiogenesis by inducing endothelial cells to release angiogenesis promoting mediators including IL-8, G-CSF/CSF-3 and VEGF-A. This study shows the changes in the composition of EVs from infected cells resulted from *Leishmania* infection and suggests that EVs from infected cells could promote the vascularization in *Leishmania* infections [32].

3.1.5 *Host manipulation and immunomodulatory properties of* Leishmania *EVs associated with parasite virulence factors*

Protozoan parasites have developed numerous effective strategies to improve their protective mechanisms to escape from the immune system by modulation of the hosts' immune response and signaling pathways, as well as virulence factor secretion [20, 25, 71, 75, 82–84]. Moreover, they secrete EVs containing various parasitic factors and signaling molecules to modify the hostile microenvironment of their hosts to their benefit [26, 29, 33, 52]. By secreting EVs with proteases, parasites suppress the initial immune response raised at the point of infections for long enough to establish a foothold in their hosts [26, 29, 33, 52].

Leishmania utilizes multiple virulence factors including lipophosphoglycan (LPG) and surface acid proteinase (GP63), which trigger the modulation with the activation of protein tyrosine phosphatases (PTP), inhibition on pro-inflammatory transcription factors NF-κB, AP-1 and STAT-1 as well as other signaling molecules such as JAK-2, IRAK-1 and MAP kinases to successfully deactivate and infect on their host macrophages [52].

Together with the parasite surface molecules, multiple host cell receptors (complement receptor type 1 and type 3 (CRl, CR3), mannose-fucose-receptor, fibronectin receptor, macrophage receptor for advanced glycosylation end products) play a crucial role in the attachment and uptake of promastigotes by the immune cells [25].

Leishmania metacyclic promastigotes (infective) have to evade the complement-mediated cell-lysis via parasitic virulence factors such as GP63 and LPG, before being eliminated by phagocytosis. Moreover, they are resistant to complement activation in contrast to procyclic promastigotes (non-infective) that are extremely sensitive to the complement system, explained by the role of surface LPG. The surface LPG plays a central role in the parasite's entry and survival in host cells. In the metacyclic promastigotes, LPG is longer than non-infective procyclic forms and is almost completely absent in amastigotes, resulted in inhibiting the attachment of the C5b-C9 complement system subunits to the parasite surface [85]. In addition, surface protein kinases were indicated to phosphorylate the complement system, therefore, hampering the cascade. The surface protein, gp63, a zinc-dependent metalloprotease, is 10-fold less abundant than LPG, as an important *Leishmania* virulence factor that is expressed at the surface of the parasite via a glycophosphatidylinositol (GPI) anchor, or is directly secreted to the extracellular environment. GP63 promotes parasite survival by the stimulation of immunomodulation on the macrophages, and thus, plays a crucial role in pathogenesis. Previous studies on the action of GP63 in parasitic infections reported that GP63 can protect *L. amazonensis* and *L. major* against cell-lysis by converting the C3b complement subunit into C3bi which accumulates on the surface of the parasites [85]. Fixation of C3 by the parasite increases the recognition of parasites by the macrophages' complement receptors 1 (CR1) and complement receptors 3 (CR3) allowing intracellular survival [86]. Thus, it appears that *Leishmania* not only inhibits activation of the lytic membrane attack complex (CSb-C9), but instead exploits C3 for "silent" invasion of host macrophages [25].

Experiments on mice and macrophages showed that these exosomes exhibit immunomodulatory activity, confirming the presence of parasite virulence factors in their content such as the surface metalloprotease GP63 [15, 26, 30, 33, 52, 54, 69, 87]. Hassani et al. previously showed that the contents of the macrophage exosomes undergo changes following LPS stimulation or *Leishmania* infection. Furthermore, they indicated that exosomes released from *Leishmania*-infected cells display unique signatures regarding composition and abundance of several functional groups of proteins such as plasma-membrane associated proteins, chaperons and metabolic enzymes [26]. In this study, surface metalloprotease GP63 was shown in the contents of the exosomes from *Leishmania*-infected macrophages, which could induce signaling molecules such as MAP kinases (except JNK) and immune-related gene expression like NF-kB associated with the immune system in naive macrophages [26]. The induction of phosphorylation of signaling proteins and translocation of activatory transcription factors into the nucleus was determined within 15 min and up to 1 h after treatment of exosomes isolated from LPS and *Leishmania*-induced macrophages and in particular in pro-inflammatory nuclear translocation of NF-kB and AP-1 and early tyrosine phosphorylation of MAP kinases ERK and P38. So, the overall effect of macrophage-infected exosomes in naive macrophages can be claimed as the down-regulation of pro-inflammatory genes and suppression of macrophage activation.

Another study comparing the EVs of wild-type and GP63-knockout *Leishmania* parasites showed the importance of GP63 in the modulation of macrophage responses [52]. While the wild-type EVs were capable of downregulating several genes associated with the immune response, GP63-knockout parasite EVs alteration of immune response genes occurred in a different pattern and had significantly reduced immunosuppressive capabilities. Furthermore, the lack of GP63 altered the proteome of EVs, suggesting that GP63 may have roles in the cargo-determinacy of parasite EVs [26, 52]. In addition, evidence suggests that exosomes secreted from *Leishmania*-infected cells containing GP63, may down-regulate the generation of specific host miRNAs and facilitate infection of the liver [87]. In one study, EVs secreted by *L. donovani* were shown to reduce miR-122 activity in hepatic cells, which reduced serum cholesterol levels and increased the infectivity of the parasite. The GP63 proteins of parasites EVs were suggested as the agent behind this alteration, as they could target the miRNA processor Dicer1 [87]. All these studies indicate that EVs from *Leishmania spp.* display a wide range of targets in mammalian hosts and, have an immune-hampering role.

3.2 Other protozoan parasites

3.2.1 Toxoplasma spp.

Toxoplasma gondii is a globally protozoan pathogen that uses felids (cats) as their primary host. When infecting other mammalians, the parasite infects the hosts' brain tissues, forming cysts. Infected rodents exhibit behavioral changes, such as reduced aversion of felines [88]. The effects of the parasite in humans are less understood, however, studies link *T. gondii* infection with neural diseases such as Alzheimer's [89].

T. gondii EVs carry several virulence factors that aid their infectivity. In one study, complete mRNAs of neurologically active proteins, as well as various miRNAs were found in *T. gondii* EVs, which may have the capacity to affect the neural cells that they enter. The most enriched mRNA's belonged to various neurologically active proteins, Rab-13, eukaryotic translation EF 1-α1, thymosin beta 4 and LLP homolog [90]. One mRNA observed in the study, e.g. eukaryotic translation elongation factor 1, was also reported to be present in *Leishmania* EVs and associated with autism [90, 91]. Furthermore, immunoregulatory miRNA miR23-b was observed in the EVs, which regulates the secretion of IL-17. In addition to mRNA and miRNA components, *T. gondii* EVs were also shown to carry several proteins under the excreted/secreted antigens family, such as surface antigens, microneme proteins, dense granule antigens and rhoptry proteins, which are known to regulate the immune response of their hosts [42, 92].

3.2.2 Plasmodium spp.

Malaria is one of the deathliest protozoan parasitic diseases in the world and the leading cause of mortality in sub-Saharan Africa. It is caused by the family of *Plasmodium* parasites, which are spread through infected Anopheles mosquitoes, leading to fatal conditions such as cerebral malaria or severe malarial anemia. When passed to a human, the parasite infects red blood cells, allowing it to evade the immune response and penetrate deep tissues. The infected red blood cells increase vascular permeability and cause the apoptosis of endothelial cells, which both increase the severity of the disease and facilitate the spread of the parasite throughout the body.

As with other parasites, EVs secreted by malaria parasites modulate the hosts' immune system to increase the survivability of the *Plasmodium* parasite. When parasites were blocked from secreting EVs, they had reduced virulence and lessened symptoms in models of cerebral malaria [93]. Secretion of EVs continues after the infection of red blood cells. Studies show that the parasite hijacks the EV secretion in infected red blood cells, modifying their cargo. Infected red blood cells secrete EVs enriched in parasite surface antigens, and contain proteins associated with immunosuppression [94]. One study observed 120 plasmodial RNAs in infected red blood cells, which coded for proteins involved in drug resistance, as well regulatory small RNAs. The presence of these modified EVs can be used as a marker for the diagnosis of malaria [31]. In another study, infected red blood cells were shown to secrete EVs with parasite-specific proteins and RNA. Furthermore, proteins and miRNA that can alter gene expressions in endothelial cells, such as Ago2, were observed in these EVs. These infected EVs may explain malaria-associated vascular dysfunction [95].

3.2.3 Trypanosoma spp.

Trypanosomatids are insect-borne parasites that cause fatal diseases such as Chagas' disease [96] or African trypanosomiasis, "the sleeping sickness" [97]. EVs secreted by trypanosomes were shown to increase virulence in various studies. Proteins associated with metabolism, parasite survival and virulence were observed in parasite EVs [45]. In one study, EVs of *Trypanosoma brucei rhodesiense* were shown to carry serum resistance-associated protein – a key protein for human infectivity- as well as flagellar proteins that increase virulence. Furthermore, the parasite EVs were shown to have the capacity to induce rapid erythrocyte clearance and anemia, suggesting a parasite-free pathogenesis pathway [44]. Another study observed that the parasite uses EVs to increase infectivity and survivability. Secreted vesicles enhanced parasite cyclogenesis, and lead to up to five times increased infection rates on susceptible cells [46].

4. EVs as diagnostic and therapeutic tools for protozoan parasitic infections

EVs offer exciting clinical opportunities in many diseases as diagnostic tools, drug delivery vehicles, or therapeutic agents – and parasitic infections are no exception. Both protozoan and host cell EVs are used in clinical applications against parasitic diseases. Moreover, immune cells infected with parasites also produce EVs that can induce inflammatory responses through the secretion of cytokines and chemokines *in vitro* and *in vivo* [21, 22, 54, 98, 99]. Considering their immunomodulatory effects, EVs could be potential vaccine candidates as components for infectious diseases [100–106].

EVs take part in the complex web of interactions that happen between immune cells. In particular, EV secreted by regulator immune cells like dendritic or T cells mimic the actions of their parental cell and prime the immune system against pathogens. When antigens of *L. major* are given to DCs, when administered, EVs secreted by those DCs were observed to protect mice from the parasite to great effect [100]. The EVs reduced footpad swelling and were capable of inducing antigen-specific T-cell responses [100]. A similar approach was also successful in inducing antigen-specific T-cell response against *T. gondii* [101, 102]. Using EVs instead of whole cells has several advantages, such as increased stability in freeze-thaw situations, and cannot alter their antigen-presentation, which may sometimes be the case with freeze-thawed DCs [103].

In addition to pulsing immune cells with protozoan antigens, protozoan EVs can also be used to induce the immune system, similar to vaccines. EVs from *Plasmodium yoelii*- infected reticulocytes were found to be capable of immunizing mice against the protozoan. Immunized mice were capable of producing IgG antibodies that could target the infected reticulocytes [39]. Similarly, EVs isolated from *L. amazonensis*-infected macrophages induce the production of the proinflammatory cytokines IL-12, IL-1b and TNF-α by neighboring macrophages, which contributes to modulate the immune system in favor of a Th1 immune response as well as the elimination of the *Leishmania*, and therefore, control of the infection [23].

As an image of the secreting cell, EVs have considerable potential as a diagnostic tool against parasitic diseases. The protein and miRNA cargo of EVs can allow a non-invasive biopsy of the parasite and may allow the determination of any drug resistance [104]. Regrettably, there are few examples of the use of EVs for the diagnosis of parasitic infections. One study of *Trigonoscuta cruzi* EV proteome revealed enrichment of antigen proteins used for the diagnosis of the parasite. Moreover, one category of proteins, retrotransposon hot spot proteins, do not cause any cross-reactivity with parasites of other diseases such as malaria, leishmaniasis or others, and may allow a definitive diagnosis of Chagas disease [105].

The natural ability of EVs to deliver cargo between cells gives makes them an attractive candidate for drug delivery applications. It has been shown that encapsulating drugs within EVs may grant them cell-specific targeting, reduced toxicity, increased circulation times and increased biodistribution with the ability to pass through tissue barriers such as the blood-brain barrier. However, the field of EV-mediated drug delivery is still at its infancy [106], with few studies done on delivering anti-protozoan drugs. The one study available to the field showed that antimalarial drugs atovaquone and tafenoquine were more effective in inhibiting the growth of *P. falciparum* when loaded into vesicles isolated from malaria-infected red blood cells [38].

5. Conclusion

With the expansion of knowledge in parasitic diseases, the critical function of EVs became more evident in the development of the diseases. EVs applies many strategies not only to provide the survival and reproduction of *Leishmania* parasites inside the host, but also to enable the invasion by means of immune strategies including change in host antigens, development of self-tolerance, immune inactivation, immunosuppression and intervention of molecule-mimetic mechanisms between parasites and host antigens [16, 24, 25]. Recent studies propose that the parasites actually utilize the EVs as one infection strategy [18, 20, 21, 26–31], where the questions are arisen on how EVs modulate the host immune system and ultimately cause the infection. Based on the cell of origin, the release mechanisms of EVs from different protozoan parasites, including Apicomplexa and Kinetoplastids such as *Leishmania* species (spp.) [22, 23, 26, 32–35], *Plasmodium spp.* [31, 36–41], *Toxoplasma spp.* [36, 42, 43] and *Trypanosoma spp.* [44–49] were described, where the parasitic infections were studied in detail for leishmaniasis, malaria, toxoplasmosis and Chagas disease independently.

Several studies indicated that *Leishmania* exosomes can modulate monocyte cytokine production in response to *Leishmania* infection by influencing the innate and adaptive immune systems using parasitic virulence factors [22, 26, 30, 52, 54, 61]. Silverman and colleagues found that *L. donovani* exosomes could be predominantly immunosuppressive regarding cytokine responses on IFN-γ inhibition and IL-10 production by human moDCs [54]. In another study, macrophage-infected exosomes in naive macrophages were shown to downregulate the pro-inflammatory genes and suppression of macrophage activation [26]. Similarly, EVs secreted by the malaria parasite modulate the hosts' immune system to increase the survivability of the *Plasmodium* parasite. When parasites were blocked from secreting EVs, they had reduced virulence and lessened symptoms in models of cerebral malaria [93].

In addition to cytokine response, studies indicated that EVs can involve in the pathogenesis by modulating the microenvironment of the mammalian hosts which is at a high temperature and a low pH than the midgut of the sandfly and thus causing the disease [30, 61, 69]. Up-regulation of EV secretion induced by infection-like temperatures suggested that these vesicles were released into the extracellular environment, before the invasion of a host such as macrophage, neutrophil or DC occurs.

While EVs play such a multifaceted role in immunomodulation and disease development at protozoan diseases, the application potential of EVs as therapeutic agents or drug delivery vehicles in therapy or as a biomarker at diagnostics attracts the researchers' attention working on these fields. Considering their immunomodulatory effects, EVs could be potential vaccine candidates as components for infectious diseases [100–106] and the application of protozoan EVs in the clinic may be expected in the near future.

Acknowledgements

The authors thank Department of Genetics and Bioengineering Yeditepe University for financial support.

Conflict of interest

No conflict of interest was declared by the authors.

Abbreviations

CRl	Complement receptor type 1
CR3	Complement receptor type 3
CL	Cutaneous leishmaniasis
DC	Dendritic cell
EF 1-α	Elongation factor 1-alpha
EVs	Extracellular vesicles
HSAPB	Hydrophilic acylated surface protein B
GFP	Green fluorescent protein
IL	Interleukin
IFN-γ	Interferon-gamma
L.	*Leishmania*
Leishmania spp.	*Leishmania species*
LPG	Lipophosphoglycan
MHC class II	Major histocompatibility complex class II
moDCs	monocyte-derived dendritic cells
MCL	Mucocutaneous leishmaniasis
NO	Nitric oxide
PTP	Protein tyrosine phosphatases
ROS	Radical oxygen species
Th1	T helper 1
Th17	T helper 17
TNF-α	Tumor necrosis factor-alpha
T. gondii	Toxoplasma gondii
VL	Visceral Leishmaniasis

Author details

Zeynep Islek[1,2], Batuhan Turhan Bozkurt[1], Mehmet Hikmet Ucisik[1] and Fikrettin Sahin[1*]

1 Department of Genetics and Bioengineering, Faculty of Engineering, Yeditepe University, Ataşehir/Istanbul, Turkey

2 Department of Pharmaceutical Technology, Faculty of Pharmacy, Istanbul Health and Technology University, Zeytinburnu/Istanbul, Turkey

*Address all correspondence to: fsahin@yeditepe.edu.tr

References

[1] Simons M, Raposo G. Exosomes – Vesicular carriers for intercellular communication. Current Opinion in Cell Biology. 2009;**21**(4):575-581

[2] Twu O, Johnson PJ. Parasite extracellular vesicles: Mediators of intercellular communication. PLoS Pathogens. 2014;**10**(8):E1004289

[3] Colombo M, Raposo G, Théry C. Biogenesis, secretion, and intercellular interactions of exosomes and other extracellular vesicles. Annual Review of Cell and Developmental Biology. 2014;**30**(1):255-289

[4] Raposo G, Stoorvogel W. Extracellular vesicles: Exosomes, microvesicles, and friends. Journal of Cell Biology. 2013;**200**:373-383

[5] Abels ER, Breakefield XO. Introduction to extracellular vesicles: Biogenesis, RNA cargo selection, content, release, and uptake. Cellular and Molecular Neurobiology. 2016; **36**(3):301-312

[6] Lötvall J, Hill AF, Hochberg F, Buzás EI, Di VD, Gardiner C, et al. Minimal experimental requirements for definition of extracellular vesicles and their functions: A position statement from the International Society for Extracellular Vesicles. Journal of Extracellular Vesicles. 2014;**3**:26913

[7] You JY, Kang SJ, Rhee WJ. Isolation of cabbage exosome-like nanovesicles and investigation of their biological activities in human cells. Bioactive Materials. 2021;**6**(12):4321-4332. DOI: 10.1016/j.bioactmat.2021.04.023

[8] Savci Y, Kirbas OK, Bozkurt BT, Abdik EA, Tasli PN, Sahin F, et al. Grapefruit-derived extracellular vesicles as a promising cell-free therapeutic tool for wound healing. Food and Function. 2021;**12**(11):5144-5156

[9] Akuma P, Okagu OD, Udenigwe CC. Naturally occurring exosome vesicles as potential delivery vehicle for bioactive compounds. Frontiers in Sustainable Food Systems. 2019;**3**:23

[10] Kim KW. Visualization of extracellular vesicles of prokaryotes and eukaryotic microbes. Applied Microscopy. 2018;**48**:96-101

[11] Deatheragea BL, Cooksona BT. Membrane vesicle release in bacteria, eukaryotes, and archaea: A conserved yet underappreciated aspect of microbial life. Infection and Immunity. 2012;**80**: 1948-1957

[12] Théry C, Witwer KW, Aikawa E, Alcaraz MJ, Anderson JD, Andriantsitohaina R, et al. Minimal information for studies of extracellular vesicles 2018 (MISEV2018): A position statement of the international society for extracellular vesicles and update of the MISEV2014 guidelines. Journal of Extracellular Vesicles. 2018;7(1):1535750

[13] Coakley G, Maizels RM, Buck AH. Exosomes and other extracellular vesicles: The new communicators in parasite infections. Trends in Parasitology. 2015;**31**:477-489

[14] Lambertz U, Silverman JM, Nandan D, McMaster WR, Clos J, Foster LJ, et al. Secreted virulence factors and immune evasion in visceral leishmaniasis. Journal of Leukocyte Biology. 2012;**91**(6):887-899

[15] Silverman JM, Reiner NE. Exosomes and other microvesicles in infection

biology: Organelles with unanticipated phenotypes. Cellular Microbiology. 2011;**13**:1-9

[16] Barteneva NS, Maltsev N, Vorobjev IA. Microvesicles and intercellular communication in the context of parasitism. Vol. 4. Frontiers in Cellular and Infection Microbiology. 2013. p. 1-11

[17] Schorey JS, Cheng Y, Singh PP, Smith VL. Exosomes and other extracellular vesicles in host–pathogen interactions. EMBO Reports. 2015;**16**(1):24-43

[18] Bhatnagar S, Shinagawa K, Castellino FJ, Schorey JS. Exosomes released from macrophages infected with intracellular pathogens stimulate a proinflammatory response in vitro and in vivo. Blood. 2007;**110**(9):3234-3244

[19] Deolindo P, Evans-Osses I, Ramirez MI. Microvesicles and exosomes as vehicles between protozoan and host cell communication. In: Biochemical Society Transactions. 2013. p. 252-257

[20] Marcilla A, Martin-Jaular L, Trelis M, de Menezes-Neto A, Osuna A, Bernal D, et al. Extracellular vesicles in parasitic diseases. Vol. 3. Journal of Extracellular Vesicles. 2014. p. 1-15

[21] Atayde VD, Hassani K, da Silva Lira Filho A, Borges AR, Adhikari A, Martel C, et al. Leishmania exosomes and other virulence factors: impact on innate immune response and macrophage functions. Cell Immunol. 2016;**309**:7-18

[22] Nogueira PM, de Menezes-Neto A, Borges VM, Descoteaux A, Torrecilhas AC, Xander P, et al. Immunomodulatory properties of Leishmania extracellular vesicles during host-parasite interaction: Differential activation of TLRs and NF-κB translocation by dermotropic and viscerotropic species. Frontiers in Cellular and Infection Microbiology. 2020;**10**:1-9

[23] Cronemberger-Andrade A, Aragão-França L, de Araujo CF, Rocha VJ, Borges-Silva M d C, Figueiras CP, et al. Extracellular vesicles from Leishmania-infected macrophages confer an anti-infection cytokine-production profile to naïve macrophages. PLoS Neglected Tropical Diseases. 2014;**8**(9)

[24] Miller LH, Scott P. Immunity to protozoa. Current Opinion in Immunology. 1990;**2**(3):368-374

[25] Bogdan C, Röllinghoff M. The immune response to Leishmania: Mechanisms of parasite control and evasion. International Journal for Parasitology. 1998;**28**:121-134

[26] Hassani K, Olivier M. Immunomodulatory impact of Leishmania-induced macrophage exosomes: A comparative proteomic and functional analysis. PLoS Neglected Tropical Diseases. 2013;**7**(5)

[27] Montaner S, Galiano A, Trelis M, Martin-Jaular L, del Portillo HA, Bernal D, et al. The role of extracellular vesicles in modulating the host immune response during parasitic infections. Frontiers in Immunology. 2014;**5**(AUG)

[28] Soto-Serna LE, Diupotex M, Zamora-Chimal J, Ruiz-Remigio A, Delgado-Domínguez J, Cervantes-Sarabia RB, et al. Leishmania mexicana: Novel insights of immune modulation through amastigote exosomes. Journal of Immunology Research. 2020;**2020**:1-12

[29] Dong G, Filho AL, Olivier M. Modulation of host-pathogen communication by extracellular vesicles

(EVs) of the protozoan parasite Leishmania. Vol. 9. Frontiers in Cellular and Infection Microbiology. 2019. p. 1-9

[30] Silverman JM, Clos J, De'Oliveira CC, Shirvani O, Fang Y, Wang C, et al. An exosome-based secretion pathway is responsible for protein export from Leishmania and communication with macrophages. Journal of Cell Science. 2010;**123**(6):842-852

[31] Babatunde KA, Mbagwu S, Hernández-Castañeda MA, Adapa SR, Walch M, Filgueira L, et al. Malaria infected red blood cells release small regulatory RNAs through extracellular vesicles. Scientific Reports. 2018;**8**(1): 1-15

[32] Gioseffi A, Hamerly T, Van K, Zhang N, Dinglasan RR, Yates PA, et al. Leishmania-infected macrophages release extracellular vesicles that can promote lesion development. Life Science Alliance. 2020;**3**(12):1-16

[33] Marshall S, Kelly PH, Singh BK, Pope RM, Kim P, Zhanbolat B, et al. Extracellular release of virulence factor major surface protease via exosomes in Leishmania infantum promastigotes. Parasites and Vectors. 2018;**11**(1):1-10

[34] Santarém N, Racine G, Silvestre R, Cordeiro-da-Silva A, Ouellette M. Exoproteome dynamics in Leishmania infantum. Journal of Proteomics. 2013;**84**:106-118

[35] Pérez-Cabezas B, Santarém N, Cecílio P, Silva C, Silvestre R, AM Catita J, et al. More than just exosomes: Distinct Leishmania infantum extracellular products potentiate the establishment of infection. Journal of Extracellular Vesicles. 2019;**8**(1). 1541708. DOI: 10.1080/20013078.2018. 1541708

[36] de Souza W, Barrias ES. Membrane-bound extracellular vesicles secreted by parasitic protozoa: Cellular structures involved in the communication between cells. Parasitology Research. 2020;**119**(7): 2005-2023

[37] Correa R, Caballero Z, De LLF, Spadafora C, Marti M. Extracellular vesicles could carry an evolutionary footprint in interkingdom communication. Frontiers in Cellular and Infection Microbiology. 2020;**10**(March):1-11

[38] Borgheti-Cardoso LN, Kooijmans SAA, Chamorro LG, Biosca A, Lantero E, Ramirez M, et al. Extracellular vesicles derived from Plasmodium-infected and non-infected red blood cells as targeted drug delivery vehicles. International Journal of Pharmaceutics. 2020;**587**:119627

[39] Martin-Jaular L, Nakayasu ES, Ferrer M, Almeida IC, Del Portillo HA. Exosomes from Plasmodium yoelii-infected reticulocytes protect mice from lethal infections. PLoS One. 2011;**6**(10): e26588

[40] Szempruch AJ, Dennison L, Kieft R, Harrington JM, Hajduk SL. Sending a message: Extracellular vesicles of pathogenic protozoan parasites. Nature Reviews Microbiology. 2016;**14**(11): 669-675

[41] Regev-Rudzki N, Wilson DW, Carvalho TG, Sisquella X, Coleman BM, Rug M, et al. Cell-cell communication between malaria-infected red blood cells via exosome-like vesicles. Cell. 2013;**153**(5):1120-1133

[42] Silva VO, Maia MM, Torrecilhas AC, Taniwaki NN, Namiyama GM, Oliveira KC, et al. Extracellular vesicles isolated from Toxoplasma gondii induce host immune response. Parasite Immunology. 2018;**40**(9):e12571

[43] Długońska H, Gatkowska J. Exosomes in the context of Toxoplasma gondii – Host communication. Vol. 62. Annals of Parasitology. 2016. p. 169-174

[44] Szempruch AJ, Sykes SE, Kieft R, Dennison L, Becker AC, Gartrell A, et al. Extracellular vesicles from Trypanosoma brucei mediate virulence factor transfer and cause host anemia. Cell. 2016;**164**(1-2):246-257

[45] Bayer-Santos E, Aguilar-Bonavides C, Rodrigues SP, Cordero EM, Marques AF, Varela-Ramirez A, et al. Proteomic analysis of Trypanosoma cruzi secretome: characterization of two populations of extracellular vesicles and soluble proteins. Journal of Proteome Research. 2013;**12**(2):883-897

[46] Garcia-Silva MR, RFC d N, Cabrera-Cabrera F, Sanguinetti J, Medeiros LC, Robello C, et al. Extracellular vesicles shed by Trypanosoma cruzi are linked to small RNA pathways, life cycle regulation, and susceptibility to infection of mammalian cells. Parasitology Research. 2014;**113**(1):285-304

[47] Lovo-Martins MI, Malvezi AD, Zanluqui NG, Lucchetti BFC, Hideko Tatakihara VL, Mörking PA, et al. Extracellular vesicles shed By Trypanosoma cruzi potentiate infection and Elicit Lipid body formation and PGE2 production in murine macrophages. Frontiers in Immunology. 2018;**9**(APR):1-16

[48] de Souza W. Exosomes in the Pathogenic Protozoan Trypanosoma Cruzi. International Journal of Pathology Clinical Research. 2017;**3**(1):1-9

[49] Castillo C, Carrillo I, Liempi A, Medina L, Navarrete A, López P, et al. Trypanosoma cruzi exosomes increases susceptibility to parasite infection in human placental chorionic villi explants. Placenta. 2017;**51**:123-124

[50] Organization WH. Sustaining the drive to overcome the global impact of neglected tropical diseases. World Health Organization. 2013. 1-138 p.

[51] Schorey JS, Bhatnagar S. Exosome function: From tumor immunology to pathogen biology. Traffic. 2008;**9**: 871-881

[52] Hassani K, Shio MT, Martel C, Faubert D, Olivier M. Absence of metalloprotease GP63 alters the protein content of leishmania exosomes. PLoS One. 2014;**9**(4)

[53] Atayde VD, Aslan H, Townsend S, Hassani K, Kamhawi S, Olivier M. Exosome secretion by the parasitic protozoan Leishmania within the sand fly midgut. Cell Reports. 2015;**13**(5): 957-967

[54] Silverman JM, Clos J, Horakova E, Wang AY, Wiesgigl M, Kelly I, et al. Leishmania exosomes modulate innate and adaptive immune responses through effects on monocytes and dendritic cells. Journal of Immunology. 2010;**185**(9): 5011-5022

[55] Théry C, Zitvogel L, Amigorena S. Exosomes: Composition, biogenesis and function. Nature Reviews Immunology. 2002;**2**(8):569-579

[56] Arenas R, Torres-Guerrero E, Quintanilla-Cedillo MR, Ruiz-Esmenjaud J. Leishmaniasis: A review. Vol. 6. F1000Research. 2017. p. 1-15

[57] Liew FY, Li Y, Millott S. Tumour necrosis factor (TNF-α) in leishmaniasis. II. TNF-α-induced macrophage leishmanicidal activity is mediated by nitric oxide from L-arginine. Immunology. 1990;**71**(4):556-559

[58] Samant M, Sahu U, Pandey SC, Khare P. Role of cytokines in experimental and human visceral Leishmaniasis. Vol. 11. Frontiers in Cellular and Infection Microbiology. 2021. p. 1-18

[59] Bacellar O, D'Oliveira A, Jerônimo S, Carvalho EM. IL-10 and IL-12 are the main regulatory cytokines in visceral leishmaniasis. Cytokine. 2000;**12**(8):1228-1231

[60] Novais FO, Scott P. Immunology of Leishmaniasis. In: Encyclopedia of Immunobiology. 2016. p. 114-124

[61] Silverman JM axwel., Reiner NE. Leishmania exosomes deliver preemptive strikes to create an environment permissive for early infection. Vol. 1. Frontiers in cellular and infection microbiology. 2011. p. 26

[62] Forrest DM, Batista M, Marchini FK, Tempone AJ, Traub-Csekö YM. Proteomic analysis of exosomes derived from procyclic and metacyclic-like cultured Leishmania infantum chagasi. Journal of Proteomics. 2020;**227**:1-10

[63] Figuera L, Acosta H, Gómez-Arreaza A, Dávila-Vera D, Balza-Quintero A, Quiñones W, et al. Plasminogen binding proteins in secreted membrane vesicles of Leishmania mexicana. Molecular and Biochemical Parasitology. 2013;**187**(1):14-20

[64] Maxwell MJ, Chan SK, Robinson DP, Dwyer DM, Nandan D, Foster LJ, et al. Proteomic analysis of the secretome of Leishmania donovani. Genome Biology. 2008;**9**(2):R35

[65] Lin WC, Tsai CY, Huang JM, Wu SR, Chu LJ, Huang KY. Quantitative proteomic analysis and functional characterization of Acanthamoeba castellanii exosome-like vesicles. Parasites and Vectors. 2019;**12**(1):1-12. DOI: 10.1186/s13071-019-3725-z

[66] Biyani N, Madhubala R. Quantitative proteomic profiling of the promastigotes and the intracellular amastigotes of Leishmania donovani isolates identifies novel proteins having a role in Leishmania differentiation and intracellular survival. Biochim Biophys Acta – Proteins Proteomics. 2012; **1824**(12):1342-1350. DOI: 10.1016/j.bbapap.2012.07.010

[67] Douanne N, Dong G, Douanne M, Olivier M, Fernandez-Pradaid C. Unravelling the proteomic signature of extracellular vesicles released by drug-resistant leishmania infantum parasites. PLoS Neglected Tropical Diseases. 2020;**14**:1-32

[68] Ward OP. Proteases. In: Comprehensive Biotechnology. 2011:571

[69] Hassani K, Antoniak E, Jardim A, Olivier M. Temperature-induced protein secretion by leishmania mexicana modulates macrophage signalling and function. PLoS One. 2011;**6**(5)

[70] Kushawaha PK, Gupta R, Sundar S, Sahasrabuddhe AA, Dube A. Elongation factor-2, a Th1 stimulatory protein of Leishmania donovani, generates strong IFN-γ and IL-12 response in cured Leishmania-infected patients/hamsters and protects hamsters against Leishmania challenge. Journal of Immunology. 2011;**187**(12):6417-6427

[71] Gabriel ÁM, Galué-Parra A, Pereira WLA, Pedersen KW, da Silva EO. Leishmania 360°: Guidelines for exosomal research. Vol. 9. Microorganisms. 2021. p. 1-29

[72] Liévin-Le Moal V, Loiseau PM. Leishmania hijacking of the macrophage intracellular compartments. FEBS Journal. 2016;**283**:598-607

[73] Berman JD, Dwyer DM, Wyler DJ. Multiplication of Leishmania in human macrophages in vitro. Infection and Immunity. 1979;**26**(1):375-379

[74] Bates PA. Transmission of Leishmania metacyclic promastigotes by phlebotomine sand flies. International Journal for Parasitology. 2007;**37**: 1097-1106

[75] Martínez-López M, Soto M, Iborra S, Sancho D. Leishmania Hijacks myeloid cells for immune escape. Vol. 9. Frontiers in Microbiology. 2018. p. 1-16

[76] Laskay T, Van Zandbergen G, Solbach W. Neutrophil granulocytes – Trojan horses for Leishmania major and other intracellular microbes? Trends in Microbiology. 2003;**11**: 210-214

[77] Pedraza-Zamora CP, Delgado-Domínguez J, Zamora-Chimal J, Becker I. Th17 cells and neutrophils: Close collaborators in chronic Leishmania mexicana infections leading to disease severity. Parasite Immunology. 2017;**39**(4)

[78] Boaventura VS, Santos CS, Cardoso CR, De Andrade J, Dos Santos WLC, Clarêncio J, et al. Human mucosal leishmaniasis: Neutrophils infiltrate areas of tissue damage that express high levels of Th17-related cytokines. European Journal of Immunology. 2010;40(10):2830-2836

[79] Kaye P, Scott P. Leishmaniasis: Complexity at the host-pathogen interface. Nature Reviews Microbiology. 2011;**9**:604-615

[80] León B, López-Bravo M, Ardavín C. Monocyte-derived dendritic cells formed at the infection site control the induction of protective T Helper 1 responses against Leishmania. Immunity. 2007;**26**(4): 519-531

[81] Barbosa FMC, Dupin TV, Toledo M d S, Reis NF d C, Ribeiro K, Cronemberger-Andrade A, et al. Extracellular vesicles released by Leishmania (Leishmania) amazonensis promote disease progression and induce the production of different cytokines in macrophages and B-1 Cells. Frontiers in Microbiology. 2018;**9**:3056

[82] Rossi M, Fasel N. How to master the host immune system? Leishmania parasites have the solutions! International Immunology. 2018;**30**(3):103-111

[83] Gupta G, Oghumu S, Satoskar AR. Mechanisms of Immune evasion in Leishmaniasis. Advances in Applied Microbiology. 2013. p. 155-184

[84] Walker DM, Oghumu S, Gupta G, McGwire BS, Drew ME, Satoskar AR. Mechanisms of cellular invasion by intracellular parasites. Cellular and Molecular Life Sciences. 2013;**71**(7): 1245-1263

[85] Puentes SM, Da Silva RP, Sacks DL, Hammer CH, Joiner KA. Serum resistance of metacyclic stage Leishmania major promastigotes is due to release of C5b-9. Journal of Immunology. 1990;**145**(12):4311-4316

[86] Mosser DM, Edelson PJ. The third component of complement (C3) is responsible for the intracellular survival of Leishmania major. Nature. 1987;**327**: 329-331. DOI: 10.1038/327329b0

[87] Ghosh J, Bose M, Roy S, Bhattacharyya SN. Leishmania donovani targets dicer1 to downregulate miR-122, lower serum cholesterol, and facilitate murine liver infection. Cell Host & Microbe. 2013;**13**(3):277-288

[88] Tenter AM, Heckeroth AR, Weiss LM. Toxoplasma gondii: From animals to humans. International Journal

for Parasitology. 2000;**30**(12-13): 1217-1258

[89] Kusbeci OY, Miman O, Yaman M, Aktepe OC, Yazar S. Could Toxoplasma gondii have any role in Alzheimer disease? Alzheimer Disease and Associated Disorders. 2011;**25**(1):1-3

[90] Pope SM, Lässer C. Toxoplasma gondii infection of fibroblasts causes the production of exosome-like vesicles containing a unique array of mRNA and miRNA transcripts compared to serum starvation. Journal of Extracellular Vesicles. 2013;**2**(1):22484

[91] McLachlan F. Investigating the role of eukaryotic translation elongation factor eEF1A2 in autism, epilepsy and intellectual disability. The University of Edinburgh; 2020

[92] Quiarim TM, Maia MM, da Cruz AB, Taniwaki NN, Namiyama GM, Pereira-Chioccola VL. Characterization of extracellular vesicles isolated from types I, II and III strains of Toxoplasma gondii. Acta Tropica. 2021;**219**:105915

[93] Debs S, Cohen A, Hosseini-Beheshti E, Chimini G, Hunt NH, Grau GER. Interplay of extracellular vesicles and other players in cerebral malaria pathogenesis. Biochim Biophys Acta (BBA)-General Subj. 2019; **1863**(2):325-331

[94] Mantel P-Y, Hoang AN, Goldowitz I, Potashnikova D, Hamza B, Vorobjev I, et al. Malaria-infected erythrocyte-derived microvesicles mediate cellular communication within the parasite population and with the host immune system. Cell Host & Microbe. 2013;**13**(5):521-534

[95] Mantel P-Y, Hjelmqvist D, Walch M, Kharoubi-Hess S, Nilsson S, Ravel D, et al. Infected erythrocyte-derived extracellular vesicles alter vascular function via regulatory Ago2-miRNA complexes in malaria. Nature Communications. 2016;**7**(1):1-15

[96] Rassi A Jr, Rassi A, Marin-Neto JA. Chagas disease. Lancet. 2010;**375**(9723): 1388-1402

[97] Brun R, Blum J, Chappuis F, Burri C. Human african trypanosomiasis. Lancet. 2010;**375**(9709):148-159

[98] Castelli G, Bruno F, Saieva L, Alessandro R, Galluzzi L, Diotallevi A, et al. Exosome secretion by Leishmania infantum modulate the chemotactic behavior and cytokinic expression creating an environment permissive for early infection. Experimental Parasitology. 2019

[99] Gupta A, Pulliam L. Exosomes as mediators of neuroinflammation. Journal of Neuroinflammation. 2014;**11**:1-10

[100] Schnitzer JK, Berzel S, Fajardo-Moser M, Remer KA, Moll H. Fragments of antigen-loaded dendritic cells (DC) and DC-derived exosomes induce protective immunity against Leishmania major. Vaccine. 2010;**28**(36): 5785-5793

[101] Aline F, Bout D, Amigorena S, Roingeard P, Dimier-Poisson I. Toxoplasma gondii antigen-pulsed-dendritic cell-derived exosomes induce a protective immune response against T. gondii infection. Infection and Immunity. 2004;**72**(7):4127-4137

[102] Beauvillain C, Ruiz S, Guiton R, Bout D, Dimier-Poisson I. A vaccine based on exosomes secreted by a dendritic cell line confers protection against T. gondii infection in syngeneic and allogeneic mice. Microbes and Infection. 2007;**9**(14-15):1614-1622

[103] Lener T, Gimona M, Aigner L, Börger V, Buzas E, Camussi G, et al. Applying extracellular vesicles based therapeutics in clinical trials–An ISEV position paper. Journal of Extracellular Vesicles. 2015;**4**(1):30087

[104] Wu Z, Wang L, Li J, Wang L, Wu Z, Sun X. Extracellular vesicle-mediated communication within host-parasite interactions. Frontiers in Immunology. 2019;**9**:3066

[105] Bautista-López NL, Ndao M, Camargo FV, Nara T, Annoura T, Hardie DB, et al. Characterization and diagnostic application of Trypanosoma cruzi trypomastigote excreted-secreted antigens shed in extracellular vesicles released from infected mammalian cells. Journal of Clinical Microbiology. 2017;**55**(3):744-758

[106] Elsharkasy OM, Nordin JZ, Hagey DW, de Jong OG, Schiffelers RM, Andaloussi S EL, et al. Extracellular vesicles as drug delivery systems: Why and how? Vol. 159. Advanced Drug Delivery Reviews. 2020. p. 332-343

Section 4

Extracellular Vesicles and Cancer

Chapter 10

Exosomes in Cancer Diagnosis and Radiation Therapy

Ai Nakaoka, Kana Kobayashi, Mennaallah Hassan and Ryohei Sasaki

Abstract

Exosomes are a subgroup of extracellular vesicles that are released by all types of cells, including tumor cells, and mediate intercellular communication via the transport of various intracellular components, including microRNAs, messenger RNAs, and proteins. Radiation produces reactive oxygen species and induces DNA double-strand break in cancer cells and normal cells. Cancer cells have severe damage and die by irradiation, but normal cells can keep proliferation with their high DNA repair ability. Irradiated cells generate communication signals and cause biological changes in neighboring or distant non-irradiated cells. This review outlines the role of exosomes in radiation therapy. In the tumor microenvironment, exosomes are considered to regulate cell survival, migration, and resistance to therapy by interacting with vascular endothelial cells and various types of immune cells. Nowadays, radiation therapy is typically combined with immunotherapy. Regulation of the activity of exosomes may overcome the problem of resistance to immunotherapy. Furthermore, exosomes can attenuate resistance to chemotherapy by transporting certain types of microRNA. The current evidence suggests that exosomes may be useful in the diagnosis and treatment of cancer in the future.

Keywords: exosomes, cancer, microRNAs, liquid biopsy, radiation therapy

1. Introduction

Approximately half of all patients with cancer receive radiation therapy as part of their treatment making it critical in the treatment of cancer [1]. Radiation therapy is used in curative and palliative regimens to achieve locoregional control [2]. Furthermore, radiation therapy is often combined with surgery, chemotherapy, and more recently, immunotherapy [3]. Despite the progress made in approaches to deliver radiation, precision medicine and combined therapies, resistance to treatment, and recurrence continue to occur in the clinical setting. According to recent studies, components of exosomes such as miRNAs and lncRNAs perpetuate drug resistance. In gastric cancer, it is reported that exosomes from M2-macrophage-induced miR-21 mediated upregulation of PI3K/Akt signaling and reduced apoptosis and cisplatin resistance [4]. In breast cancer, exosomal miR-221/222 modulated p27 and estrogen receptor (ER) for tamoxifen resistance, and exosomal cargo-lncRNA

UCA1 mediated tamoxifen resistance [5, 6]. In prostate cancer, cancer-associated fibroblast-derived exosomes conferred gemcitabine resistance via Snail and miR-146a [7].

Biologically, radiation induces damage to DNA and other host cell structures by oxidative stress [8]. Following exposure to X-rays, free radical production caused by the interaction of ionizing radiation with water molecules and redox-mediated biological pathways are responsible for oxidative DNA damage and cell death. Interaction of free radicals with DNA causes the formation of different types of DNA oxidation in both nucleus and mitochondria. Oxidation of DNA and also cell death through necrosis or apoptosis can stimulate inflammatory responses and oxidative stress, leading to further DNA damage. Oxidized cell-free DNA is elevated in cancer patients and also patients undergoing radiation therapy for their malignancies [9, 10].

Irradiated cells generate communication signals and cause biological changes in neighboring or distant non-irradiated cells. This phenomenon is known as the radiation-induced bystander effect (RIBE) [11, 12]. Although, the exact mechanism of RIBE remains unclear, there is increasing evidence to suggest that exosomal microRNAs (miRNAs) support various cellular regulatory roles in the response to radiation [13–15].

Radiation therapy has recently been shown to have systemic immune-modulating effects in patients with cancer, including an abscopal effect whereby local radiation elicits a systemic immune response and alleviates the tumor burden in untreated areas [16]. Although, the mechanism of the abscopal effect remains to be fully elucidated, there is accumulating evidence indicating a strong association between this effect and tumor-derived exosomes.

Exosomes are membrane vesicles with a diameter of 30–150 nm that are released by the fusion of an organelle in the endocytic pathway, the multivesicular body, and

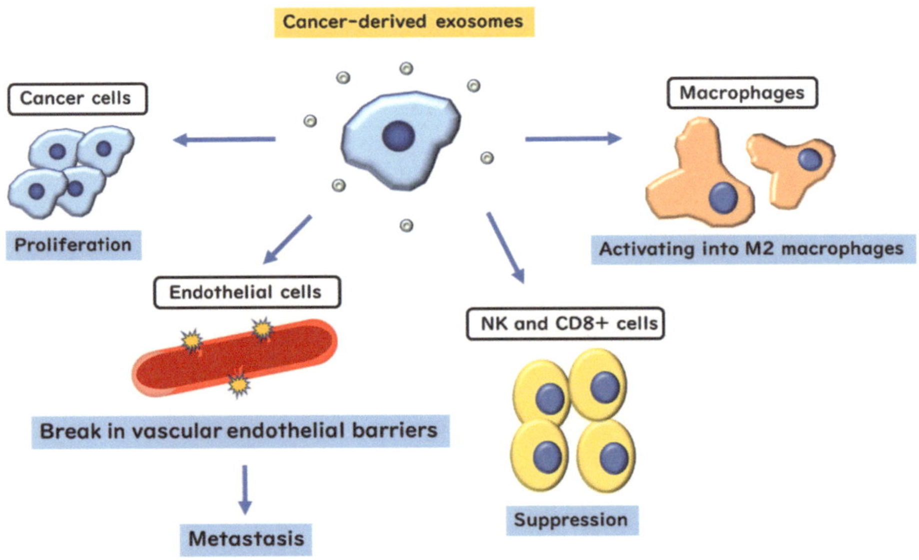

Figure 1.
Cancer-derived exosomes mediate various intercellular communications and induce the proliferation of cancer cells. Through communication with endothelial cells, vascular endothelial barriers are destroyed and metastasis is induced by cancer-derived exosomes. NK and CD8-positive cells are suppressed by cancer-derived exosomes. Macrophages are activated into M2 macrophages by cancer-derived exosomes. NK, natural killer.

the plasma membrane [17]. Exosomes are released in a four-stage process, namely, initiation, endocytosis, formation of multivesicular bodies, and release of exosomes. The exosome with its cargo, consisting of messenger RNA (mRNA), miRNA, mitochondrial DNA (mtDNA), single-stranded DNA, double-stranded DNA, retrotransposons, and various proteins (MHC class I and II molecules, cytokines, adhesion molecules, transmembrane molecules, integrins, and tetraspanin) are released by the host cells and transferred to the recipient cells [18–20]. Exosomes are a subgroup of extracellular vesicles that are released by all types of cells, including tumor cells. Exosomes mediate intercellular communication by transporting various intracellular components, including cargos [21] and induce a variety of phenomena, including the proliferation of cancer cells [22], metastasis [23], and resistance to treatment [24]. Tumor-derived exosomes affect not only cancer cells themselves but also vascular endothelial cells and various types of immune cells (**Figure 1**). Exosomes released from tumor cells may affect proximal tumor cells and stromal cells in the local microenvironment and can also have systemic effects via their functional components, such as microRNAs and proteins, via the circulation. Emerging evidence suggests that tumor-derived exosomes have both immunostimulatory and immunosuppressive activity that depends on the molecules inside these structures and the status of immune cells in the tumor microenvironment [24, 25].

2. Diagnosis of exosome-mediated cancer using liquid biopsies

Liquid biopsy is a term generally used to describe the collection of body fluid and has been explored as a non-invasive complementary tool for the diagnosis of cancer. Non-invasive measurement of cancer biomarkers using liquid biopsy allows for patient stratification, screening, monitoring of response to treatment, and detecting minimal residual disease following recurrence. Extracellular vesicles are considered to be significant biomarkers in the liquid biopsy-based diagnosis of cancer, and profiling of these vesicles has the potential to improve the early detection of cancer.

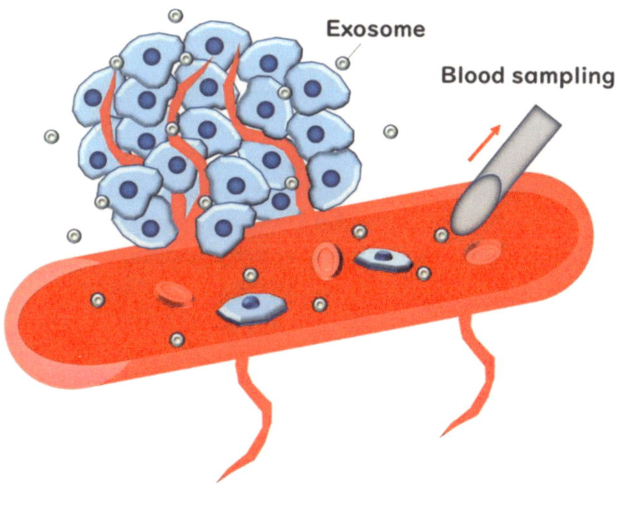

Figure 2.
Schema showing the liquid biopsy technique. Exosomes are released from cancer cells and enter blood vessels. Early diagnosis can be assisted by minimally invasive analysis of exosomes in blood vessels.

Liquid biopsy can provide more comprehensive information on the genetic landscape and track genomic evolution during disease progression in patients. The main biological biomarkers used in liquid biopsy include circulating tumor cells, circulating tumor DNA, and exosomes originating from healthy tissue or tumor tissue [26]. Given that many tissues are difficult or impossible to biopsy or resect, conventional biopsies cannot provide information on the efficacy of treatment. Exosomes can be easily acquired from most types of body fluid; therefore, they are attracting attention as biomarkers in liquid biopsy for diagnosis (**Figure 2**).

3. Roles of exosomes in common malignant diseases

3.1 Breast cancer

Early diagnosis of breast cancer increases the chances of cure and survival. Although, mammography is widely used, it is difficult to detect cancers in women with dense mammary gland tissue when using this method [27], and unnecessary radiated exposure should be avoided. Blood, nipple suction fluid, sweat, urine, and tears are being investigated as alternative diagnostic methods; among these, urine and tear tests have relatively high sensitivity and specificity. For example, the urinary miRNA profile in patients with primary breast cancer is different from that in healthy controls [28–30]. Cala et al. identified cancer-specific patterns and constructed models for the early diagnosis of breast cancer. They found that a combination of succinic acid and dimethylheptanoyl carnitine was able to separate patients with breast cancer (n = 31) from healthy controls (n = 29) with a sensitivity of 93.5% and a specificity of 86.2% [29].

Currently, reported urinary biomarkers of breast cancer are still in the discovery stage, and their specificity and sensitivity need to be verified in cohort studies. There are also multiple reports of biomarkers in tears [31–35], which are relatively easy to obtain using non-invasive methods. Inubushi et al. compared the exosomes in tears from five women with metastatic breast cancer with those from eight healthy volunteers and found higher amounts of exosome markers in tears than in serum and higher expression of breast cancer-specific miR-21 and miR-200c in tears from the patients with metastatic breast cancer [35]. Therefore, exosomes in tears could also be a useful diagnostic and prognostic biomarker.

Drug resistance is also a factor that affects prognosis. The role of exosomes in drug resistance has received a great deal of attention. There are numerous reports on the association between exosomes and drug resistance in breast cancer [36–43]. Biomarkers can be used to predict the response of a tumor to a particular treatment and may reflect tumor susceptibility and drug resistance. Although, few methods are currently recommended for routine clinical application, blood-based monitoring of the therapeutic response is a minimally invasive and promising technique. Based on those methods, liquid biopsy analysis may bring novel information as biomarkers of breast cancer [44, 45].

3.2 Prostate cancer

Prostate cancer is a tumor with a high mortality rate, and early diagnosis has a significant effect on the prognosis. Serum and urinary exosomes are promising non-invasive biomarkers that could aid in early diagnosis [46–48]. Prostate cancer has a good prognosis when hormone therapy is effective, but castrate-resistant prostate cancer has a markedly worse prognosis. Hessvik et al. identified 36 exosomal miRNAs as candidate biomarkers for prostate cancer in clinical studies [49]. Moreover, plasma miR-1290 and miR-375 levels correlate with reduced overall survival and have been identified as potential prognostic biomarkers of castrate-resistant prostate cancer [50, 51]. Expression of AR-V7 RNA in circulating tumor cells was identified as a predictor of the response to enzalutamide (an anti-androgen) and abiraterone (an anti-androgen and CYP17 inhibitor) in prostate cancer [52]. Del Re et al. detected the AR-V7 transcript preferentially in patients resistant to treatment with enzalutamide or abiraterone and proposed that the level of the AR-V7 transcript measured in extracellular vesicles in plasma could serve as a biomarker [53].

3.3 Lung cancer

Screening with low-dose CT is currently used for early diagnosis of lung cancer but is associated with radiation exposure. miRNAs derived from exosomes in body fluids are stable and relatively easy to obtain. Therefore, they are expected to be useful biomarkers for the early diagnosis of lung cancer [54–61]. Asakura et al. reported that serum miR-1268b and miR-6075 expression levels showed 99% sensitivity and 99% specificity for the diagnosis of lung cancer regardless of histological type or TNM stage [61]. This finding may lead to significant improvements in the results of screening for lung cancer and selection of more effective treatment for non-small cell lung cancer (NSCLC). Lebanony et al. reported that expression of exosomal miR-205 can distinguish between squamous and non-flat epithelial lung cancer even in poorly differentiated tumors [60]. Biomarkers that can predict the progression of lung cancer are also being investigated. Downregulation of miR-503 relative to non-malignant lung tissue has been observed in NSCLC tissue. For example, Liu et al. reported a link between the miR-503 level and advanced tumor stage and a poor prognosis [62]. These findings indicate that miR-503 may be a useful biomarker of survival in patients with NSCLC.

4. Exosome in radio-resistance

Radio-resistance is induced by certain conditions or pathways, and eventually limits the efficacies of radiation therapy. Molecular oxygen status is recognized as one of the most influential factors regulating radio-resistance. In general, cancer cells may be radio-resistant under hypoxic conditions in solid tumors [63–65]. Adapting to the hypoxic conditions, tumor cells acquire a hypoxia-resistant phenotype with the characteristic alterations in signaling, gene expression and metabolism. It is widely known that tumor cells in the hypoxia microenvironment induced radio-resistance by hypoxia-inducible factor 1 (HIF-1) and several pathways such as phosphatidylinositol 3-kinase Akt/mammalian target of rapamycin (mTOR), nuclear factor-κB (NF-κB) [66–68]. Tumor-derived exosomes promote angiogenesis by suppressing the expression of factor-inhibiting hypoxia-inducible factor 1 (HIF-1) and by transporting numerous pro-angiogenic biomolecules like vascular endothelial growth factor (VEGF), matrix metalloproteinases (MMPs), and microRNAs. The hypoxic condition induces angiogenesis. Uptake of the tumor-derived exosomes by normal endothelial cells activates angiogenic signaling pathways in endothelial cells and stimulates new vessel formation [69]. Exosomes can affect T cells through direct receptor-ligand interactions, but in ECs, exosomes usually use the internalization pathway [70].

The NF-κB is known as the most important transcription factors regulating the expression of a large number of genes involved in cellular processes, such as inflammatory responses, cellular growth, developmental processes and apoptosis. The inducers of NF-κB activity are highly variable, including ROS, TNFα, IL-1β, LPS, isoproterenol, ionizing radiation, viruses, and chemotherapeutic reagents [71]. Active NF-κB enters the nucleus and up-regulates the transcription of Bcl-2, Bcl-xL, XIAP, survivin and Akt leading to chemo- and radio-resistance. Yoshida and coworkers demonstrated that NF-κB is the strongest indicator of radio-resistance in laryngeal squamous cell carcinoma cases treated with radiation therapy alone without surgery nor combined chemotherapy [72]. As for the relationship between exosomes and NF-κB, Zeng et al. elaborated that exosomal miR-183-5p shuttled by M2-TAM mediated Akt/NF-κB pathway to accelerate colon cancer progression through targeting thioesterase superfamily member 4 (THEM4) [73]. Li J et al. demonstrated that hypoxic colorectal cancer-derived extracellular vesicles deliver microRNA-361-3p to facilitate cell proliferation by targeting TRAF3 via the noncanonical NF-κB pathways [74].

Farias et al. reported that exosomes derived from irradiated MSCs (mesenchymal stem cells) delay tumor growth and reduce metastasis after treatment with MSC plus radiation therapy [75]. In this report, the authors indicate that the effect of MSCs on the tumor is associated with ANXA1; recognized as an anti-inflammatory mediator which regulates migration and cellular responses. They also showed MSCs plus radiation therapy administered on an experimental murine model had a positive effect on the tumor-volume reduction of the contralateral, untreated tumor [75]. According to these reports, exosomes derived from MSC are expected to be able to improve the efficacy of irradiation for radiation-resistant cancers.

5. Perspectives on the future use of exosomes in radiation therapy

Radiation induces oxidative stress in host cells [8], and the RIBE induced by exosomes has been investigated in several studies. It has been reported that irradiated cells release more exosomes than non-irradiated cells [12, 76, 77]. Arscott et al. identified a relationship between increased secretion of exosomes and overexpression or mutation of p53 in glioblastoma cells [78]. Furthermore, it has been found that exosomes derived from irradiated cells are taken up by neighboring cells in greater numbers than those derived from non-irradiated cells [79]. Hazawa et al. suggested that one reason for this difference could be the enhancement of cellular attachment to exosomes via augmented formation of the CD29/CD81 complex induced by radiation [80]. Thus, the influence of irradiation is transferred to neighboring cells via interaction with exosomal cargo. Although, the mechanism via which cargo is included in exosomes is not clearly understood, exosomal expression of miRNA influences the effect of irradiation on cells [79, 81]. Nakaoka et al. also investigated this effect and found that exosomes derived from irradiated MIAPaCa-2 human pancreatic cancer cells induced a radiosensitive effect on neighboring cells through an increase in levels of reactive oxygen species in cells [79], which they attributed to a reduction in expression of antioxidant enzymes via changes in the miRNA profile in exosomes. A summary of the molecules involved in RIBE is presented in **Table 1**.

Radiation therapy can have profound immune-stimulatory effects and is increasingly viewed as a promising partner in combination therapy for patients receiving immunotherapy. Intrinsic events in cancer cells induced by DNA damage are central to the immune-modifying effects of radiation therapy [82]. These events have been investigated in studies of the DNA damage response, which focuses on the tumor microenvironment. In addition to its ability to destroy cancer cells by damaging DNA,

Cargo	Molecule	Cell line	Interactions	Ref.
miRNAs	miR-7-5p	BEP2D	Induction of autophagy	13
	miR-208a	Non-small cell lung cancer cells	Promotion of cell proliferation Induction of radioresistance	14
	miR-301a	Glioblastoma cells	Promotion of radiation resistance	15
	miR-6823-5p	MIAPaCa-2	Enhancement of radiosensitive effect	79
Proteins	Connective tissue growth factor Insulin-like growth factor-binding protein 2	LN18, U87MG and U251	Enhancement of tumorigenesis and migration	78

Table 1.
Radiation-induced bystander effect induced by exosomes.

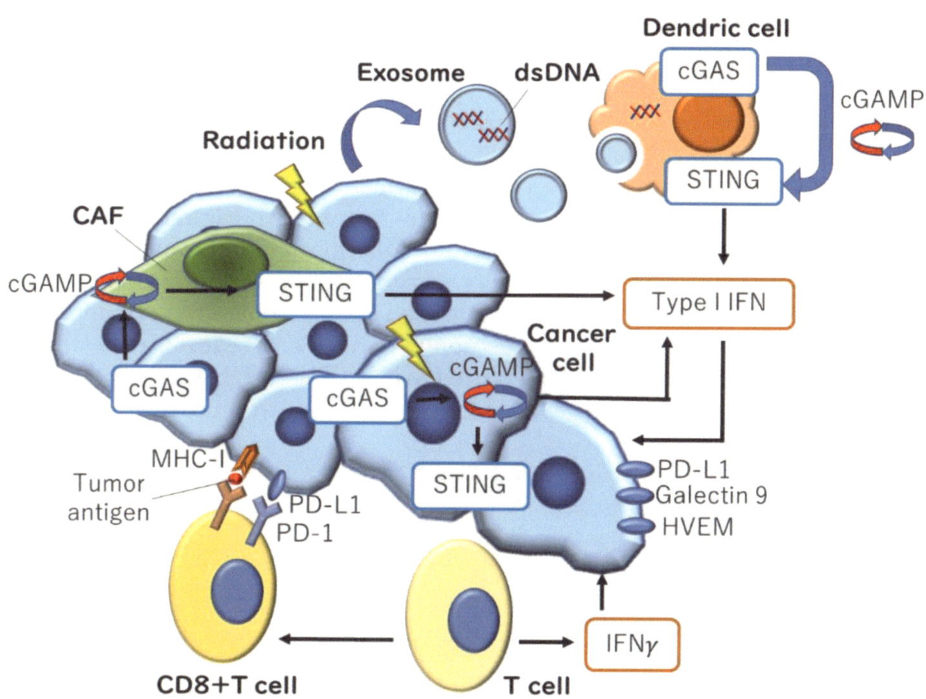

Figure 3.
Radiation and immune response via cancer-derived exosomes in the tumor microenvironment. The pathway related to activation of cytoplasmic nucleic acid sensors. Tumor-derived exosomes can transport type I IFN-stimulatory dsDNA from irradiated cancer cells to dendritic cells via the STING pathway. T-cells produce IFN-γ. Inhibition of activation of CD8-positive T-cells by PD-1/L1 binding. cGAS cytoplasmic DNA-sensing cyclic GMP-AMP synthase; dsDNA, double-stranded DNA; IFN, interferon; STING, stimulator of interferon genes.

radiation therapy can modulate both the immunotherapy and adjuvant therapy of tumors by triggering the release of pro-inflammatory mediators, increasing tumor-infiltrating immune-stimulatory cells. and enhancing the expression of neoantigens [82–84] and immune-stimulatory signaling by cyclic dinucleotide cyclic GMP-AMP. Understanding the mechanistic basis of radiation therapy as an anticancer treatment has been transformed by the recent discovery that DNA damage in cycling cancer cells can activate cytoplasmic nucleic acid sensors [82, 83]. The cGAS (cytoplasmic DNA-sensing cyclic GMP–AMP synthase)-STING (stimulator of interferon genes) pathway is involved in this process [85, 86]. Using this pathway, tumor-derived exosomes can shuttle TREX1-sensitive type I interferon-stimulatory double-stranded DNA from irradiated cancer cells to dendritic cells (**Figure 3**) [87].

6. Exosomes in chemotherapy and other therapy

Chemo-resistance should be solved for effective cancer therapy. Several studies indicated that signal transducer and activator of transcription 3 (STAT3), focal adhesion kinase (FAK) and epithelial-mesenchymal transition (EMT) contribute to the development of chemotherapeutic resistance [88–91]. The anti-apoptotic signal which is activated by the STAT3 pathway is an important factor in causing initial drug resistance [88]. FAK is reported to promote invasive tumor growth with β-catenin [89]. EMT is popular with the induction of cancer metastasis [90], but it is also deeply related to drug resistance. Mani et al. reported that EMT may cause cancer cells to acquire epithelial stem cell properties [91]. EMT formation induces chemo-resistance [92] and it is induced by various signaling pathways, such as TGFβ, Wnt signal, etc. [93, 94].

Several studies have demonstrated that exosomal activity might be involved in chemo-resistance. In EMT, Shan et al. indicated that downregulation of exosomal miR-148b-3p may offer opportunities in the treatment of bladder cancer by increasing chemosensitivity via inhibition of the Wnt/β-catenin pathway and promoting expression of PTEN [95]. In addition to it, Liu et al. reported that exosome-transmitted miR-128-3p increased chemosensitivity of oxaliplatin-resistant colorectal cancer by suppressing epithelial-mesenchymal transition and inducing drug accumulation in cancer cells [96]. Furthermore, exosomes derived from cancer-associated fibroblasts and cancer cells are often reported to contribute to chemo-resistance [97, 98]. According to these reports, exosomes are also important to the regulation of chemotherapy.

7. Conclusions

Exosomes play a critical role in cancer progression, including cell–cell communication, tumor-stromal interactions, activation of signaling pathways, and immunomodulation. Emerging data indicate that radiation-derived exosomes increase tumor burden, decrease survival, cause radiation-induced bystander effects and promote radio-resistance. Exosomes show abilities as diagnostic and predictive biomarkers in various malignancies. Exosomes play a role in radio- and chemo-resistance through multiple pathways. The mechanism of the intercellular communication by exosomes, the transport of exosomal cargo, the secretion mechanism, cell dependence, immune response and radiation response have been intensively explored. Specifically, research on exosomes about radiation therapy has been expanding. Radiation-exposed cells release altered exosomes, and those exosomes play a role as cargo to bring multiple messages to their recipient cells leading to various radiation responses. Further understanding of the mechanisms of exosome-mediated radio-resistance might ultimately lead to the development of novel treatment strategies.

Acknowledgements

This research was supported by Grants-in-Aid for Exploratory Research (21 J21801 to AN, and 21 K07642 to KK) from the Ministry of Education, Culture, Sports, Science, and Technology of Japan.

Conflict of interest

The authors declare no conflict of interest.

Author details

Ai Nakaoka[1], Kana Kobayashi[1], Mennaallah Hassan[1,2] and Ryohei Sasaki[1*]

1 Division of Radiation Oncology, Kobe University Graduate School of Medicine, Kobe, Japan

2 Faculty of Medicine, Department of Clinical Oncology, Sohag University, Sohag, Egypt

*Address all correspondence to: rsasaki@med.kobe-u.ac.jp

References

[1] Delaney G, Jacob S, Featherstone C, Barton M. The role of radiotherapy in cancer treatment: estimating optimal utilization from a review of evidence-based clinical guidelines. Cancer. 2005;**104**:1129-1137. DOI: 10.1002/cncr.21324

[2] Grass GD, Krishna N, Kim S. The immune mechanisms of abscopal effect in radiation therapy. Current Problems in Cancer. 2016;**40**:10-24. DOI: 10.1016/j.currproblcancer.2015.10.003

[3] Yaromina A, Krause M, Baumann M. Individualization of cancer treatment from radiotherapy perspective. Molecular Oncology. 2012;**6**:211-221. DOI: 10.1016/j.molonc.2012.01.007

[4] Liao J, Liu R, Shi YJ, Yin LH, Pu YP. Exosome-shuttling microRNA-21 promotes cell migration and invasion-targeting PDCD4 in esophageal cancer. International Journal of Oncology. 2016;**48**:2567-2579. DOI: 10.3892/ijo.2016.3453

[5] Wei Y, Lai X, Yu S, Chen S, Ma Y, Zhang Y, et al. Exosomal miR-221/222 enhances tamoxifen resistance in recipient ER-positive breast cancer cells. Breast Cancer Research and Treatment. 2014;**147**:423-431. DOI: 10.1007/s10549-014-3037-0

[6] Xu CG, Yang MF, Ren YQ, Wu CH, Wang LQ. Exosomes mediated transfer of lncRNA UCA1 results in increased tamoxifen resistance in breast cancer cells. European Review for Medical and Pharmacological Sciences. 2016;**20**:4362-4368

[7] Richards KE, Zeleniak AE, Fishel ML, Wu J, Littlepage LE, Hill R. Cancer-associated fibroblast exosomes regulate survival and proliferation of pancreatic cancer cells. Oncogene. 2017;**36**:1770-1778. DOI: 10.1038/onc.2016.353

[8] Ward JF, Evans JW, Limoli CL, Calabro-Jones PM. Radiation and hydrogen peroxide induced free radical damage to DNA. British Journal of Cancer. 1987;**55**:105-112

[9] Cheng C, Omura-Minamisawa M, Kang Y, Hara T, Koike I, Inoue T. Quantification of circulating cell-free DNA in the plasma of cancer patients during radiation therapy. Cancer Science. 2009;**100**:303-309. DOI: 10.1111/j.1349-7006.2008.01021.x

[10] Deligezer U, Eralp Y, Akisik EZ, Akisik EE, Saip P, Topuz E, et al. Effect of adjuvant chemotherapy on integrity of free serum DNA in patients with breast cancer. Annals of the New York Academy of Sciences. 2008;**113**:175-179. DOI: 10.1196/annals.1448.010

[11] Al-Mayah AHJ, Irons SL, Pink RC, Carter DRF, Kadhim MA. Possible role of exosomes containing RNA in mediating nontargeted effect of ionizing radiation. Radiation Research. 2012;**177**:539-545. DOI: 10.1667/rr2868.1

[12] Al-Mayah A, Bright S, Chapman K, Irons S, Luo P, Carter D, et al. The non-targeted effects of radiation are perpetuated by exosomes. Mutation Research. 2015;**772**:38-45. DOI: 10.1016/j.mrfmmm.2014.12.007

[13] Song M, Wang Y, Shang ZF, Liu XD, Xie DF, Wang Q, et al. Bystander autophagy mediated by radiation-induced exosomal miR-7-5p in non-targeted human bronchial epithelial cells. Scientific Reports. 2016;**6**:30165. DOI: 10.1038/srep30165

[14] Tang Y, Cui Y, Li Z, Jiao Z, Zhang Y, He Y, et al. Radiation-induced miR-208a increases the proliferation and radioresistance by targeting p21 in human lung cancer cells. Journal of Experimental & Clinical Cancer Research. 2016;**35**:7. DOI: 10.1186/s13046-016-0285-3

[15] Yue X, Lan F, Xia T. Hypoxic glioma cell-secreted exosomal miR-301a activates Wnt/β-catenin signaling and promotes radiation resistance by targeting TCEAL7. Molecular Therapy. 2019;**27**:1939-1949. DOI: 10.1016/j.ymthe.2019.07.011

[16] Hu ZI, McArthur HL, Ho AY. The abscopal effect of radiation therapy: what is it and how can we use it in breast cancer? Current Breast Cancer Reports. 2017;**9**:45-51. DOI: 10.1007/s12609-017-0234-y

[17] Hessvik NP, Llorente A. Current knowledge on exosome biogenesis and release. Cellular and Molecular Life Sciences. 2018;**75**:193-208. DOI: 10.1007/s00018-017-2595-9

[18] Théry C, Zitvogel L, Amigorena S. Exosomes: composition, biogenesis and function. Nature Reviews. Immunology. 2002;**2**:569-579. DOI: 10.1038/nri855

[19] Zhang C, Ji Q, Yang Y, Li Q, Wang Z. Exosome: Function and role in cancer metastasis and drug resistance. Technology in Cancer Research & Treatment. 2018;**17**:1-12. DOI: 10.1177/1533033818763450

[20] McAndrews KM, Kalluri R. Mechanisms associated with biogenesis of exosomes in cancer. Molecular Cancer. 2019;**18**:52. DOI: 10.1186/s12943-019-0963-9

[21] Valadi H, Ekstrom K, Bossios A, Sjostrand M, Lee JJ, Lotvall JO. Exosome mediated transfer of mRNAs and microRNAs is a novel mechanism of genetic exchange between cells. Nature Cell Biology. 2007;**9**:654-659. DOI: 10.1038/ncb1596

[22] Wu H, Zhou J, Mei S, Wu D, Mu Z, Chen B, et al. Circulating exosomal microRNA-96 promotes cell proliferation, migration and drug resistance by targeting LMO7. Journal of Cellular and Molecular Medicine. 2017;**21**:1228-1236. DOI: 10.1111/jcmm

[23] Zhou W, Fong MY, Min Y, Somlo G, Liu L, Palomares MR, et al. Cancer secreted miR-105 destroys vascular endothelial barriers to promote metastasis. Cancer Cell. 2014;**25**:501-515. DOI: 10.1016/j.ccr.2014.03.007

[24] Li I, Nabet BY. Exosomes in the tumor microenvironment as mediators of cancer therapy resistance. Molecular Cancer. 2019;**18**:32. DOI: 10.1186/s12943-019-0975-5

[25] Whiteside TL. Exosomes and tumor-mediated immune suppression. Journal of Clinical Investigation. 2016;**126**:1216-1223. DOI: 10.1172/JCI81136

[26] De Rubis G, Krishnan SR, Bebawy M. Circulating tumor DNA – Current state of play and future perspectives. Pharmacological Research. 2018;**136**:35-44. DOI: 10.1016/j.phrs.2018.08.017

[27] Bakker MF, de Lange SV, Pijnappel RM, Mann RM, PHM P, Monninkhof EM, et al. Supplemental MRI screening for women with extremely dense breast tissue. The New England Journal of Medicine. 2019;**381**:2091-2102. DOI: 10.1056/NEJMoa1903986

[28] Ando W, Kikuchi K, Uematsu T, Yokomori H, Takaki T, Sogabe M, et al.

Novel breast cancer screening: Combined expression of miR-21 and MMP-1 in urinary exosomes detects 95% of breast cancer without metastasis. Scientific Reports. 2019;**9**:13595. DOI: 10.1038/s41598-019-50084-5

[29] Cala M, Aldana J, Sanchez J, Guio J, Meesters RJW. Urinary metabolite and lipid alterations in Colombian Hispanic women with breast cancer: A pilot study. Journal of Pharmaceutical and Biomedical Analysis. 2018;**152**:234-241. DOI: 10.1016/j.jpba.2018.02.009

[30] Dinges SS, Hohm A, Vandergrift LA, Nowak J, Habbel P, Kaltashov IA, et al. Cancer metabolomic markers in urine: Evidence, techniques and recommendations. Nature Reviews. Urology. 2019;**16**:339-362. DOI: 10.1038/s41585-019-0185-3

[31] Lebrecht A, Boehm D, Schmidt M, Koelbl H, Grus FH. Surface-enhanced laser desorption/ionisation time-of-flight mass spectrometry to detect breast cancer markers in tears and serum. Cancer Genomics and Proteomics. 2009;**6**:75-83

[32] Lebrecht A, Boehm D, Schmidt M, Koelbl H, Schwirz RL, Grus FH. Diagnosis of breast cancer by tear proteomic pattern. Cancer Genomics and Proteomics. 2009;**6**:177-182

[33] Böhm D, Keller K, Pieter J, Boehm N, Wolters D, Siggelkow W, et al. Comparison of tear protein levels in breast cancer patients and healthy controls using a de novo proteomic approach. Oncology Reports. 2012; **28**:429-438. DOI: 10.3892/or.2012.1849

[34] Kim S, Kim TG, Lee SH, Kim W, Bang A, Moon SW, Song J, Shin JH, Yu JS. Choi S. Label-free surface-enhanced Raman spectroscopy biosensor for on-site breast cancer detection using human tears. ACS Applied Materials & Interfaces 2020; **12**:7897-7904. DOI: 10.1021/acsami.9b19421

[35] Inubushi S, Kawaguchi H, Mizumoto S, Kunihisa T, Baba M, Kitayama Y, et al. Oncogenic miRNAs identified in tear exosomes from metastatic breast cancer patients. Anticancer Research. 2020;**40**:3091-3096. DOI: 10.21873/anticanres.14290

[36] Wang X, Pei X, Guo G, Qian X, Dou D, Zhang Z, et al. Exosome-mediated transfer of long noncoding RNA H19 induces doxorubicin resistance in breast cancer. Journal of Cellular Physiology. 2020;**235**:6896-6904. DOI: 10.1002/jcp.29585

[37] Wang B, Zhang Y, Ye M, Wu J, Ma L, Chen H. Cisplatin-resistant MDA-MB-231 cell-derived exosomes increase the resistance of recipient cells in an exosomal miR-423-5p-dependent manner. Current Drug Metabolism. 2019;**20**:804-814. DOI: 10.2174/1389200220666190819151946

[38] Yang SJ, Wang DD, Li J, Xu HZ, Shen HY, Chen X, et al. Predictive role of GSTP1-containing exosomes in chemotherapy-resistant breast cancer. Gene. 2017;**623**:5-14. DOI: 10.1016/j.gene.2017.04.031

[39] Ning K, Wang T, Sun X, Zhang P, Chen Y, Jin J, et al. UCH-L1-containing exosomes mediate chemotherapeutic resistance transfer in breast cancer. Journal of Surgical Oncology. 2017;**115**:932-940. DOI: 10.1002/jso.24614

[40] Li XJ, Ren ZJ, Tang JH, Yu Q. Exosomal microRNA MiR-1246 promotes cell proliferation, invasion and drug resistance by targeting CCNG2 in breast cancer. Cellular Physiology and

Biochemistry. 2017;**44**:1741-1748. DOI: 10.1159/000485780

[41] Muluhngwi P, Klinge CM. Identification of miRNAs as biomarkers for acquired endocrine resistance in breast cancer. Molecular and Cellular Endocrinology. 2017;**456**:76-86. DOI: 10.1016/j.mce.2017.02.004

[42] Semina SE, Scherbakov AM, Vnukova AA, Bagrov DV, Evtushenko EG, Safronova VM, et al. Exosome-mediated transfer of cancer cell resistance to antiestrogen drugs. Molecules. 2018;**23**:829. DOI: 10.3390/molecules23040829

[43] Martinez VG, O'Neill S, Salimu J, Breslin S, Clayton A, Crown J, et al. Resistance to HER2-targeted anti-cancer drugs is associated with immune evasion in cancer cells and their derived extracellular vesicles. Oncoimmunology. 2017;**6**:e1362530. DOI: 10.1080/2162402X.2017.1362530

[44] Braden AM, Stankowski RV, Engel JM, Onitilo AA. Breast cancer biomarkers: risk assessment, diagnosis, prognosis, prediction of treatment efficacy and toxicity, and recurrence. Current Pharmaceutical Design. 2014;**20**:4879-4898. DOI: 10.2174/1381612819666131125145517

[45] Bardelli A, Pantel K. Liquid biopsies, what we do not know (yet). Cancer Cell. 2017;**31**:172-179. DOI: 10.1016/j.ccell.2017.01.002

[46] Campos-Carrillo A, Weitzel JN, Sahoo P, Rockne R, Mokhnatkin JV, Murtaza M, et al. Circulating tumor DNA as an early cancer detection tool. Pharmacology & Therapeutics. 2020;**207**:107458. DOI: 10.1016/j.pharmthera.2019.107458

[47] Vlaeminck-Guillem V. Extracellular vesicles in prostate cancer carcinogenesis, diagnosis, and management. Frontiers in Oncology. 2018;**8**:222. DOI: 10.3389/fonc.2018.00222

[48] Øverbye A, Skotland T, Koehler CJ, Thiede B, Seierstad T, Berge V, et al. Identification of prostate cancer biomarkers in urinary exosomes. Oncotarget. 2015;**6**:30357-30376. DOI: 10.18632/oncotarget.4851

[49] Hessvik NP, Sandvig K, Llorente A. Exosomal miRNAs as biomarkers for prostate cancer. Frontiers in Genetics. 2013;**4**:36. DOI: 10.3389/fgene.2013.00036

[50] Huang X, Yuan T, Tschannen M, Sun Z, Jacob H, Du M, et al. Characterization of human plasma-derived exosomal RNAs by deep sequencing. BMC Genomics. 2013;**14**:319. DOI: 10.1186/1471-2164-14-319

[51] Huang X, Yuan T, Liang M, Du M, Xia S, Dittmar R, et al. Exosomal miR-1290 and miR-375 as prognostic markers in castration-resistant prostate cancer. European Urology. 2015;**67**:33-41. DOI: 10.1016/j.eururo.2014.07.035

[52] Antonarakis ES, Lu C, Wang H, Luber B, Nakazawa M, Roeser JC, et al. AR-V7 and resistance to enzalutamide and abiraterone in prostate cancer. The New England Journal of Medicine. 2014;**371**:1028-1038. DOI: 10.1056/NEJMoa1315815

[53] Re MD, Biasco E, Crucitta S, Derosa L, Rofi E, Orlandini C, et al. The detection of androgen receptor splice variant 7 in plasma-derived exosomal RNA strongly predicts resistance to hormonal therapy in metastatic prostate cancer patients. European Urology. 2017;**71**:680-687. DOI: 10.1016/j.eururo.2016.08.012

[54] Kanaoka R, Iinuma H, Dejima H, Sakai T, Uehara H, Matsutani N, et al.

Usefulness of plasma exosomal microRNA-451a as a noninvasive biomarker for early prediction of recurrence and prognosis of non-small cell lung cancer. Oncology. 2018;**94**: 311-323. DOI: 10.1159/000487006

[55] Zhou X, Wen W, Shan X, Zhu W, Xu J, Guo R, et al. A six-microRNA panel in plasma was identified as a potential biomarker for lung adenocarcinoma diagnosis. Oncotarget. 2017;**8**:6513-6525. DOI: 10.18632/oncotarget.14311

[56] Liu Q, Yu Z, Yuan S, Xie W, Li C, Hu Z, et al. Circulating exosomal microRNAs as prognostic biomarkers for non-small-cell lung cancer. Oncotarget. 2017;**8**:13048-13058. DOI: 10.18632/oncotarget.14369

[57] Grimolizzi F, Monaco F, Leoni F, Bracci M, Staffolani S, Bersaglieri C, et al. Exosomal miR-126 as a circulating biomarker in non-small-cell lung cancer regulating cancer progression. Scientific Reports. 2017;**7**:15277. DOI: 10.1186/s12935-020-01653-6

[58] Jin X, Chen Y, Chen H, Fei S, Chen D, Cai X, et al. Evaluation of tumor-derived exosomal miRNA as potential diagnostic biomarkers for early-stage non-small cell lung cancer using next-generation sequencing. Clinical Cancer Research. 2017;**23**:5311-5319. DOI: 10.1158/1078-0432.CCR-17-0577

[59] Lai X, Friedman A. Exosomal miRs in lung cancer: a mathematical model. PLoS One. 2016;**11**:e0167706. DOI: 10.1371/journal.pone.0167706

[60] Lebanony D, Benjamin H, Gilad S, Ezagouri M, Dov A, Ashkenazi K, et al. Diagnostic assay based on hsa-miR-205 expression distinguishes squamous from nonsquamous non-small-cell lung carcinoma. Journal of Clinical Oncology. 2009;**27**:2030-2037. DOI: 10.1200/JCO.2008.19.4134

[61] Asakura K, Kadota T, Matsuzaki J, Yoshida Y, Yamamoto Y, Nakagawa K, et al. A miRNA-based diagnostic model predicts resectable lung cancer in humans with high accuracy. Communications Biology. 2020;**3**:134. DOI: 10.1038/s42003-020-0863-y

[62] Liu L, Qu W, Zhong Z. Down-regulation of miR-503 expression predicate advanced mythological features and poor prognosis in patients with NSCLC. International Journal of Clinical and Experimental Pathology. 2015;**8**:5609-5613

[63] Brown JM, Giaccia AJ. The unique physiology of solid tumors: opportunities (and problems) for cancer therapy. Cancer Research. 1998;**58**:1408-1416

[64] Brown JM, Wilson WR. Exploiting tumour hypoxia in cancer treatment. Nature Reviews. Cancer. 2004;**4**:437-447. DOI: 10.1038/nrc1367

[65] Kizaka-Kondoh S, Inoue M, Harada H, Hiraoka M. Tumor hypoxia: A target for selective cancer therapy. Cancer Science. 2003;**94**:1021-1028. DOI: 10.1111/j.1349-7006.2003.tb01395.x

[66] Deep G, Panigrahi GK. Hypoxia-induced signaling promotes prostate cancer progression: Exosomes role as messenger of hypoxic response in tumor microenvironment. Critical Reviews in Oncogenesis. 2015;**20**:419-434. DOI: 10.1615/CritRevOncog.v20.i5-6.130

[67] Raja R, Kale S, Thorat D, Soundararajan G, Lohite K, Mane A, et al. Hypoxia-driven osteopontin contributes to breast tumor growth through modulation of HIF1α-mediated VEGF-dependent angiogenesis. Oncogene. 2014;**33**:2053-2064. DOI: 10.1038/onc.2013.171

[68] Toulany M, Rodemann HP. Phosphatidylinositol 3-kinase/Akt

signaling as a key mediator of tumor cell responsiveness to radiation. Seminars in Cancer Biology. 2015;**35**:180-190. DOI: 10.1016/j.semcancer.2015.07.003

[69] Olejarz W, Kubiak-Tomaszewska G, Chrzanowska A, Lorenc T. Exosomes in angiogenesis and anti-angiogenic therapy in cancers. International Journal of Molecular Sciences. 2020;**21**:5840. DOI: 10.3390/ijms21165840

[70] Huang A, Dong J, Li S, Wang C, Ding H, Li H, et al. Exosomal transfer of vasorin expressed in hepatocellular carcinoma cells promotes migration of human umbilical vein endothelial cells. International Journal of Biological Sciences. 2015;**11**:961-969. DOI: 10.7150/ijbs.11943

[71] Aggarwal BB, Vijayalekshmi RV, Sung B. Targeting inflammatory pathways for prevention and therapy of cancer: short-term friend, long-term foe. Clinical Cancer Research. 2009;**15**: 425-430. DOI: 10.1158/1078-0432.CCR-08-0149

[72] Yoshida K, Sasaki R, Nishimura H, Okamoto Y, Suzuki Y, Kawabe T, et al. Nuclear factor-kappaB expression as a novel marker of radioresistance in early-stage laryngeal cancer. Head & Neck. 2010;**32**:646-655. DOI: 10.1002/hed.21239

[73] Zhang S, Li D, Zhao M, Yang F, Sang C, Yan C, et al. Exosomal miR-183-5p shuttled by M2 polarized tumor-associated macrophage promotes the development of colon cancer via targeting THEM4 mediated PI3K/AKT and NF-κB Pathways. Front Oncol. 2021;**25**(11):672684. DOI: 10.3389/fonc.2021.672684

[74] Li J, Yang P, Chen F, Tan Y, Huang C, Shen H, et al. Hypoxiccolorectal cancer-derived extracellular vesicles deliver microRNA-361-3p to facilitate cell proliferation by targeting TRAF3 via the noncanonical NF-κB pathways. Clinical and Translational Medicine. 2021;**11**:e349. DOI: 10.1002/ctm2.349

[75] Farias VA, O'Valle F, Serrano-Saenz S, Anderson P, Andrés E, López-Peñalver J, et al. Exosomes derived from mesenchymal stem cells enhance radiotherapy-induced cell death in tumor and metastatic tumor foci. Molecular Cancer. 2018;**17**:122. DOI: 10.1186/s12943-018-0867-0

[76] Jella KK, Rani S, O'Driscoll L, McClean B, Byrne HJ, Lyng FM. Exosomes are involved in mediating radiation induced bystander signaling in human keratinocyte cells. Radiation Research. 2014;**181**:138-145. DOI: 10.1667/RR13337.1

[77] Lehmann BD, Paine MS, Brooks AM, McCubrey JA, Renegar RH, Wang R, et al. Senescence-associated exosome release from human prostate cancer cells. Cancer Research. 2008;**68**:7864-7871. DOI: 10.1158/0008-5472.CAN-07-6538

[78] Arscott WT, Tandle AT, Zhao S, Shabason JE, Gordon IK, Schlaff CD, et al. Ionizing radiation and glioblastoma exosomes: implications in tumor biology and cell migration. Translational Oncology. 2013;**6**:638-648. DOI: 10.1593/tlo.13640

[79] Nakaoka A, Nakahana M, Inubushi S, Akasaka H, Salah M, Fujita Y, et al. Exosome-mediated radiosensitizing effect on neighboring cancer cells via increase in intracellular levels of reactive oxygen species. Oncology Reports. 2021;**45**:13. DOI: 10.3892/or.2021.7964

[80] Hazawa M, Tomiyama K, Saotome-Nakamura A, Obara C, Yasuda T, Gotoh T, et al. Radiation increases the cellular uptake of exosomes

through CD29/CD81 complex formation. Biochemical and Biophysical Research Communications. 2014;**446**:1165-1171. DOI: 10.1016/j.bbrc.2014.03.067

[81] Niemoeller OM, Niyazi M, Corradini S, Zehentmayr F, Li M, Lauber K, et al. MicroRNA expression profiles in human cancer cells after ionizing radiation. Radiation Oncology. 2011;**6**:29. DOI: 10.1186/1748-717X-6-29

[82] Harding SM, Benci JL, Irianto J, Discher DE, Minn AJ, Greenberg RA. Mitotic progression following DNA damage enables pattern recognition within micronuclei. Nature. 2017;**548**:466-470. DOI: 10.1038/nature23470

[83] Mackenzie KJ, Carroll P, Martin CA, Murina O, Fluteau A, Simpson DJ, et al. cGAS surveillance of micronuclei links genome instability to innate immunity. Nature. 2017;**548**:461-465. DOI: 10.1038/nature23449

[84] Dillon MT, Bergerhoff KF, Pedersen M, Whittock H, Crespo-Rodriguez E, Patin EC, et al. ATR inhibition potentiates the radiation-induced inflammatory tumor microenvironment. Clinical Cancer Research. 2019;**25**:3392-3403. DOI: 10.1158/1078-0432.CCR-18-1821

[85] Deng L, Liang H, Xu M, Yang X, Burnette B, Arina A, et al. STING-dependent cytosolic DNA sensing promotes radiation-induced type I interferon-dependent antitumor immunity in immunogenic tumors. Immunity. 2014;**41**:843-852. DOI: 10.1016/j.immuni.2014.10.019

[86] Vance RE. Cytosolic DNA sensing: the field narrows. Immunity. 2016;**45**:227-228. DOI: 10.1016/j.immuni.2016.08.006

[87] Diamond JM, Vanpouille-Box C, Spada S, Rudqvist NP, Chapman JR, Ueberheide BM, et al. Exosomes shuttle TREX1-sensitive IFN-stimulatory dsDNA from irradiated cancer cells to DCs. Cancer Immunology Research. 2018;**6**:910-920. DOI: 10.1158/2326-6066.CIR-17-0581

[88] Lee H, Herrmann A, Deng JH, Kujawski M, Niu G, Li Z, et al. Persistently activated STAT3 maintains constitutive NF-κB activity in tumors. Cancer Cell. 2009;**15**:283-293. DOI: 10.1016/j.ccr.2009.02.015

[89] Diaz Osterman CJ, Ozmadenci D, Kleinschmidt EG, Taylor KN, Barrie AM, Jiang S, et al. FAK activity sustains intrinsic and acquired ovarian cancer resistance to platinum chemotherapy. eLife. 2019;**8**:e47327. DOI: 10.7554/eLife.47327

[90] Micalizzi D, Farabaugh S, Ford H. Epithelial-mesenchymal transition in cancer: parallels between normal development and tumor progression. Journal of Mammary Gland Biology and Neoplasia. 2010;**15**:117-134. DOI: 10.1007/s10911-010-9178-9

[91] Mani SA, Guo W, Liao MJ, Eaton EN, Ayyanan A, Zhou AY, et al. The epithelial-mesenchymal transition generates cells with properties of stem cells. Cell. 2008;**133**:704-715. DOI: 10.1016/j.cell.2008.03.027

[92] Fischer KR, Durrans A, Lee S, Sheng J, Li F, Wong STC, et al. Epithelial-to-mesenchymal transition is not required for lung metastasis but contributes to chemoresistance. Nature. 2015;**527**:472-476. DOI: 10.1038/nature15748

[93] Papageorgis P. TGFβ signaling in tumor initiation, epithelial-to-mesenchymal transition, and metastasis.

Journal of Oncology. 2015;**2015**:587193. DOI: 10.1155/2015/587193

[94] Dissanayake SK, Wade M, Johnson CE, O'Connell MP, Leotlela PD, French AD, et al. The Wnt5A/protein kinase C pathway mediates motility in melanoma cells via the inhibition of metastasis suppressors and initiation of an epithelial to mesenchymal transition. The Journal of Biological Chemistry. 2007;**282**:17259-17271. DOI: 10.1074/jbc.M700075200

[95] Shan G, Zhou X, Gu J, Zhou D, Cheng W, Wu H, et al. Downregulated exosomal microRNA-148b-3p in cancer associated fibroblasts enhance chemosensitivity of bladder cancer cells by downregulating the Wnt/β-catenin pathway and upregulating PTEN. Cellular Oncology (Dordr). 2021;**44**:45-59. DOI: 10.1007/s13402-020-00500-0

[96] Liu T, Zhang X, Du L, Wang Y, Liu X, Tian H, et al. Exosome-transmitted miR-128-3p increase chemosensitivity of oxaliplatin-resistant colorectal cancer. Molecular Cancer. 2019;**18**:43. DOI: 10.1186/s12943-019-0981-7

[97] Shan G, Gu J, Zhou D, Li L, Cheng W, Wang Y, et al. Cancer-associated fibroblast-secreted exosomal miR-423-5p promotes chemotherapy resistance in prostate cancer by targeting GREM2 through the TGF-β signaling pathway. Experimental & Molecular Medicine. 2020;**52**:1809-1822. DOI: 10.1038/s12276-020-0431-z

[98] Wang X, Zhang H, Yang H, Bai M, Ning T, Deng T, et al. Exosome-delivered circRNA promotes glycolysis to induce chemoresistance through the miR-122-PKM2 axis in colorectal cancer. Molecular Oncology. 2020;**14**:539-555. DOI: 10.1002/1878-0261.12629

Chapter 11

Diversity of Extracellular Vesicles (EV) in Plasma of Cancer Patients

Theresa L. Whiteside and Soldano Ferrone

Abstract

Extracellular vesicles (EVs) are produced by all cells and are found in all body fluids. They function as intercellular messengers that carry and deliver signals regulating cellular interactions in health and disease. EVs are emerging as potential biomarkers of diseases and responses to therapies, and much attention is being devoted to understanding their role in physiological as well as pathological events. EVs are heterogenous in their origin, size, molecular characteristics, genetic content and functions. Isolation of EV subsets from plasma and characterization of their distinct properties have been a limiting factor in ongoing efforts to understand their biological importance. Here, we discuss the immunoaffinity-based strategies that are available for isolating distinct subsets of EVs from plasma and provide a road-map to their successful immunocapture and molecular profiling, with special attention to tumor-derived EVs or TEX.

Keywords: extracellular vesicles (EV), exosomes, tumor derived exosomes (TEX), immunoaffinity capture from plasma

1. Introduction

Recent progress in the understanding of the role tumor microenvironment (TME) plays in cancer development has identified intercellular communication within and outside the TME as one of the major mechanisms driving tumor progression. A detailed characterization of the crosstalk of the tumor with various immune and tissue cells has become a major goal in cancer research. For years, many soluble factors, including cytokines and chemokines, have been postulated to play a major role in the regulation of cellular interactions in healthy and pathological tissues. The recognition of extracellular vesicles, EVs, as major players in the intercellular communication network occurred only a few years ago [1]. Since then, EVs produced by cancer cells and by immune as well as non-immune cells residing in the TME have become the topic of numerous studies evaluating their involvement in the regulation of tumor progression on the one hand and of the host anti-tumor immune responses on the other. This double role of EVs in cancer as well as other diseases emphasizes their potential as reporters or markers of changes that preface or accompany the emergence of disease or its outcome.

2. The origin and characteristics of extracellular vesicles (EVs)

Extracellular vesicles (EVs) are produced and released into the extracellular space by all cells. EVs are classified based on differences in their biogenesis, size and functions [2]. The current EV nomenclature recognizes exosomes (30–150 nm), microvesicles (MVs; 150–1000 nm) and apoptotic bodies (>1000 nm). However, within these EV categories, there is considerable heterogeneity. Thus, although all exosomes, now referred to as small EVs (sEVs), originate in multivesicular bodies (MVBs) and thus share the endocytic origin [3], they are divided into tiny exomeres (<35 nm), small exosomes (Exo-S, 50–100 nm) and large exosomes (Exo-L, 100–150 nm) [2]. In our studies of EVs, we consider sEVs derived from MVBs and sized from 50 to 150 nm as exosomes [4] and intermittently refer to them as either exosomes or sEVs. Unlike exosomes, MVs bud off from the surface of parent cells, differ broadly in size and molecular content from exosomes and are called ectosomes or, if they carry oncogenes, oncosomes [2].

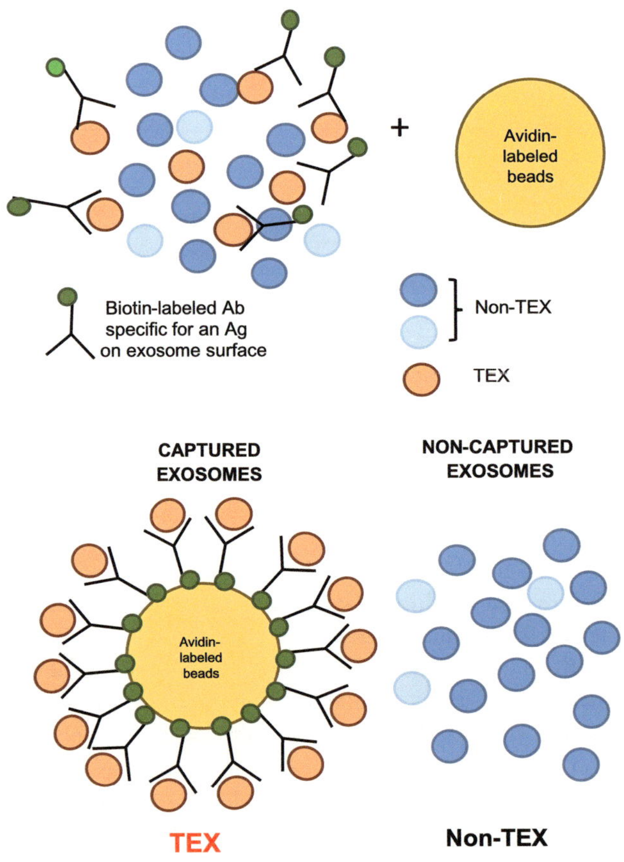

Figure 1.
Schematic view of an immune capture of tumor-derived exosomes (TEX) using an antibody (capture Ab) specific for an antigen epitope present on the surface of TEX. The capture Ab is labeled with biotin and is co-incubated with exosomes isolated by size exclusion chromatography (SEC) from plasma. These exosomes contain TEX and Non-TEX, which are produced by various non-malignant cells. TEX carry the target Ag and are captured by the Ab. Non-TEX do not carry the target Ag and are not recognized by the capture Ab. Streptavidin-labeled magnetic beads are added to the mix of exosomes and biotinylated Ab. TEX are captured on streptavidin beads and recovered using a magnet. Non-TEX are not captured and remain in solution. Captured TEX are then evaluated for their molecular content by on-bead flow cytometry.

During exosome biogenesis, when MVBs filled with intraluminal vesicle fuse with the cellular plasma membrane, exosomes are released into the extracellular space. Due to their endosomal origin, exosomes carry endocytic markers, such as TSG101, ALIX, syntenin-1, flotillin and others but do not contain cytoplasmic proteins, such as calnexin or GRPp94. Importantly, the topography of exosome molecular surface as well as molecular and genetic contents of exosomes resemble those of their parent cells [5]. This similarity of molecular and genetic signatures of tumor-derived exosomes (TEX) to parent tumor cells is the main reason for considering TEX as a "liquid tumor biopsy" [6]. Tumor cells produce large numbers of exosomes, ranging, from 10^{10} to 10^{12}/mL plasma, and plasma of cancer patients is thus variably but significantly enriched in TEX [7]. However, not only tumor cells but also non-malignant cells in the TME, e.g., immune cells, endothelial cells or cancer-associated fibroblasts (CAFs), produce exosomes, which account for a considerable fraction of total EVs present in cancer patients' plasma [8, 9]. Thus, cancer plasma contains a diverse mix of small and large EVs of various cellular origins and presumably with different molecular content in the vesicle lumen and on the vesicle surface membrane. TEX represent a fraction of all circulating EVs that differs broadly among cancer patients; in patients with malignant melanoma, TEX constitute 20–80% of total plasma EVs (our data), and the ratio of TEX/non-TEX increases with melanoma progression **Figure 1** [7].

3. Up-take of EVs and reprogramming of recipient cells

Once exosomes are released from MVBs into the extracellular space, they are disseminated throughout all tissues, enter the circulation and freely cross the blood brain barrier (BBB) and all tissue barriers [10]. Exosomes are taken up and internalized by recipient cells in the circulation and tissues by a variety of mechanisms ranging from endocytosis, phagocytosis or membrane fusion to receptor-ligand mediated entry and signaling as reviewed elsewhere [11] and deliver their cargos to recipient cells, which may be either near or distantly located. Through this mechanism, "*exosome-releasing cells*" can profoundly alter the phenotype and function of "*exosome-receiving cells*". For example, immune cells found in patients with cancer (tumor-infiltrating as well as circulating cells) become polarized or reprogrammed through interactions with TEX, so that from effector cells, they turn into cells facilitating tumor growth [12]. Further, these reprogrammed immune cells now release exosomes that are equipped with immunosuppressive molecules and effectively suppress tumor antigen-specific immune responses. This process of the immune cell "corruption" or "subversion" by TEX is orchestrated by the tumor which utilizes TEX to initiate a "cascade" of secondary EVs, thereby changing the TME into one promoting tumor growth and suppressing anti-tumor functions of immune cells [13].

Body fluids are a diverse mixture of various EV subsets, and in cancer patients, TEX constitute a substantial and functionally important EV subset that is engaged in immune regulation. Mechanistically, TEX-mediated immune suppression involves activation in recipient immune cells of numerous inhibitory pathways, leading to a loss of anti-tumor functions [9, 13]. The result of TEX-driven reprogramming of immune cells is that not only TEX but also immune cell-derived exosomes in the plasma of cancer patients are enriched in immunosuppressive proteins and upon co-incubation with primary normal immune cells or upon injection into experimental animals, these exosomes mediate immune suppression [14]. Although, TEX carry tumor-associated antigens (TAAs) and thus could be immunogenic, TEX interactions with reprogrammed antigen-presenting cells (APCs) in the TME do not support antigen processing/presentation which normally culminates in T cell responses [15]. Instead, T cells cross talking with TEX are suppressed or induced to acquire a suppressive phenotype (i.e., develop into Treg or myeloid-derived suppressor cells). Suppressive activities of TEX appear to be the major mechanism underlying negative regulation that prevails in the TME.

4. Isolation of exosomes from plasma

An existing barrier that has impeded the progress in EV research has been the lack of methods for their isolation from body fluids in a relatively "pure" form, i.e., without non-specifically attached plasma proteins, and in quantities sufficient for further studies. The current "gold standard" for the isolation of EVs has been the density gradient ultracentrifugation of pre-cleared plasma at 100,000x g for periods of time ranging from 12-24h [16]. Ultracentrifugation using iodixanol density gradients (24 spin time) is currently the preferred isolation method. However, for many reasons, including an inadequate recovery, vesicle aggregation and potential vesicle damage during prolonged ultracentrifugation as well as the isolation platform that does not lend itself to a high throughput required for clinical assessments, ultracentrifugation is being slowly replaced by other methods. The literature is replete in listing various technological advances for EV isolation from body fluids, including microfluidics and sophisticated ultrafiltration systems [17, 18]. Many factors need to be taken into account when selecting an EV isolation procedure, such as the volume of available fluid, desired recovery and purity of EVs and processing time. Among these various methods, size exclusion chromatography (SEC) emerges as the most efficient technique for the isolation of "purified" exosomes or sEVs from plasma [4, 19]. SEC is a readily applicable separation method based on differences in protein size, which removes unwanted soluble proteins from precleared plasma and allows for the recovery of partially "cleaned" exosomes in early fractions [4, 19]. Data from various studies indicate that upon sEV isolation by SEC, glycoprotein aggregates, albumin and other plasma proteins elute in the late fractions, while partly purified, tetraspanin-positive vesicles elute in the early fractions, allowing for a relatively simple, one-step separation of exosomes from most of "contaminating" plasma components. SEC outperforms various precipitation protocols which co-isolate contaminating plasma proteins [19]. It has been suggested that SEC has drawbacks, including relatively poor yield; however, as this results from removal of protein aggregates not a loss of vesicles, the lower yield is counterbalanced by increased sample purity. With relatively minor adjustments, SEC can be used for high throughput isolation of sEVs from serially collected body fluids, yielding partially purified sEVs in early fractions [4]. These exosomes retain their vesicular morphology and phenotypic as well as functional attributes, such as the ability to modify responses of recipient cells to exogenous signaling [4].

The use of SEC has facilitated the: (i) "cleaning" of exosomes from most, although not all, "contaminating' plasma proteins; (ii) separation of soluble Ags that might be weakly "associated" with exosomes from those embedded in or carried on the exosome membrane and (iii) recovery of morphologically intact, non-aggregated exosomes that retain their functional activity [4]. The isolation of non-aggregated vesicles is especially critical for the subsequent immunoaffinity capture of vesicles, because vesicle aggregation is likely to interfere with Ab-based capture. Equally important is the fact that the recovered vesicles retain their functional activity, e.g., are able to induce apoptosis following a brief co-incubation with activated T cells, after removal of soluble plasma proteins. It is for these reasons, that immune capture of EVs from body fluids should be preceded by SEC and not be used for direct EV isolation from plasma.

5. Rationale for sEV fractionation into TEX and non-TEX

In cancer patients, total exosomes isolated from plasma by SEC contain various proportions of TEX. In patients with melanoma, 20–70% of total plasma exosomes are tumor cell-derived [7]. While total plasma exosomes with a high content of TEX might largely reflect the TEX characteristics, non-TEX present in the mix might influence the estimates of effects plasma exosomes to exert in recipient cells. Thus, the separation of TEX from a mix of other vesicles in plasma is a necessary step to evaluate their unique phenotypic, molecular and functional characteristics. This step may be especially important when TEX account for only a small fraction of total sEVs in plasma. Immunoaffinity capture of TEX from plasma has been introduced as an approach to the pulldown of TEX based on the use of Abs specific for the antigens selectively expressed or markedly overexpressed by cancer cells and carried by TEX [20]. Immunocapture-based exosome isolation from body fluids has been extensively used in diseases other than cancer, including neurological diseases, where Ab-based capture is broadly used for the isolation of neuron-derived L1CAM bearing EVs (NDEVs) [21]. In cancer, TEX separated from non-TEX are expected to serve as a liquid tumor biopsy that faithfully recapitulates molecular and genetic features of parental cancer cells.

6. Immunocapture of TEX from body fluids

In principle, EVs which resemble parent tumor cells and carry on their surface the antigens expressed by tumor cells, should be readily recognized and captured by Abs specific for these tumor-associated antigens (TAAs). There are two major components to this approach that are critical for immunocapture success: one concerns the general strategy used for capture and the other one is the selection of capture Abs.

6.1 Immunobead-based EV capture

The immunocapture-based methods generally use beads coated with selected Abs for EV pulldown. In the simplest approach, beads coated with Abs are added directly to plasma diluted in phosphate-buffered saline (PBS) with the expectation that all EVs bearing the target Ag on the surface will bind to the Ab coated beads. This strategy for capture may not be and usually is not very effective, because EVs in plasma carry a variety of soluble plasma proteins on the surface, such as albumin, immunoglobulins (Igs) and other "contaminating" plasma proteins. These plasma proteins form a protein "corona" associated with the sEV surface membrane which is likely to block the access of capture Abs to targeted Ags, leading to an incomplete pulldown or even lack of pulldown. Also, if the target Ag is present in soluble form in plasma, it might compete with the counterpart carried on the EV membrane, binding to the Ab coated beads either specifically as a soluble protein or as a "contaminating" EV surface-associated protein. The soluble target Ag present in plasma (especially when its abundance is high) could non-specifically associate with proteins/glycoproteins decorating EV surfaces. As a result, the target Ag could mediate the pulldown of EVs that do not constitutively express the Ag. The result will be a pulldown of EVs carrying a soluble Ag in addition to EVs genuinely endowed with the target Ag embedded in the EV membrane. Such capture will not distinguish between these two types of EVs, and thus the strategy is useless for selective capture of EVs carrying the targeted tumor-specific Ag. This capture strategy has been also used with EVs isolated from plasma by ExoQuick, which concentrates rather than "purifies" EVs, with the same unsatisfactory results [21].

Perhaps a good example of this strategy is immune capture from cancer patients' plasma of EVs carrying PD-L1 as recently reported [22]. This protein, commonly carried by TEX in most cancers, is also present in the plasma of cancer patients both as a soluble protein derived from malignant and various non-malignant cells and as the integral membrane protein of non-TEX released by macrophages or other immune cells [23]. Therefore, beads coated with anti-PD-L1 Abs cannot be used for selective capture of PD-L1-positive TEX, because such beads will capture soluble PD-L1, non-TEX carrying PD-L1 as well as TEX carrying PD-L1, thus making it impossible to distinguish which EV subset delivers inhibitory signals to PD-1-positive recipient cells. The contribution of soluble PD-L1 to negative signaling by the captured vesicles may not be disregarded, because in addition to its specific binding to Ab-coated beads, soluble PD-L1 might non-specifically "associate" with all EVs in plasma, similar to albumin or other plasma proteins. Data in the literature [24] and the protein content of the ExoCarta data base [25] confirm that EVs isolated from plasma carry numerous non-specific plasma-derived proteins and suggest that the discrimination of "true" EV proteins from plasma "contaminants" is a major challenge in the field.

6.2 Selection of abs for immune capture of TEX

The selection of Abs for TEX immune capture depends on the convincingly demonstrated ability of such Abs to selectively bind to tumor cells expressing the target Ag on the cell surface, with the exclusion of any binding to non-malignant cells which do not express the Ag. This is a rigorous requirement and one that may be difficult to implement, because few tumor-specific antigens are known, except for mutated epitopes in cancer cells. To emphasize, the capture Abs selected for TEX capture must be specific for an antigen (or an epitope) present only in parent cells and in EVs these parent cells produce but not in any other cells or tissues. Even if such tumor Ag- specific Ab is available, it is necessary to ensure that its binding affinity for the target Ag is high and that the target antigen is not present in soluble forms in body fluids. Low-affinity Abs will not be effective in pulldowns, and the presence of a target Ag in soluble forms might not only interfere with Ab binding to TEX but upon its non-specific association with EVs, as described above, will lead to the capture of EVs derived from non-malignant cells, thus interfering with selective TEX capture. As most Ags expressed on the surface of cancer cells are enzymatically cleaved and are present in plasma, this requirement may not be readily addressed. Importantly, adding Ab coated beads directly to plasma without prior attention to these restrictions will jeopardize the selectivity and efficiency of TEX immune capture.

Clearly, the use of the best capture Abs is by far the most critical aspect of TEX immune capture from plasma. In the absence of such Abs, immune capture using a mix of Abs specific for Ags highly overexpressed on cancer cells relative to non-malignant cells and on the EVs these cells produce could be utilized for immune capture, and this approach has been successful [26]. It is possible to perform immune capture of TEX with a cocktail of Abs carefully selected for specificity to proteins overexpressed on tumor cells and weakly expressed on non-malignant cells. The complexity of immune capture increases with the use of Ab cocktails largely due to `Ab titration requirements and the need for extensive controls as well as the limitations imposed by the presence in plasma of the Ags recognized by the Abs used for immune capture, in soluble form.

7. A successful TEX capture from plasma of patients with melanoma

Compelling evidence indicates the immune capture method to isolate TEX from patients' plasma yields excellent results when essential requirements are met. For example, we have reported separation of TEX from non-TEX in plasma of patients with melanoma [20] that have allowed for extensive characterization of the molecular cargo and functional repertoire of these sEV fractions [27]. The immune capture of TEX was performed using chondroitin sulfate peptidoglycan 4 (CSPG4)-specific mAbs developed by one of us [28]. These mAbs recognize CSPG4 which is selectively expressed on melanoma cells (and on the EVs these cells produce) but is not detectable on any other non-malignant cells in the body except for activated pericytes in the TME [29, 30]. Immunohistochemical staining with mAbs of more than 2000 melanoma lesions has showed that CSPG4 is expressed on about 80% of all investigated melanoma specimens [29, 30]. In melanoma tissues, CSPG4-specific mAbs decorate the surface of malignant cells; flow cytometric analysis of EVs stained with mAbs visualizes CSPG4 on their membrane [31]. Monoclonal Abs recognizing distinct CSPG4 epitopes are available to be selectively used for the capture of TEX, which is isolated and "purified" by SEC, and for subsequent antigen detection by flow cytometry confirmed that CSPG4 is expressed on TEX but is not detectable on non-TEX [7]. As with all immune capture experiments, titrations of the capture and detection Abs are critical for success as are the vesicle/Ab ratios, and these must be determined a priori and strictly adhered to during capture. As described in detail elsewhere [7, 20], all immunocaptured melanoma TEX are positive for CSPG4 and for melanoma-associated antigens (MAA), while non-TEX are negative. The exosome recovery ranged from 60 to 100 μg protein/mL plasma, and the ratio of TEX/total exosomes in plasma varied among melanoma patients from 0.2 to 0.6. The separation by immune capture of melanoma TEX from non-TEX yielded sufficient numbers of both exosome fractions for studies of their protein content by on-bead flow cytometry and by high-resolution mass spectrometry (HRMS) as well as of their function in co-incubation assays with immune cell subsets [7]. TEX were enriched in immunosuppressive and non-TEX in immunostimulatory proteins, and co-incubation of the fractionated exosomes with immune cells confirmed their distinct immunoregulatory functions [7]. Melanoma TEX carrying CD39, CD73, FasL, PD-L1, TGF-β and TRAIL, among other suppressive proteins consistently inhibited functions of immune cells, while non-TEX were stimulatory in co-incubation assays. Using LC–MS/MS-based proteomics, we identified a profile of 16 proteins highly overexpressed in TEX which discriminated TEX from non-TEX. These proteins were components of molecular pathways mediating cellular events such as vesicle transport, immune reactivity, signal transduction, and disease activity [27]. Further, by dividing the analyzed melanoma patients into two groups of 7 patients with no evident disease (NED) and 8 with progressive disease (PD) at the time of phlebotomy for exosome isolation from plasma, we were able to identify a signature of 12 proteins significantly and consistently overexpressed in TEX of patients with PD relative to TEX of patients with NED [27]. This ability of TEX bearing the signature of 5/12 most significantly ($p < 0.0003$) overexpressed proteins to discriminate melanoma patients with PD from those with NED within a very small patient cohort emphasizes the potential of TEX to serve as a biomarker of disease activity in melanoma [27]. In addition, this immunocapture-based study was the first to show that melanoma TEX, which are especially abundant in plasma of patients with advanced disease, are largely responsible for immune suppression that potentially promotes immune escape and tumor progression.

8. Immune capture of T cell-derived sEV from plasma

The T cell receptor (TCR) is expressed only on T lymphocytes, and an Ab specific for CD3, a protein component of the TCR complex, decorates the surface of sEV produced exclusively by T cells. A high-affinity Ab specific for CD3 proved to be an excellent candidate for immunocapture and subsequent characterization of sEV in plasma of patients with cancer or HDs [32]. T cell-derived sEV account for a considerable proportion of total plasma exosomes in cancer patients, and their phenotypic profiles examined by on-bead flow cytometry recapitulate those expressed by various T cell subsets as we reported [32]. CD3-based immune capture allowed for isolation from plasma of CD3(−) sEV fraction enriched in TEX and CD3(+) sEV fraction that was useful for evaluations of phenotypic and functional changes induced in the cancer-reprogrammed T cells [33]. Thus, this type of immune capture allowed for simultaneous analysis of molecular profiles in tumor-derived and immune cell-derived sEVs in the same plasma sample. The procedure for immune capture of CD3+ EVs from the plasma of cancer patients is described in Current Protocols in Immunology [34].

9. Conclusions

TEX are rapidly emerging as the major component of immunoinhibitory signaling that prevails in the TME. TEX present in body fluids a subset of circulating EVs. The large quantity and enormous diversity of circulating EVs in plasma of patients with cancer with respect to cellular origin, molecular characteristics, genetic content and functions imposes a need for the isolation of TEX and their separation from non-malignant vesicles. This approach allows for studies of impact TEX exert on cells in the TME and of TEX value as potential cancer biomarkers. To dissect the EV diversity in body fluids, strategies are necessary for their capture, isolation from body fluids and separation of various EV subsets without interference with EV molecular identities and functions. Among various isolation strategies, immune capture with Abs specific for proteins carried on the EV surface has been most frequently utilized with a variable level of success. While in principle, immune capture is the rational strategy for EV pulldown from plasma, its application to nanovesicles requires an understanding of EV characteristics and EV biology. As the latter is still largely lacking, all EV immune capture strategies may backfire, for reasons that may not be anticipated, such as high levels of a soluble target Ag in plasma or the presence on vesicles of a "corona" of contaminating plasma proteins. The set of stringent requirements for vesicle immune capture from plasma that we have discussed allows for overcoming some, but probably not all, of the barriers we might face in the future while attempting to study the diversity EVs in body fluids. As this diversity is of key importance in understanding the role EV subsets such as, e.g., TEX, play in health and disease, immunoaffinity EV capture is likely to remain the method of choice for selective TEX pulldown. However, only when performed correctly, immunoaffinity capture of TEX yields valuable insights into their potential as cancer biomarkers and as markers of immune competence.

Author details

Theresa L. Whiteside[1,2*] and Soldano Ferrone[3,4]

1 Department of Pathology, University of Pittsburgh School of Medicine and UPMC Hillman Cancer Center, Pittsburgh, PA, USA

2 Department of Immunology, University of Pittsburgh School of Medicine and UPMC Hillman Cancer Center, Pittsburgh, PA, USA

3 Department of Surgery, Massachusetts General Hospital, Harvard Medical School, Boston, MA, USA

4 Department of Orthopedic Surgery, Massachusetts General Hospital, Harvard Medical School, Boston, MA, USA

*Address all correspondence to: whitesidetl@upmc.edu

References

[1] Wortzel I, Dror S, Kenific CM, Lyden D. Exosome-mediated metastasis: Communication from a distance. Developmental Cell. 2019;**49**(3):347-360. DOI: 10.1016/j.devcel.2019.04.011

[2] Zijlstra A, Di Vizio D. Size matters in nanoscale communication. Nature Cell Biology. 2018;**20**(3):228-230. DOI: 10.1038/s41556-018-0049-8

[3] Mathieu M, Martin-Jaular L, Lavieu G, Thery C. Specificities of secretion and uptake of exosomes and other extracellular vesicles for cell-to-cell communication. Nature Cell Biology. 2019;**21**(1):9-17. DOI: 10.1038/s41556-018-0250-9

[4] Hong CS, Funk S, Muller L, Boyiadzis M, Whiteside TL. Isolation of biologically active and morphologically intact exosomes from plasma of patients with cancer. Journal of Extracellular Vesicles. 2016;**5**:29289. DOI: 10.3402/jev.v5.29289

[5] Czystowska-Kuzmicz M, Whiteside TL. The potential role of tumor-derived exosomes in diagnosis, prognosis, and response to therapy in cancer. Expert Opinion on Biological Therapy. 2021;**21**(2):241-258. DOI: 10.1080/14712598.2020.1813276

[6] Whiteside TL. The potential of tumor-derived exosomes for noninvasive cancer monitoring. Expert Review of Molecular Diagnostics. 2015;**15**(10): 1293-1310. DOI: 10.1586/14737159.2015.1071666

[7] Sharma P, Diergaarde B, Ferrone S, Kirkwood JM, Whiteside TL. Melanoma cell-derived exosomes in plasma of melanoma patients suppress functions of immune effector cells. Scientific Reports. 2020;**10**(1):92. DOI: 10.1038/s41598-019-56542-4

[8] Whiteside TL. Exosomes and tumor-mediated immune suppression. The Journal of Clinical Investigation. 2016;**126**(4):1216-1223. DOI: 10.1172/JCI81136

[9] Whiteside TL. The effect of tumor-derived exosomes on immune regulation and cancer immunotherapy. Future Oncology. 2017;**13**(28):2583-2592. DOI: 10.2217/fon-2017-0343

[10] Banks WA, Sharma P, Bullock KM, Hansen KM, Ludwig N, Whiteside TL. Transport of extracellular vesicles across the blood-brain barrier: Brain pharmacokinetics and effects of inflammation. International Journal of Molecular Sciences. 2020;**21**(12). DOI: 10.3390/ijms21124407

[11] Mulcahy LA, Pink RC, Carter DR. Routes and mechanisms of extracellular vesicle uptake. Journal of Extracellular Vesicles. 2014;**3**. DOI: 10.3402/jev.v3.24641

[12] Azambuja JH, Ludwig N, Yerneni S, Rao A, Braganhol E, Whiteside TL. Molecular profiles and immunomodulatory activities of glioblastoma-derived exosomes. Neuro-Oncology Advances. 2020;**2**(1):vdaa056. DOI: 10.1093/noajnl/vdaa056

[13] Whiteside TL. Exosome and mesenchymal stem cell cross-talk in the tumor microenvironment. Seminars in Immunology. 2018;**35**:69-79. DOI: 10.1016/j.smim.2017.12.003

[14] Razzo BM, Ludwig N, Hong CS, et al. Tumor-derived exosomes promote carcinogenesis of murine oral squamous

cell carcinoma. Carcinogenesis. 2020;**41**(5):625-633. DOI: 10.1093/carcin/bgz124

[15] Ludwig S, Floros T, Theodoraki MN, et al. Suppression of lymphocyte functions by plasma exosomes correlates with disease activity in patients with head and neck cancer. Clinical Cancer Research. 2017;**23**(16):4843-4854. DOI: 10.1158/1078-0432.CCR-16-2819

[16] Thery C, Amigorena S, Raposo G, Clayton A. Isolation and characterization of exosomes from cell culture supernatants and biological fluids. Current Protocols in Cell Biology. Chapter 3:Unit 3 22. DOI: 10.1002/0471143030.cb0322s30

[17] Chen Y, Zhu Q, Cheng L, et al. Exosome detection via the ultrafast-isolation system: EXODUS. Nature Methods. 2021;**18**(2):212-218. DOI: 10.1038/s41592-020-01034-x

[18] Royo F, Thery C, Falcon-Perez JM, Nieuwland R, Witwer KW. Methods for separation and characterization of extracellular vesicles: Results of a worldwide survey performed by the ISEV Rigor and standardization subcommittee. Cell. 2020;**9**(9). DOI: 10.3390/cells9091955

[19] Lobb RJ, Becker M, Wen SW, et al. Optimized exosome isolation protocol for cell culture supernatant and human plasma. Journal of Extracellular Vesicles. 2015;**4**:27031. DOI: 10.3402/jev.v4.27031

[20] Sharma P, Ludwig S, Muller L, et al. Immunoaffinity-based isolation of melanoma cell-derived exosomes from plasma of patients with melanoma. Journal of Extracellular Vesicles. 2018;**7**(1):1435138. DOI: 10.1080/20013078.2018.1435138

[21] Norman M, Ter-Ovanesyan D, Trieu W, et al. L1CAM is not associated with extracellular vesicles in human cerebrospinal fluid or plasma. Nature Methods. 2021;**18**(6):631-634. DOI: 10.1038/s41592-021-01174-8

[22] Chen G, Huang AC, Zhang W, et al. Exosomal PD-L1 contributes to immunosuppression and is associated with anti-PD-1 response. Nature. 2018;**560**(7718):382-386. DOI: 10.1038/s41586-018-0392-8

[23] Peng Q, Qiu X, Zhang Z, et al. PD-L1 on dendritic cells attenuates T cell activation and regulates response to immune checkpoint blockade. Nature Communications. 2020;**11**(1):4835. DOI: 10.1038/s41467-020-18570-x

[24] Abramowicz A, Marczak L, Wojakowska A, et al. Harmonization of exosome isolation from culture supernatants for optimized proteomics analysis. PLoS One. 2018;**13**(10):e0205496. DOI: 10.1371/journal.pone.0205496

[25] Keerthikumar S, Chisanga D, Ariyaratne D, et al. ExoCarta: A web-based Compendium of Exosomal Cargo. Journal of Molecular Biology. 2016;**428**(4):688-692. DOI: 10.1016/j.jmb.2015.09.019

[26] Whiteside TL, Diergaarde B, Hong CS. Tumor-Derived Exosomes (TEX) and their role in immuno-oncology. International Journal of Molecular Sciences. 2021;**22**(12). DOI: 10.3390/ijms22126234

[27] Pietrowska M, Zebrowska A, Gawin M, et al. Proteomic profile of melanoma cell-derived small extracellular vesicles in patients' plasma: A potential correlate of melanoma progression. Journal of Extracellular Vesicles. 2021;**10**(4):e12063. DOI: 10.1002/jev2.12063

[28] Wang X, Katayama A, Wang Y, et al. Functional characterization of an scFv-Fc antibody that immunotherapeutically targets the common cancer cell surface proteoglycan CSPG4. Cancer Research. 2011;**71**(24):7410-7422. DOI: 10.1158/0008-5472.CAN-10-1134

[29] Campoli M, Ferrone S, Wang X. Functional and clinical relevance of chondroitin sulfate proteoglycan 4. Advances in Cancer Research. 2010;**109**:73-121. DOI: 10.1016/B978-0-12-380890-5.00003-X

[30] Wang Y, Sabbatino F, Wang X, Ferrone S. Detection of chondroitin sulfate proteoglycan 4 (CSPG4) in melanoma. Methods in Molecular Biology. 2014;**1102**:523-535. DOI: 10.1007/978-1-62703-727-3_28

[31] Ferrone S, Whiteside TL. Targeting CSPG4 for isolation of melanoma cell-derived exosomes from body fluids. HNO. 2020;**68**(2):100-105. CSPG4 im Fokus zur Isolierung von aus Melanomzellen stammenden Exosomen aus Korperflussigkeiten. DOI: 10.1007/s00106-019-00811-1

[32] Theodoraki MN, Hoffmann TK, Whiteside TL. Separation of plasma-derived exosomes into CD3((+)) and CD3((−)) fractions allows for association of immune cell and tumour cell markers with disease activity in HNSCC patients. Clinical and Experimental Immunology. 2018;**192**(3):271-283. DOI: 10.1111/cei.13113

[33] Theodoraki MN, Matsumoto A, Beccard I, Hoffmann TK, Whiteside TL. CD44v3 protein-carrying tumor-derived exosomes in HNSCC patients' plasma as potential noninvasive biomarkers of disease activity. OncoImmunology. 2020;**9**(1):1747732. DOI: 10.1080/2162402X.2020.1747732

[34] Ludwig N, Hong CS, Ludwig S, et al. Isolation and analysis of tumor-derived exosomes. Current Protocols in Immunology. 2019;**127**(1):e91. DOI: 10.1002/cpim.91

Chapter 12

Extracellular Vesicles as Biomarkers and Therapeutic Targets in Cancers

Prince Amoah Barnie, Justice Afrifa, Eric Ofori Gyamerah and Benjamin Amoani

Abstract

Extracellular vesicles refer to exosomes, apoptotic bodies, microvesicles and large oncosomes, which are membrane bound structures secreted by cells including cancer cells. The pathological role and translational potential of extracellular vesicles (EVs) in cancers are receiving research attention recently. The cargoes of cancer-derived EVs retain the molecular properties of their sources and cancer cells actively release EVs into body fluids that are easy to access. EVs released from cancer cells not only promote cancer progression through the delivery of cancer-associated molecules but also reflect alterations in the state of cancers during therapy. They are considered promising biomarkers for therapeutic response evaluation, especially resistance to therapy and diagnostics. This chapter discusses the various roles of extracellular vesicles in cancers and their potential as therapeutic targets.

Keywords: extracellular vesicles, biomarkers, cancers, therapeutics

1. Introduction

Cancer is described as one of the challenging diseases globally, which accounts for 19 million newly diagnosed cases and over 10 million deaths annually making it the leading cause of death [1]. The burden of cancer incidence and mortality is rapidly growing worldwide [1]. Cancer development in humans is a multistep process, which involves various genetic or epigenetic changes and results in the malignant transformation of the normal cells [2]. Recently the use of proteomics, genomics and bioinformatic techniques has unraveled the intricate interplay of numerous cellular genes and regulatory genetic elements that account for the cancerous phenotypes. Higher mortalities associated with cancers are as a result of the absence of very reliable cancer biomarkers, which could be used to diagnose early cancers, predict prognostics and treatment response as well as detection of biomarkers for drug resistance [3]. The unavailability of good biomarkers is a major hindrance for cancer treatment. Cancer biomarkers are not only important for diagnostic purposes but can also be of great prognostic value. With the identification of the right biomarker the cancer progression and effect of chemotherapeutic drugs can be evaluated in great detail [4]. Again, the presence of resistance to therapy, disease relapse, and individual differences continue to reduce the survival chances of cancer patients and makes the disease impossible to cure [5]. It is predicted that therapeutic response assessment, especially treatment response prediction, is valuable to guide treatment strategy determinations and provide responsive therapy for better survival [6]. The identification of reliable cancer biomarkers in the management of cancers may play a crucial role in reducing cancer-related mortality.

Cancer biomarkers are biological molecules that suggest the presence of cancer in a patient. They are either produced by the cancer cells or by other non-cancer cells in response to cancer [7]. Cancer biomarkers may be used to identify the presence of cancer and also help determine its stage, subtype, and whether they will respond to therapy [8]. Cancer biomarkers identified from serum are the most desirable form of the biomarkers that can be used for regular personalized screening, diagnosis, establishing prognosis, monitoring treatment, and detecting relapse. Cancer biomarkers can be classified into three main categories: prognostic biomarkers, which allow prediction of the disease course and survival chances; predictive biomarkers; to assess if a patient benefits from a certain treatment; and pharmacodynamic biomarkers, which are used in the clinics to guide personalized drug dosing and response assessment. In recent years, a group of biological molecules receiving research attention due to their potential utility as circulating biomarkers for cancer are the extracellular vesicles.

Extracellular vesicles (EVs) are small, lipid-bound particles containing nucleic acid and protein cargo which are excreted from cells under a variety of normal and pathological conditions [9]. Recent studies indicated that cancer-associated EVs play pivotal roles in constructing favorable microenvironments for cancer cells. They are therefore considered as new and promising biomarkers for many cancer types. EVs secreted from a variety of cancer types, including pancreatic cancer, ovarian cancer, prostate cancer, breast cancer, colorectal cancer, glioblastoma multiforme (GBM) are reported to contain cancer-associated protein markers [10]. The EVs play important roles in the regulation of intercellular communication and cell microenvironment homeostasis and again as important biomarkers of various cancers. As EVs are increasingly revealed to play important roles in cancer development and to carry specific information related to cancer state. In cancer research, growing evidence indicates that EVs possess the ability to promote tumor growth, metastasis, and angiogenesis [11] mediate tumor immune responses [12]; and stimulate chemotherapeutic resistance, Identification and modification of cancer cell-derived extracellular vesicles may allow for the development of novel diagnostic, preventive and therapeutic approaches in cancers. This chapter summarizes the functions of EV's in cancers, their potential as biomarkers and therapeutic targets. It further emphasizes the roles of EV's in cancer prognosis, treatment response and drug resistance.

2. Overview and biogenesis of extracellular vesicles

As membranous vesicles, many cell types in the human body release EVs and cancer cells actively secrete EVs even during the early phase of the disease. Another interesting characteristic of EVs is that its contents are protected from degradative enzymes in body fluids [13]. Based on various characteristics, ranging from size, biogenesis, content, cell of origin, morphology, EV are categorized into four main classes: endosomal-derived small exosomes (Exo) (30–150 nm), plasma membrane-derived middle-sized microvesicles (MV) (100–1000 nm), and large oncosomes (LO) (1000–10,000 nm), as well as apoptotic bodies (500–4000 nm) that are released from dying cells [14].

2.1 Exosomes

Typically, exosomes are about 30–150 nm in diameter and are generated via an endosomal route [15–17]. Exosomes are generated through the endosomal network. This is a compartment, which is membranous in nature and aids in the sorting and direction of intraluminal vesicles such as cell surface membranes and lysosomes to their specific destinations. It is known that exosomal vesicles are formed during an inward budding of early endosomal limiting membrane, which develops into multivesicular bodies in the process [17–19]. When late endosomal membranes invaginate, intraluminal vesicles (ILVs) are formed within the larger multivesicular bodies [20]. It is during this process that the molecules carried by exosomes including proteins, lipids and nucleic acids are incorporated into the invaginating membrane whiles the components of the cytosol are engulfed by the ILVs [21]. They are liberated into the surrounding body fluids when the multivesicular bodies fuse with plasma membrane. The general function of these early endosomes and multivesicular bodies are endocytic and transportation of the cell's material. These include storage, recycling, transport, protein sorting and release of these materials [16].

2.2 Apoptotic cell-derived extra cellular vesicles

Apoptotic cell-derived extracellular vesicles (ApoEVs) are subcellular and membrane bound in nature. They are produced when cells are undergoing senescence. Further they can be derived from various cell types including endothelial cells, osteoblasts, precursor cells, stem cells and immunocytes [22]. Basically, three major steps are involved in ApoEVs formation. Firstly, there is a prerequisite step which involves cell surface membrane blebbing [23], which is then followed by projections of apoptotic membrane such as apoptopodia, beaded apoptopodia and microtubule spikes which releases 10–20 ApoEVs [24] and lastly the final formation of ApoEVs. Several factors have been shown to present the regulatory function on the generation of ApoEVs, these includes Rho-associated kinase (ROCK1) [25, 26] and Myosin-Light Chain Kinase (MLCK) [27]. Specifically, MLCK is known to enhance nuclear material packaging into ApoEVs, thus molecules that could inhibit caspases, MLCK and ROCK1 are also able to downregulate the production of ApoEVs [28]. Orlando et al., report that formation of blebs which is the first stage in ApoEVs formation are mediated by the presence of actomyosin which increases cell contraction leading to elevated hydrostatic pressure [29]. Researchers have unraveled that ApoEVs are key messengers released by dying cells to regulate processes including cell clearance, tissue homeostasis, pathogen dissemination and immunity thereby implicating them as therapeutic targets and diagnostic purposes.

2.3 Microvessicles

Microvascular vesicles are derived from myriad cell types surfaces [19]. Unlike ApoEVs, which are generated via indiscriminate surface blebbing or exosomes, which are derived intracellularly within MVBs, microvesicles are formed through active interaction between cytoskeletal protein contraction and the redistribution of phospholipids. Aminophospholipid translocases closely regulate an uneven distribution of the phospholipids in the plasma membrane leading to the formation of micro-domains [30–32]. Specifically, the plasma membrane budding process is induced by translocation of phosphatidylserine to the outer-membrane leaflet [33, 34]. The process is completed via actin–myosin interactions which cause cytoskeletal structures to contract. This ensures the release of nascent microvesicles into the extracellular space via the direct outward blebbing and breaking off of the plasma membrane [35, 36]. After blebbing, there is a distinct localization of plasma membrane lipids and proteins which informs the rigidity and curvature of the membrane [37, 38]. In addition to the redistribution of membrane lipids and proteins, there is a selective redistribution of the components of microvesicles' cargo for specific microvesicles enrichment [39]. MVs carry proteins, such as enzymes, growth factors, growth factor receptors, cytokines and chemokines. They also carry lipids, and nucleic acids, including mRNA, miRNA, ncRNA, and genomic DNA [40]. MVs have been detected in the circulation of patients with several cancers, such as lung, breast, prostate ovarian, gastric cancer and colorectal cancer [41, 42]. They have been identified to contribute to tumorigenesis, progression of cancer cells, evasion of apoptosis by tumor cells, and induction of angiogenesis. The tumor-promoting role of MV in tumor mediated exosome communication largely depends on their bioactive cargo. It is believed that the shuttling of tumor-specific proteins to the surrounding cells influence tumor growth. This is achieved through the transfer of oncogenic traits between tumor cells, which result in enhanced tumor growth, and progression [43]. MVs are recently receiving research attention as potential biomarkers because tumor cells are able to constitutively release large amounts of MVs bearing tumor-specific antigens into the bloodstream and other bodily fluids [35]. Researchers have proposed many uses of MV in cancers. Others believe that MVs can be useful for disease staging as well as evaluate the response to therapy by permitting an accurate assessment of a patient's responsiveness and personalization of treatment [44].

2.4 Large oncosomes

Large oncosomes (LO) are atypically large (1–10 μm diameter) cancer-derived extracellular vesicles (EVs), originating from the shedding of membrane blebs and associated with advanced disease [45]. They contain proteins and nucleic acids [46]. Proteins such as caveolin-1 and metalloproteinases 2–9 (MMP2, 9) and GTPase ADP-ribosylation factor 6 (ARF6) are reported to be contained in LO [45]. LO contain miRNA, mRNA and DNA, which transmit signaling complexes between cell and tissue compartments. They can propagate oncogenic information, including transfer of signal transduction complexes, across tissue spaces. Compared to other EVs such as exosomes and MV, LO remains a poorly characterized EV type. LO exerts some functional effects varying on different cells from a direct proteolytic activity to the activation of pro-tumorigenic signals into different types of target cells including other tumor cells or cells of tumor microenvironment [47]. LO has been identified in highly migratory and invasive prostate cancer cells [48]. Recent studies have found that LO can contribute to tumor progression because they are able to degrade directly ECM in vitro [45]. Other researchers again have revealed that they have the ability to establish a tumor growth-supporting environment. This they believe is through the export of specific oncogenic cargo to other tumor or stromal cells [49]. Prostate cancer cell–derived oncosomes contain bioactive MMP9 and MMP2 and exhibit proteolytic activity on gelatin. This suggests that they could be a means to focally concentrate proteases that facilitate migration of tumor cells, thus promoting metastasis [50]. Considering their atypical size and their specific release from cancer cells, LO are promising source of both diagnostic and prognostic markers in cancers.

3. Extracellular vesicles in the pathology of cancers

The importance and the role played by the tumor microenvironment on tumor development and progression has been established in recent years [51]. EVs are known to influence the tumor microenvironment either through a direct impact on the tumor or from a distant site which promote future metastasis of circulating cancer cells [51]. Due to these characteristics, key processes involved in cancer developments such as angiogenesis, thrombosis, oncogenic transfer, immune modulation and pre-metastatic niche formation have seen an up-regulation of EVs [52–57]. Compared to non-malignant cells, tumor cells are known to release higher amounts of EVs. In this regard increased levels ESCRT components as well as heparanase and syntenin have been expressed in various cancers [58–60]. Specifically, in colorectal cancer and pancreatic carcinoma, hyperactivity of RalB has been observed and in non-small-cell lung cancer YKT6 overexpression coupled with elevated Rho-ROCK signaling expressed in various type of cancers may contribute to EVs generation in tumor cells [61–64]. On the basis that tumorigenesis occurs due to accumulation of genetic alterations, the metastatic traits of EVs are expressed through the transfer of their oncogenic cargo. Tumor derived EVs through the co-transfer of protein crosslinking enzymes (tissue transglutaminase) and fibronectin, are able to import transformed characteristics of cancer cells on to recipients endothelial cells and fibroblast [57]. Both the cell-intrinsic and environmental signals may influence EV release in tumor cells. EVs production in tumor cells may be induced by the activation of H-RAS^{v12} and EGFRvIII oncogenic signal pathways [65–67]. Again, the level (de)regulation of the machinery, which aid in plasma membrane fusion could also influence the release of EVs in tumor cells. For example, it has been demonstrated that EV secretion could be enhanced when PKM2 (a glycolytic enzyme associated with the Warburg effect) is over expressed leading to phosphorylating tSNARE SNAP23 [68]. Also SRC, a proto-oncogene, through the phosphorylation of the cytosolic domains of syntenin and syndecan is able to stimulate the syntenin exosome biogenesis pathway [69]. On the other hand, in some cancers such as colon cancer cells, mutant proto-oncogene, KRAS could be transferred via EVs to increase the population of recipients colon cancer cells expressing the wild-type KRAS [70]. Further an increase in levels of tissue factor (TF) bearing EVs are known to mediate thrombosis occurrence in cancer patients. Available evidence indicates a possible role of tumor-derived EVs in thrombosis occurrence among cancer subjects [54]. Specifically, P-selectin glycoprotein ligand-1 (PSGL-1) and TF have been implicated in cancer associated thrombosis [71]. In mice with induced pancreatic tumor, formation of thrombosis was high compared to cancer free mice [72]. A major hallmark of tumor growth and development is increased angiogenesis. That is to say for the development of the tumor beyond its minute size an adequate supply of oxygen and nutrients is essential for its survival. Thus, numerous studies have established that besides the cell's intrinsic mechanisms, the release and regulation of exosomes and microvesicles could be due to enhanced prevailing hypoxic microenvironmental conditions [73–75]. In hypoxic glioma cells, an induction of a pro-angiogenic process mediated by derived EVs was able to influence the vasculature surrounding cell [55]. Another specific example where EVs promotes angiogenesis is reported in squamous carcinoma. It was reported that in A431 squamous carcinoma cells, angiogenesis was induced as a results of a direct transfer oncogenic epidermal growth factor receptor (EGFR) from the derived EVs to endothelial cells [76].

RNAs are a major important cargo incorporated into EVs. Cancer cells promote an increase in the release of EVs containing varying amount and types of proteins and RNAs compared to normal cells [77, 78]. There exist an EV-RNA mediated crosstalk within tumors and also between tumors and stroma which could modify the malignant behavior of cancer cells [79]. EV-RNAs derived from tumor may be implicated in the devolvement of oncogenic, pro-angiogenic, and pro-metastatic processes as well as stromal cell differentiation in the tumor microenvironment. Also it is known that normal and tumor cells subpopulations are likely to be driven towards malignant phenotypes aided by tumor derived EVs [79]. Some EV-RNAs are known to actively mediate proliferation, migration, invasion, apoptosis, dormancy and therapy resistance of cancer cells. There seems to be a dual function of EV-RNAs in cancer pathology. Whiles some are known to promote the malignant characteristics of cancer cells, it also possible for some EV-RNAs to inhibit the malignant characteristics of cancer cells. In this regard various studies have reported the ability of EV-RNAs to inhibit mechanisms that favor tumor growth. In order to establish homeostasis, various non tumor cells can produce miRNAs which could suppress the malignant phenotypes of adjacent cancerous cells [80]. This is due to the fact that a natural competition exists between cancerous and adjacent non-cancerous cells during the development of cancer [81, 82]. In hepatocellular carcinoma, EV-miRNAs released from liver stem cells were able to promote apoptosis whiles inhibiting cell proliferation *in vitro and in vivo* [83]. Again, in pancreatic ductal adenocarcinoma cells, tumor-associated stroma cells derived EV-miR-145 inhibited cancer cell viability whiles promoting apoptosis [84]. Similarly, EV-miE-145 derived from adipose tissue-derived mesenchymal stem cells promoted apoptosis and inhibited proliferation in prostate cancer cells [85]. Another important tumor modulatory role influenced by EVs is the immune system modulation [86–88]. Examples of cancer derived EVs in immune-modulation have been reported in the peripheral circulation of oral squamous carcinoma patients in which Fas ligand positive EVs were able to induce apoptosis of effector cytotoxic T cells [89]. Other studies have demonstrated that various Treg regulatory mechanisms such as Treg expansion promotion, Treg induction, Treg suppressor functional upregulation and others have been promoted by cancer-derived EVs (**Figure 1**) [90].

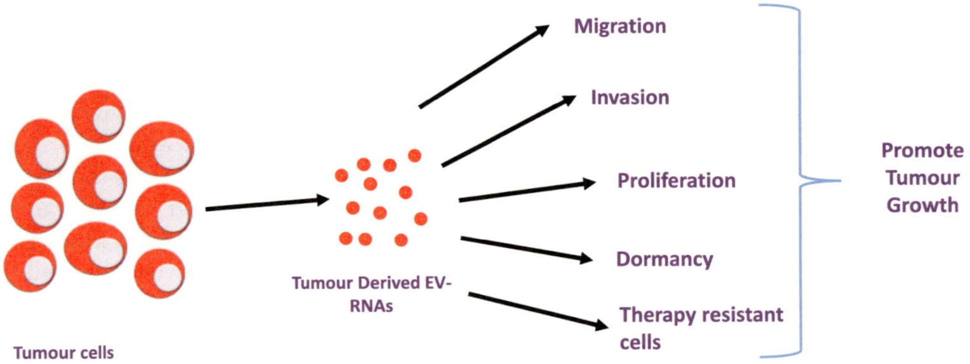

Figure 1.
Tumor cells release EV-RNAs. These EV-RNAs mediate many functions including sustaining proliferation, migration, invasion and metastasis, evading growth suppression, dormancy and therapy resistance of tumor cells, which promote growth.

4. Extracellular vesicles in cancer metastasis

The release and (de)regulation of cancer EVs and their cargo critically influence the crosstalk between tumor and stromal in the tumor microenvironment, adjacent normal cells and even distant (pre-) metastatic areas. Various stages of cancer metastasis especially the epithelial-mesenchymal transition (EMT) stage are influenced by cancer stromal cell derived EVs [91]. Mesenchymal stromal cells (MSC) are very important in the cancer stromal EMT induction [92]. There is ample evidence to show that certain components of the cargo carried by MSC-derived EVs could promote cancer metastasis by stimulating, inducing and promoting EMT. Specifically, it was shown that in breast cancer cells, EVs generated from adipose-tissue MSCs could activate the Wnt signaling pathway thus promoting cancer cell migration [93]. Again it has been demonstrated that EVs generated from human umbilical cord MSCs promoted EMT through Extracellular signal-regulated kinase (1/2) (ERK) signaling pathway with subsequent promotion of invasion and migration of breast cancer cells [94]. Also in lung cancer, EVs derived human umbilical cord MSCs promoted EMT and when TGF-β in the MSCs were knocked down EMT was inhibited [95].

5. Extracellular vesicles in biomarkers in cancer diagnosis

EVs have gained extensive attention as promising biomarkers for cancer diagnostics. Characteristics of EV include their relatively short-lived or highly labile in the cytoplasm of donor cells make them a stable biomarker cargo, be it protein, lipid, nucleic acid. In cancer these molecules can be reflective of both the tumors presence and also of cancer staging. Some studies have demonstrated that biomolecules in serum or plasma exosomes are of great value for tumor diagnosis including long non-coding RNAs (lncRNAs), miRNAs, and proteins [96]. Some other important properties of EVs which make researchers believe they represent cancer biomarkers are: (a) most of EVs populations are shed from all cell types in the organism; (b) molecular determinants contained in EVs are dependent on cells/tissues of origin; (c) however the specific EVs cargo (i.e. proteins, miRNAs) is not always coupled to the overexpression in the cells of origin; (d) molecular cargos in EVs can be affected by microenvironment conditions such as inflammation, oxygen deprivation, and metabolic balance; (e) EVs size may affect their content. In several cancers, including ovarian cancer, it has been demonstrated that the expression of a specific subset of miRNAs may potentially be used in clinical practice, for example, for screening or early diagnosis to evaluate the response to therapeutic treatments. EVs in blood and urine of prostate cancer patients contain unique prostate-cancer specific contents that are biomarkers of prostate cancer [97, 98]. EVs are proving to be valuable diagnostic biomarker in pancreatic cancer; flow cytometry coupled with mass spectrometry analysis of exosome glypican-1 can distinguish benign disease from early and late stage cancer [99]. Again, the detection of DEL-1 on circulating EVs facilitated early-stage breast cancer diagnosis and discrimination of breast cancer from benign breast disease [100]. EV-survivin is proposed to be useful in breast cancer diagnosis [101]. Kibria et al. also suggested that EV-CD47 may be a possible breast cancer biomarker [102].

6. Extracellular vesicles as therapeutic targets in cancers

Communication between cells in a tumor microorganism is largely via chemokines, cytokines, or growth factors [10]. These notwithstanding, EVs from cells in the tumor microenvironment are also noted to facilitate such communications owing to their role in tumor progression [103]. EVs are endogenous vesicles whose composition and function makes them attractive vehicles for the delivery of therapeutic agents to target cells. They have experienced increasing attention in recent years since studies into their roles demonstrated their importance as therapeutic nanomaterials. Compared to some existing synthetic or traditional carriers, EVs are considered more suitable for use as nanovesicles due to their characteristic properties of being intrinsically biocompatible, low immunogenicity and toxicity and their ability to cross-physiological barriers such as the blood-brain barrier. In addition, they have biodegradable and modification abilities and have the capability to escape clearing actions of the immune system [104, 105]. The first report of a successful therapeutic application of EVs was reported by Alvarez-Erviti et al. [106] in 2011. In that study, modified exosomes were exploited and a transfer of siRNAs was made into the brain of mice, which resulted in a knock down of the targeted gene. Supporting this hypothesis, a study by Saari et al. observed the delivery of chemotherapeutics to recipient cells and these were subsequently released into intracellular milieu to give rise to an increased cytotoxic bioactivity. The chemotherapeutics were noted to have been loaded by tumor-cell derived EVs [107]. There are three main approaches that are utilized by EVs in their role as therapeutic agents that include elimination of EVs in circulation, inhibition of secretion and disruption of the absorption of EVs.

The elimination of EVs secreted by cancer cells has been one of EV-targeting therapeutic strategies. The first report of the use of this target therapeutic approach was by Marleau et al., [108]. In the study, a hemofiltration system that was capable of targeting EVs from cancer cells by specifically aiming at human epidermal growth factor receptor 2 (HER-2) on the surface of EVs was proven [108]. This targeting of HER-2 which results in the selective elimination of cancer derived-EVs could be very valuable for cancer treatment [109].

A number of studies have focused on other strategies that block EV secretion. Inhibition of intraluminal vesicles formation and release of EVs by the fusion of MVBs to the plasma membrane have been achieve by the use of a sphingomyelinase inhibitor drug, GW4869 [110, 111]. Again, the inhibition of EV production and the transfer of miR-210-3p have reportedly been achieved by the attenuation of neutral sphingomyelinase 2 (nSMase2). nSMase2 is known to control the synthesis of ceramide and suppresses angiogenesis and metastasis in breast cancer xenograft model [112]. Conversely, EV secretion from prostate cancer cells was not inhibited by the downregulation of nSMase2. Meanwhile, nSMases have been revealed in normal neural cells [113, 114]. Their presence in these normal cells indicates the inhibition of some other fundamental pathways. Cancer specific mechanisms of EV secretion are therefore very crucial in the establishment of the role of EVs as cancer therapeutic targets. Quite recently, a group of researchers have identified a number of activators and inhibitors of EV production from prostate cancer cells [115]. This implies a clear understanding of cancer specific mechanism of EV production is required in identifying cancer-specific therapies mediated by targeted EVs.

Reports into the role of EVs have shown that the process of anti-melanoma is facilitated by EVs released by natural killer cells [116]. Similarly, the abundance of histocompatibility complex classes I and II from dendritic cells are capable of triggering other immune system cell types and also activate antitumour immune responses [117]. The use of these traditional methods in obtaining EVs for direct use as cancer therapeutic targets are not without challenges. Indistinct production mechanisms, low product yield, and the high probability of obtaining EV contents that stand the chance of mutation are a few of such challenges faced by these methods. The intrinsic properties of EVs, however, makes the engineering of these nanoparticles for the purpose of drug delivery to target cells a more favorable approach for cancer management. Engineering parental cells to shed EVs with a particular cargo or loading it directly can achieve encapsulating of therapeutic cargoes into EVs. This has been utilized in breast cancer and leukemia cell studies by Usman et al., [118] in the delivery of RNA drugs by RBC-derived EVs (RBCEVs) which showed an improved miRNA inhibition and CRISPR-Cas9 genome editing with no known cytotoxicity. Other studies on the engineering of EVs include research using mesenchymal stem cells in the overexpression of MiR-379 to obtain MiR-379-rich EVs which functions to subdue metastatic breast cancer development [119].

7. Role of extracellular vesicles in cancer prognosis and treatment response

An increasing amount of research has established that EVs are present in every human biological fluid including lymphatic and seminal fluids, bile, urine, breast milk, ascites, cerebrospinal fluid, saliva and blood, making these fluids a good source for many liquid biopsy approaches [120–122]. An increase in the rate of release of EVs on the account of cellular activation and/or during pathologic conditions may be considered an indication of a possible pathologic condition [123–125]. Real-time cancer treatment response and monitoring can be done using cancer-derived components obtained from these body fluids. The components include EVs, microRNA, circulating tumor cells (CTC), circulating cell-free tumor DNA, long non-coding RNA and EVs [126]. During the development and treatment of cancer conditions, the state of the cell is revealed by the level of active secretion of EVs, which provide timely information on the changing dynamics of the cell [18, 127]. Cancer-derived components like miRNAs obtained through liquid biopsy inhibits mRNA degradation by binding to coding sequences, 5'untranslated region (UTR), or 3'UTR of target mRNAs leading to the inhibition of mRNA degradation or translation [128]. When miRNAs bind to target mRNAs, the mRNA level as well as protein expression are essentially regulated. This means, circulating EVs are latent tools that are utilized in the quest to find a way of monitoring changes in tumor cells during treatment.

A number of studies have reported the relationship between EVs and cancer treatment response. The presence of immune checkpoints and the application of the blocking of these points by some drugs have been exploited in novel anti-cancer treatment regimens [129]. Research into the capacity of EVs as a regulating tool for checkpoint therapy has contributed immensely to the growing need of the essence of monitoring immunotherapy. Anti-tumor immunity and related expressions can be suppressed by programmed cell death 1 ligand (PDL-1) and the identification of these ligands on EVs has shown the potential for use as biomarkers in tumor patients [130]. In a syngeneic mouse melanoma model in C57BL/6 mice and B16-F10 cells experiment by Chen et al. [131]. Analysis of PDL-1 expression proved the application of EVs as a potential monitoring tool in PDL-1 therapy in melanoma patients. PDL-1 expression was either present or knocked down in these models and the levels of tumor-infiltrating CD8+ T-lymphocytes was significantly reduced in the PDL-1 expressing group compared those knocked down. A positive correlation which varied all through anti-PDL-1 therapy was observed of interferon-γ and the level of EV associated PDL-1 during the analysis of patients with metastatic melanoma [131].

In some specific cancer studies, König et al., [132] analyzed EV concentration and circulating tumor cells in breast cancer patients as a marker for the close observation, monitoring and prediction of prognosis in primary and locally advanced breast cancer. Analysis of the cells and EVs were done before and after the administration of neoadjuvant chemotherapy (NACT) prior to a surgical procedure. Patients' response to NAC is an early indication of the efficacy of subsequent systemic therapy. The overall after-NACT response is a strong prognostic factor for the risk of reoccurrence [133]. Patients with a pathologic complete response (pCR) after NACT have a significant higher overall as well as disease-free survival (OS, DFS) than their counterpart patients with residual invasive disease [134]. Studies have shown that before the administration of NACT during therapy, there is an overall an increase in EV concentration, which is linked to lymph node infiltration, while the after-NACT elevation of EV concentration is associated with reduced three-year progression-free and overall survival. This means, the analysis of EVs together with CTC analysis is a promising tool in the assessment of residual disease and the monitoring of therapy and disease outcome [132]. Other studies have used EVs in diverse ways with respect to their role in treatment response and prognosis. The first exosome-based liquid biopsy test, ExoDx™ Prostate IntelliScore (Exosome Diagnostics, Inc., Waltham, MA, USA), was approved by the Food and Administration Authority (FDA) in 2019 to analyze the exosomal RNA for the biomarkers PCA3, TMPRSS2:ERG, and SPDEF on urine specimen [135]. The prostate specific antigen available in this approach is an effective diagnostic and prognostic tool and the monitoring of this antigen together with digital rectal examination is utilized in men who have gone through a definitive therapy for localized cancer of the prostate. Again, in non-metastatic prostate cancer patients undergoing radiotherapy, a higher concentration of circulating EVs have been detected by Nano tracking analysis as a means of monitoring treatment response [136]. The study proposed a possible radiation specific induction resulting from the upregulation of hsa-miR-21-5p and hsa-let-7a-5p, both of which are specific miRNAs related to prostate cancer and radiotherapy [137]. This is further supported by the observation of altered expression of blood extracted EVs and their miRNA cargoes in the monitoring of prostate cancer radiotherapy response [138]. A high expression of some specific miRNAs before radiotherapy were noted to be an indication of better therapeutic outcomes [138]. More applications of the role of EVs in the cancer are recorded for cancer conditions such as glioblastoma [139], colorectal [140], liver [141], and non-solid cancers [142].

8. The role of extracellular vesicles in drug resistance in cancers

Due to an improved effectiveness of cancer therapies lately, there have been an increase in the survival rate of diagnosed cases [143]. Some tumors however, remain non-responsive to available treatment regimen resulting in patients going through relapse. Many cancer drugs work by causing damage in the DNA of dividing cells, which eventually result in their apoptotic death. Research has shown that some cells gain the ability to effectively repair the damage caused to them or lose the capacity to recognize apoptotic signals which renders them less capable of submitting to programmed cell death [144, 145]. Such cells become more likely to grow resistant.

The failure of treatment in general and for that matter cancer treatment could occur through various ways. It could happen through drug metabolism alterations, or changes in the efflux and/or absorption of drugs from target cells. In addition to this, the ability of drugs to induce mutations and the inhibition of cellular apoptotic pathways are all ways by which drug resistance could occur. Once acquired, the multidrug resistance of cancer drugs can lead to resistance to other drugs of different structural make-up or target.

Cancer cells have specific characteristic genetic make-up together with varying expressions of tumor suppressor genes and oncogenes. This makes them respond distinctively to various drugs. Interactions exist between host and tumor microenvironment together with changes in these genetic factors which contribute to drug resistance [146, 147]. Drug resistance represents a daunting challenge in the treatment of cancer patients. There are two types of drug resistance: de novo drug resistance, which refers to the insensitivity of cancer cells to chemotherapy before receiving drug treatment, and acquired resistance, which refers to the acquired drug resistance of cancer cells after being treated [148]. Understanding drug resistance has not been an easy task because of how complex and challenging their supporting molecular mechanisms are [149–152]. In fact, the source of resistance of a drug in a person may be very different from that of another individual because of the variations in fundamentals of different cellular processes. Playing an important role in drug resistance are extracellular vesicles. They mediate cancer drug resistance such that cells that secrete more of these vesicles show higher level of resistance than those that secrete less [153].

Recently, there are many studies concerning the role and effects of EVs in disease control and drug resistance. They have been identified to show a profound role in the development of chemo-insensitivity and drug resistance [148]. The study into their characteristics increased after they were discovered to be involved aspects of cancer progression including in proliferation, tumorigenesis, angiogenesis, and invasiveness [154–156]. In cancer drug resistance analysis, therapeutic targets are very much implicated in the development of resistance. The mediation of drug resistance by EVs takes place through a number of mechanisms. One is by the reduction of the effective concentration of cytotoxic drugs at target sites through the behavior of EVs acting as a pathway for the sequestration of such drugs. That is, resistance can arise when there is an up-regulation of vesicles that export drugs from cells or a reduction in those carriers that import drugs into the cells [157–161]. The results of such changes are the alteration in the concentrations of chemotherapeutics at the active sites. These vesicles may also act as decoys, carrying membrane proteins and capturing monoclonal antibodies intended to target receptors at the cell surface. They can also mediate cross-talk between cancer cells and stromal cells in the tumor microenvironment, leading to tumor progression and acquisition of therapeutic resistance. Apart from their role in drug resistance within a cell, EVs can transfer the resistance from a cell to another cell [162].

9. The role of EV proteins and RNA transfers in cancer drug resistance

Although cargoes of EVs are passively packaged into EVs, evidence have shown the existence of selective packaging as well [163]. Major components of EVs include protein that contribute to determining the destination of EVs and also influence the phenotype of recipient cells [163]. Vesicular protein transfers between cells constitute a potential mechanism of action of these effects. An example is the transfer of an ATP binding cassette called P-glycoprotein (P-gp), which has been reported to mediate resistance in recipient cells during their transfer between cells. In other instances, it is the expression of P-gp that becomes induced in receiver cells after a different kind of protein is delivered [164]. For example, the transfer of TrpC5 protein to recipient cells by adriamycin resistant MCF7 cells through EVs is known to stimulate the translocation of NFATc3 protein resulting in the transcriptional activation of MDR1 (ABCB1) promoter [165]. The characteristics and pathways by which drug-sensitive cells acquire resistance from EVs containing P-gp have been investigated quite extensively. The process of transfer of cancer traits from drug resistant cells to drug sensitive cells is dependent on characteristics of donor cell. While EVs from cells of leukemia transfer P-gp to malignant and non-malignant cells, those from drug resistant breast cancer cells transfer P-gp to malignant recipient cells only [166]. These findings demonstrated that P-gp transfer by EVs are potentially tissue selective and are likely associated with the cell of origin of EVs rather than their possible relation to a particular feature of recipient cell membrane [166]. In another study, de-Souza et al., [167] explored the selectivity of P-gp transfer and found no discrimination in relation to cell-type. In their study, EVs from drug-resistant leukemia cells could transfer P-gp to drug-sensitive lung and breast cancer cells. Altogether, these findings indicate the debatable issue of the selectivity of EV cargo and therefore require further investigation.

EVs can also carry non-coding RNAs such as miRNAs, IncRNAs, and circRNAs which are noted to mediate cell to cell transfer of resistance [164, 168]. These RNAs have been found to be associated with cancer progression and their deregulation is noted to support drug resistance in tumors of diverse origins [169–171]. Through the transmission of active biomolecules to neighboring cells, these various RNAs induce drug resistance in recipient cells. The transmitted biomolecules regulates certain genes together with their corresponding signaling pathways [168].

miRNAs contributes to the progress of chemoresistance by influencing the genes that are involved in cell cycle, cell proliferation and survival, apoptosis and immunity [172]. They regulate the genes by inhibiting the translation of mRNA. miRNAs have been reported in several studies to play a role in EV mediated chemoresistance. For instance, several miRNAs have been identified to be involved in the transfer of gemcitabine resistance. Gemcitabine is a chelator of DNA that gets activated by deoxycytidine kinase. EVs secreted by macrophages associated with tumors and having miR-365 cargoes have been identified to induce resistance of pancreatic ductal adenocarcinoma cells in the treatment of gemcitabine. The concentration of triphosphate nucleotides (NTPs) in the recipient cells become increased by miRNAs and the result is competitive interaction between activated gemcitabine and the increased levels of NTPs which efficiently reduces the efficacy of gemcitabine [173]. In another study, abundance of miR-1246 was observed to have been present in EVs secreted by paclitaxel resistant ovarian cancer cells. The transfer of this miRNA upregulated the expression of ABCB1 and inhibited the expression of Cav1 to facilitate paclitaxel efflux and in the process promoting drug resistance phenotype in recipient cells [174]. Conversely, EVs secreted by cancer-associated fibroblasts were found to contain miR-106 when exposed to gemcitabine. Resistance of pancreatic cancer cells (AsPC-1) against the treatment of gemcitabine has been found to be associated with the uptake of miR-106 enriched EVs [175].

Several studies have also made efforts to demonstrate the link between specific EV-transferred miRNAs and drug resistance. Through the EV-mediated transfer of miR-21, drug resistance has been found to be induced in MCF7 cells after they were co-cultured with EVs from multidrug-resistant chronic myeloid leukemia cell lines [167], Again, cisplatin resistance in lung cancer cells by miR-96 [176], Adriamycin resistance in breast cancer by miR-222 [148], and EVs' miR-155 mediated gemcitabine resistance in pancreatic cancer cells [177] have been reported. Further studies found exosomal miR-19b mediated oxaliplatin-resistance in SW480 colorectal cancer cells [178], tamoxifen-resistance in ER-positive breast cancer MCF7 cells by miR221/222 [179].

10. Conclusion

The EVs are secreted from various types of cells and are regulated by physiological conditions and other pathological conditions including cancers. EVs are considered to be attractive resources for cancer biomarker development. More research to identify potential biomarkers should be performed. This may provide more clues for elucidating the biological functions of EVs in cancer development as well as predicting the disease progression. These researches about the EVs will again offer valuable information to increase our understanding into the pathology of cancer and provide the novel ways to advance the diagnosis and prognosis of cancers.

Acknowledgements

This work was supported by China Postdoctoral Science Foundation (Grant No. 2018M642187).

Author details

Prince Amoah Barnie[1,2]*, Justice Afrifa[3]*, Eric Ofori Gyamerah[4,5]
and Benjamin Amoani[2]

1 Department of Immunology, School of Medical Science and Laboratory Medicine, Jiangsu University, Zhenjiang, China

2 Department of Biomedical Sciences, School of Biological Sciences, University of Cape Coast, Cape Coast, Ghana

3 Department of Medical Laboratory Science, School of Biological Sciences, University of Cape Coast, Cape Coast, Ghana

4 Department of Biology Education, Faculty of Science Education, University of Education, Winneba, Ghana

5 Department of Molecular Biology and Biotechnology, School of Biological Sciences, University of Cape Coast, Cape Coast, Ghana

*Address all correspondence to: pamoah-barnie@ucc.edu.gh and jafrifa@ucc.edu.gh

References

[1] Sung H et al. Global cancer statistics 2020: GLOBOCAN estimates of incidence and mortality worldwide for 36 cancers in 185 countries. CA: A Cancer Journal for Clinicians. 2021;**71**(3): 209-249

[2] Shimada S, Tanaka S. A New Era for Understanding Genetic Evolution of Multistep Carcinogenesis. New York: Springer; 2019

[3] Henry NL, Hayes DF. Cancer biomarkers. Molecular Oncology. 2012;**6**(2):140-146

[4] Costa-Pinheiro P et al. Diagnostic and prognostic epigenetic biomarkers in cancer. Epigenomics. 2015;7(6): 1003-1015

[5] Wang Z. Drug Resistance and Novel Therapies in Cancers. Basel, Switzerland: Multidisciplinary Digital Publishing Institute; 2020

[6] Tan L-L, Lyon AR. Role of biomarkers in prediction of cardiotoxicity during cancer treatment. Current Treatment Options in Cardiovascular Medicine. 2018;**20**(7):1-14

[7] Xue H, Lu B, Lai M. The cancer secretome: A reservoir of biomarkers. Journal of Translational Medicine. 2008;**6**(1):1-12

[8] Chatterjee SK, Zetter BR. Cancer Biomarkers: Knowing the Present and Predicting the Future. Future Oncology. 2005;**1**(1):37-50

[9] Lane R et al. Extracellular vesicles as circulating cancer biomarkers: Opportunities and challenges. Clinical and Translational Medicine. 2018; **7**(1):1-11

[10] Xu R et al. Extracellular vesicles in cancer—Implications for future improvements in cancer care. Nature Reviews Clinical oncology. 2018; **15**(10):617-638

[11] Żmigrodzka M et al. The biology of extracellular vesicles with focus on platelet microparticles and their role in cancer development and progression. Tumor Biology. 2016;**37**(11):14391-14401

[12] Xie F et al. Extracellular vesicles in cancer immune microenvironment and cancer immunotherapy. Advanced Science. 2019;**6**(24):1901779

[13] Katsuda T, Kosaka N, Ochiya T. The roles of extracellular vesicles in cancer biology: Toward the development of novel cancer biomarkers. Proteomics. 2014;**14**(4-5):412-425

[14] Lässer C, Jang SC, Lötvall J. Subpopulations of extracellular vesicles and their therapeutic potential. Molecular Aspects of Medicine. 2018;**60**:1-14

[15] Zaborowski MP et al. Extracellular vesicles: Composition, biological relevance, and methods of study. Bioscience. 2015;**65**(8):783-797

[16] Borges FT, Reis L, Schor N. Extracellular vesicles: Structure, function, and potential clinical uses in renal diseases. Brazilian Journal of Medical and Biological Research. 2013;**46**:824-830

[17] Bebelman MP et al. Biogenesis and function of extracellular vesicles in cancer. Pharmacology & Therapeutics. 2018;**188**:1-11

[18] Yáñez-Mó M et al. Biological properties of extracellular vesicles and

their physiological functions. Journal of Extracellular Vesicles. 2015;**4**(1):27066

[19] Raposo G, Stoorvogel W. Extracellular vesicles: Exosomes, microvesicles, and friends. Journal of Cell Biology. 2013;**200**(4):373-383

[20] Kishore R, Garikipati VNS, Gumpert A. Tiny shuttles for information transfer: Exosomes in cardiac health and disease. Journal of Cardiovascular Translational Research. 2016;**9**(3): 169-175

[21] Sahu R et al. Microautophagy of cytosolic proteins by late endosomes. Developmental Cell. 2011;**20**(1):131-139

[22] Jiang L et al. Determining the contents and cell origins of apoptotic bodies by flow cytometry. Scientific Reports. 2017;7(1):1-12

[23] Lane JD, Allan VJ, Woodman PG. Active relocation of chromatin and endoplasmic reticulum into blebs in late apoptotic cells. Journal of Cell Science. 2005;**118**(17):4059-4071

[24] Xu X, Lai Y, Hua Z-C. Apoptosis and apoptotic body: Disease message and therapeutic target potentials. Bioscience Reports. 2019;**39**(1):BSR20180992

[25] Coleman ML et al. Membrane blebbing during apoptosis results from caspase-mediated activation of ROCK I. Nature Cell Biology. 2001;**3**(4):339-345

[26] Aoki K et al. Coordinated changes in cell membrane and cytoplasm during maturation of apoptotic bleb. Molecular Biology of the Cell. 2020;**31**(8):833-844

[27] Mills J, Stone NL, Erhardt J, Pittman RN. Apoptotic membrane blebbing is regulated by myosin light chain phosphorylation. The Journal of Cell Biology. 1998;**140**:627-636

[28] Zirngibl M et al. Loading of nuclear autoantigens prototypically recognized by systemic lupus erythematosus sera into late apoptotic vesicles requires intact microtubules and myosin light chain kinase activity. Clinical & Experimental Immunology. 2015;**179**(1):39-49

[29] Orlando KA, Pittman RN. Rho kinase regulates phagocytosis, surface expression of GlcNAc, and Golgi fragmentation of apoptotic PC12 cells. Experimental Cell Research. 2006; **312**(17):3298-3311

[30] Leventis PA, Grinstein S. The distribution and function of phosphatidylserine in cellular membranes. Annual Review of Biophysics. 2010;**39**:407-427

[31] Bevers EM et al. Lipid translocation across the plasma membrane of mammalian cells. Biochimica et Biophysica Acta (BBA)-Molecular and Cell Biology of Lipids. 1999;**1439**(3): 317-330

[32] Zwaal RF, Schroit AJ. Pathophysiologic implications of membrane phospholipid asymmetry in blood cells. Blood, The Journal of the American Society of Hematology. 1997;**89**(4):1121-1132

[33] McConnell RE et al. The enterocyte microvillus is a vesicle-generating organelle. Journal of Cell Biology. 2009;**185**(7):1285-1298

[34] Muralidharan-Chari V et al. ARF6-regulated shedding of tumor cell-derived plasma membrane microvesicles. Current Biology. 2009;**19**(22):1875-1885

[35] Muralidharan-Chari V et al. Microvesicles: Mediators of extracellular communication during cancer progression. Journal of Cell Science. 2010;**123**(10):1603-1611

[36] Ohno Y et al. Analysis of substrate specificity of human DHHC protein acyltransferases using a yeast expression system. Molecular Biology of the Cell. 2012;**23**(23):4543-4551

[37] McMahon HT, Boucrot E. Membrane curvature at a glance. Journal of Cell Science. 2015;**128**(6):1065-1070

[38] Kozlov MM et al. Mechanisms shaping cell membranes. Current Opinion in Cell Biology. 2014;**29**:53-60

[39] D'Souza-Schorey C, Clancy JW. Tumor-derived microvesicles: Shedding light on novel microenvironment modulators and prospective cancer biomarkers. Genes & Development. 2012;**26**(12):1287-1299

[40] Tricarico C, Clancy J, D'Souza-Schorey C. Biology and biogenesis of shed microvesicles. Small GTPases. 2017;**8**(4):220-232

[41] Galindo-Hernandez O et al. Elevated concentration of microvesicles isolated from peripheral blood in breast cancer patients. Archives of Medical Research. 2013;**44**(3):208-214

[42] Wang W et al. Peripheral blood microvesicles are potential biomarkers for hepatocellular carcinoma. Cancer Biomarkers. 2013;**13**(5):351-357

[43] Ghasemi R et al. Tumor-derived microvesicles: The metastasomes. Medical Hypotheses. 2013;**80**(1):75-82

[44] Panfoli I et al. Microvesicles as promising biological tools for diagnosis and therapy. Expert Review of Proteomics. 2018;**15**(10):801-808

[45] Minciacchi VR, Freeman MR, Di Vizio D. Extracellular vesicles in cancer: Exosomes, microvesicles and the emerging role of large oncosomes. In: Seminars in Cell & Developmental Biology. Amsterdam, Netherlands: Elsevier; 2015

[46] Minciacchi VR et al. Large oncosomes contain distinct protein cargo and represent a separate functional class of tumor-derived extracellular vesicles. Oncotarget. 2015;**6**(13):11327

[47] Ciardiello C et al. Large extracellular vesicles: Size matters in tumor progression. Cytokine & Growth Factor Reviews. 2020;**51**:69-74

[48] Ciardiello C et al. Large oncosomes overexpressing integrin alpha-V promote prostate cancer adhesion and invasion via AKT activation. Journal of Experimental & Clinical Cancer Research. 2019;**38**(1): 1-16

[49] Wendler F et al. Extracellular vesicles swarm the cancer microenvironment: From tumor–stroma communication to drug intervention. Oncogene. 2017;**36**(7):877-884

[50] Di Vizio D et al. Large oncosomes in human prostate cancer tissues and in the circulation of mice with metastatic disease. The American Journal of Pathology. 2012;**181**(5):1573-1584

[51] Yuan Y et al. Role of the tumor microenvironment in tumor progression and the clinical applications. Oncology Reports. 2016;**35**(5):2499-2515

[52] Costa-Silva B et al. Pancreatic cancer exosomes initiate pre-metastatic niche formation in the liver. Nature Cell Biology. 2015;**17**(6):816-826

[53] Chalmin F et al. Membrane-associated Hsp72 from tumor-derived exosomes mediates STAT3-dependent immunosuppressive function of mouse and human myeloid-derived suppressor cells. The Journal of Clinical Investigation. 2010;**120**(2):457-471

[54] Geddings JE, Mackman N. Tumor-derived tissue factor–positive microparticles and venous thrombosis in cancer patients. Blood, The Journal of the American Society of Hematology. 2013;**122**(11):1873-1880

[55] Kucharzewska P et al. Exosomes reflect the hypoxic status of glioma cells and mediate hypoxia-dependent activation of vascular cells during tumor development. Proceedings of the National Academy of Sciences. 2013;**110**(18):7312-7317

[56] Xiao D et al. Melanoma cell–derived exosomes promote epithelial–mesenchymal transition in primary melanocytes through paracrine/autocrine signaling in the tumor microenvironment. Cancer Letters. 2016;**376**(2):318-327

[57] Antonyak MA et al. Cancer cell-derived microvesicles induce transformation by transferring tissue transglutaminase and fibronectin to recipient cells. Proceedings of the National Academy of Sciences. 2011;**108**(12):4852-4857

[58] Toyoshima M et al. Inhibition of tumor growth and metastasis by depletion of vesicular sorting protein Hrs: Its regulatory role on E-cadherin and β-catenin. Cancer Research. 2007;**67**(11):5162-5171

[59] Oh K et al. Tsg101 is upregulated in a subset of invasive human breast cancers and its targeted overexpression in transgenic mice reveals weak oncogenic properties for mammary cancer initiation. Oncogene. 2007;**26**(40): 5950-5959

[60] Liu R-T et al. Overexpression of tumor susceptibility gene TSG101 in human papillary thyroid carcinomas. Oncogene. 2002;**21**(31):4830-4837

[61] Ruiz-Martinez M et al. YKT6 expression, exosome release, and survival in non-small cell lung cancer. Oncotarget. 2016;7(32):51515

[62] Morgan-Fisher M, Wewer UM, Yoneda A. Regulation of ROCK activity in cancer. Journal of Histochemistry & Cytochemistry. 2013;**61**(3):185-198

[63] Martin TD, Der CJ. Differential involvement of RalA and RalB in colorectal cancer. Small GTPases. 2012;**3**(2):126-130

[64] Lim K-H et al. Divergent roles for RalA and RalB in malignant growth of human pancreatic carcinoma cells. Current Biology. 2006;**16**(24):2385-2394

[65] Takasugi M et al. Small extracellular vesicles secreted from senescent cells promote cancer cell proliferation through EphA2. Nature Communications. 2017;**8**(1):1-11

[66] Lee TH et al. Oncogenic ras-driven cancer cell vesiculation leads to emission of double-stranded DNA capable of interacting with target cells. Biochemical and Biophysical Research Communications. 2014;**451**(2):295-301

[67] Al-Nedawi K et al. Intercellular transfer of the oncogenic receptor EGFRvIII by microvesicles derived from tumour cells. Nature Cell Biology. 2008;**10**(5):619-624

[68] Wei Y et al. Pyruvate kinase type M2 promotes tumour cell exosome release via phosphorylating synaptosome-associated protein 23. Nature Communications. 2017;**8**(1):1-12

[69] Imjeti NS et al. Syntenin mediates SRC function in exosomal cell-to-cell communication. Proceedings of the National Academy of Sciences. 2017;**114**(47):12495-12500

[70] Beckler MD et al. Proteomic analysis of exosomes from mutant KRAS colon cancer cells identifies intercellular transfer of mutant KRAS. Molecular & Cellular Proteomics. 2013;**12**(2): 343-355

[71] Thomas GM et al. Cancer cell–derived microparticles bearing P-selectin glycoprotein ligand 1 accelerate thrombus formation in vivo. Journal of Experimental Medicine. 2009;**206**(9): 1913-1927

[72] Thomas G et al. Tissue factor expressed by circulating cancer cell-derived microparticles drastically increases the incidence of deep vein thrombosis in mice. Journal of Thrombosis and Haemostasis. 2015;**13**(7):1310-1319

[73] King HW, Michael MZ, Gleadle JM. Hypoxic enhancement of exosome release by breast cancer cells. BMC Cancer. 2012;**12**(1):1-10

[74] Li L et al. Exosomes derived from hypoxic oral squamous cell carcinoma cells deliver miR-21 to normoxic cells to elicit a prometastatic phenotype. Cancer Research. 2016;**76**(7):1770-1780

[75] Wang T et al. Hypoxia-inducible factors and RAB22A mediate formation of microvesicles that stimulate breast cancer invasion and metastasis. Proceedings of the National Academy of Sciences. 2014;**111**(31):E3234-E3242

[76] Al-Nedawi K. Intercellular transfer of the oncogenic EGFRv III via tumor cell derived microvesicles. Nature Cell Biology. 2008;**10**:619-624

[77] Griffiths SG et al. Differential proteome analysis of extracellular vesicles from breast cancer cell lines by chaperone affinity enrichment. Proteomes. 2017;**5**(4):25

[78] Tűzesi Á et al. Pediatric brain tumor cells release exosomes with a miRNA repertoire that differs from exosomes secreted by normal cells. Oncotarget. 2017;**8**(52):90164

[79] Hu W et al. Comprehensive landscape of extracellular vesicle-derived RNAs in cancer initiation, progression, metastasis and cancer immunology. Molecular Cancer. 2020;**19**:1-23

[80] Kosaka N et al. Competitive interactions of cancer cells and normal cells via secretory microRNAs. Journal of Biological Chemistry. 2012;**287**(2): 1397-1405

[81] Kanada M, Bachmann MH, Contag CH. Signaling by extracellular vesicles advances cancer hallmarks. Trends in Cancer. 2016;**2**(2):84-94

[82] Kosaka N et al. Versatile roles of extracellular vesicles in cancer. The Journal of Clinical Investigation. 2016;**126**(4):1163-1172

[83] Fonsato V et al. Human liver stem cell-derived microvesicles inhibit hepatoma growth in SCID mice by delivering antitumor microRNAs. Stem Cells. 2012;**30**(9):1985-1998

[84] Han S et al. Stroma-derived extracellular vesicles deliver tumor-suppressive miRNAs to pancreatic cancer cells. Oncotarget. 2018;**9**(5):5764

[85] Takahara K et al. microRNA-145 mediates the inhibitory effect of adipose tissue-derived stromal cells on prostate cancer. Stem Cells and Development. 2016;**25**(17):1290-1298

[86] Greening DW et al. Exosomes and their roles in immune regulation and cancer. In: Seminars in cell & developmental biology. Amsterdam, Netherlands: Elsevier; 2015

[87] Teng Y et al. MVP-mediated exosomal sorting of miR-193a promotes colon cancer progression. Nature Communications. 2017;**8**(1):1-16

[88] Whiteside T. Advances in clinical chemistry. Vol. 74. Amsterdam, Netherlands: Academic Press; 2016. pp. 103-141

[89] Taylor DD et al. T-cell apoptosis and suppression of T-cell receptor/CD3-ζ by Fas ligand-containing membrane vesicles shed from ovarian tumors. Clinical Cancer Research. 2003;**9**(14):5113-5119

[90] Wieckowski EU et al. Tumor-derived microvesicles promote regulatory T cell expansion and induce apoptosis in tumor-reactive activated CD8+ T lymphocytes. The Journal of Immunology. 2009;**183**(6):3720-3730

[91] Zhao H et al. The key role of extracellular vesicles in the metastatic process. Biochimica et Biophysica Acta (BBA)-Reviews on Cancer. 2018; **1869**(1):64-77

[92] Dong C et al. Loss of FBP1 by Snail-mediated repression provides metabolic advantages in basal-like breast cancer. Cancer Cell. 2013;**23**(3):316-331

[93] Commisso C et al. Macropinocytosis of protein is an amino acid supply route in Ras-transformed cells. Nature. 2013;**497**(7451):633-637

[94] Zhou X et al. Mesenchymal stem cell-derived extracellular vesicles promote the in vitro proliferation and migration of breast cancer cells through the activation of the ERK pathway. International Journal of Oncology. 2019;**54**(5):1843-1852

[95] Zhao X et al. Knockdown of TGF-β1 expression in human umbilical cord mesenchymal stem cells reverts their exosome-mediated EMT promoting effect on lung cancer cells. Cancer Letters. 2018;**428**:34-44

[96] Kalluri R. The biology and function of exosomes in cancer. The Journal of Clinical Investigation. 2016;**126**(4): 1208-1215

[97] Pang B et al. Extracellular vesicles: The next generation of biomarkers for liquid biopsy-based prostate cancer diagnosis. Theranostics. 2020;**10**(5): 2309

[98] Dhondt B et al. Unravelling the proteomic landscape of extracellular vesicles in prostate cancer by density-based fractionation of urine. Journal of extracellular vesicles. 2020;**9**(1):1736935

[99] Yee NS et al. Extracellular vesicles as potential biomarkers for early detection and diagnosis of pancreatic cancer. Biomedicine. 2020;**8**(12):581

[100] Moon P-G et al. Identification of developmental endothelial locus-1 on circulating extracellular vesicles as a novel biomarker for early breast cancer detection. Clinical Cancer Research. 2016;**22**(7):1757-1766

[101] Khan S et al. Early diagnostic value of survivin and its alternative splice variants in breast cancer. BMC Cancer. 2014;**14**(1):1-10

[102] Kibria G et al. A rapid, automated surface protein profiling of single circulating exosomes in human blood. Scientific Reports. 2016;**6**(1):1-9

[103] Naito Y et al. How cancer cells dictate their microenvironment: Present roles of extracellular vesicles. Cellular and Molecular Life Sciences. 2017;**74**(4):697-713

[104] Andaloussi SE et al. Extracellular vesicles: Biology and emerging

therapeutic opportunities. Nature Reviews Drug Discovery. 2013; **12**(5):347-357

[105] Colao IL et al. Manufacturing exosomes: A promising therapeutic platform. Trends in Molecular Medicine. 2018;**24**(3):242-256

[106] Alvarez-Erviti L et al. Delivery of siRNA to the mouse brain by systemic injection of targeted exosomes. Nature Biotechnology. 2011;**29**(4):341-345

[107] Shedden K et al. Expulsion of small molecules in vesicles shed by cancer cells: Association with gene expression and chemosensitivity profiles. Cancer Research. 2003;**63**(15):4331-4337

[108] Marleau AM et al. Exosome removal as a therapeutic adjuvant in cancer. Journal of Translational Medicine. 2012;**10**(1):1-12

[109] Ciravolo V et al. Potential role of HER2-overexpressing exosomes in countering trastuzumab-based therapy. Journal of Cellular Physiology. 2012;**227**(2):658-667

[110] Kosaka N et al. Secretory mechanisms and intercellular transfer of microRNAs in living cells*♦. Journal of Biological Chemistry. 2010;**285**(23): 17442-17452

[111] Middleton RC et al. Newt cells secrete extracellular vesicles with therapeutic bioactivity in mammalian cardiomyocytes. Journal of Extracellular Vesicles. 2018;**7**(1):1456888

[112] Kosaka N et al. Neutral sphingomyelinase 2 (nSMase2)-dependent exosomal transfer of angiogenic microRNAs regulate cancer cell metastasis. Journal of Biological Chemistry. 2013;**288**(15):10849-10859

[113] Yuyama K et al. Sphingolipid-modulated exosome secretion promotes clearance of amyloid-β by microglia. Journal of Biological Chemistry. 2012;**287**(14):10977-10989

[114] Phuyal S et al. Regulation of exosome release by glycosphingolipids and flotillins. The FEBS Journal. 2014;**281**(9):2214-2227

[115] Datta A et al. High-throughput screening identified selective inhibitors of exosome biogenesis and secretion: A drug repurposing strategy for advanced cancer. Scientific Reports. 2018; **8**(1):1-13

[116] Zhu L et al. Exosomes derived from natural killer cells exert therapeutic effect in melanoma. Theranostics. 2017;**7**(10):2732

[117] Chulpanova D et al. Therapeutic prospects of extracellular vesicles in cancer treatment. Frontiers in Immunology. 2018;**9**:1534

[118] Usman WM et al. Efficient RNA drug delivery using red blood cell extracellular vesicles. Nature Communications. 2018;**9**(1):1-15

[119] O'brien K et al. Employing mesenchymal stem cells to support tumor-targeted delivery of extracellular vesicle (EV)-encapsulated microRNA-379. Oncogene. 2018; **37**(16):2137-2149

[120] Pant S, Hilton H, Burczynski ME. The multifaceted exosome: Biogenesis, role in normal and aberrant cellular function, and frontiers for pharmacological and biomarker opportunities. Biochemical Pharmacology. 2012;**83**(11):1484-1494

[121] Vlassov AV et al. Exosomes: Current knowledge of their composition,

biological functions, and diagnostic and therapeutic potentials. Biochimica et Biophysica Acta (BBA)-General Subjects. 2012;**1820**(7):940-948

[122] Akers JC et al. Biogenesis of extracellular vesicles (EV): Exosomes, microvesicles, retrovirus-like vesicles, and apoptotic bodies. Journal of Neuro-Oncology. 2013;**113**(1):1-11

[123] Szajnik M et al. Exosomes in plasma of patients with ovarian carcinoma: Potential biomarkers of tumor progression and response to therapy. Gynecology & obstetrics (Sunnyvale, Calif.). 2013:3

[124] Baran J et al. Circulating tumour-derived microvesicles in plasma of gastric cancer patients. Cancer Immunology, Immunotherapy. 2010;**59**(6):841-850

[125] Rabinowits G et al. Exosomal microRNA: A diagnostic marker for lung cancer. Clinical Lung Cancer. 2009;**10**(1):42-46

[126] Babayan A, Pantel K. Advances in liquid biopsy approaches for early detection and monitoring of cancer. Genome Medicine. 2018;**10**(1):21

[127] Tai Y-L et al. Basics and applications of tumor-derived extracellular vesicles. Journal of Biomedical Science. 2019;**26**(1):35

[128] Kai K, Dittmar RL, Sen S. Secretory microRNAs as biomarkers of cancer. In: Seminars in Cell & Developmental Biology. Amsterdam, Netherlands: Elsevier; 2018

[129] Gu D et al. Soluble immune checkpoints in cancer: Production, function and biological significance. Journal for Immunotherapy of Cancer. 2018;**6**(1):1-14

[130] Ricklefs FL et al. Immune evasion mediated by PD-L1 on glioblastoma-derived extracellular vesicles. Science Advances. 2018;**4**(3):eaar2766

[131] Chen G et al. Exosomal PD-L1 contributes to immunosuppression and is associated with anti-PD-1 response. Nature. 2018;**560**(7718):382-386

[132] König L et al. Elevated levels of extracellular vesicles are associated with therapy failure and disease progression in breast cancer patients undergoing neoadjuvant chemotherapy. Oncoimmunology. 2018;**7**(1):e1376153

[133] Gralow JR et al. Preoperative therapy in invasive breast cancer: Pathologic assessment and systemic therapy issues in operable disease. Journal of Clinical Oncology. 2008;**26**(5):814-819

[134] Cortazar P et al. Pathological complete response and long-term clinical benefit in breast cancer: The CTNeoBC pooled analysis. The Lancet. 2014;**384**(9938):164-172

[135] Press B, Schulster M, Bjurlin MA. Differentiating molecular risk assessments for prostate cancer. Reviews in Urology. 2018;**20**(1):12

[136] Malla B, Aebersold DM, Dal Pra A. Protocol for serum exosomal miRNAs analysis in prostate cancer patients treated with radiotherapy. Journal of Translational Medicine. 2018;**16**(1):1-13

[137] Malla B et al. Exosomes and exosomal microRNAs in prostate cancer radiation therapy. International Journal of Radiation Oncology* Biology* Physics. 2017;**98**(5):982-995

[138] Yu Q et al. Nano-vesicles are a potential tool to monitor therapeutic efficacy of carbon ion radiotherapy in

prostate cancer. Journal of Biomedical Nanotechnology. 2018;**14**(1):168-178

[139] Skog J et al. Glioblastoma microvesicles transport RNA and proteins that promote tumour growth and provide diagnostic biomarkers. Nature Cell Biology. 2008;**10**(12): 1470-1476

[140] Willms A et al. Tumour-associated circulating microparticles: A novel liquid biopsy tool for screening and therapy monitoring of colorectal carcinoma and other epithelial neoplasia. Oncotarget. 2016;**7**(21):30867

[141] Julich-Haertel H et al. Cancer-associated circulating large extracellular vesicles in cholangiocarcinoma and hepatocellular carcinoma. Journal of Hepatology. 2017;**67**(2):282-292

[142] van Eijndhoven MA et al. Plasma vesicle miRNAs for therapy response monitoring in Hodgkin lymphoma patients. JCI Insight. 2016;**1**(19):e89631

[143] Yu S et al. Extracellular vesicles in breast cancer drug resistance and their clinical application. Tumor Biology. 2016;**37**(3):2849-2861

[144] Kim D et al. AKT/PKB signaling mechanisms in cancer and chemoresistance. Frontiers in Bioscience. 2005;**10**(14):975

[145] Hata AN, Engelman JA, Faber AC. The BCL2 family: Key mediators of the apoptotic response to targeted anticancer therapeutics. Cancer Discovery. 2015;**5**(5):475-487

[146] Gottesman MM. Mechanisms of cancer drug resistance. Annual Review of Medicine. 2002;**53**(1):615-627

[147] Broxterman HJ, Lankelma J, Hoekman K. Resistance to cytotoxic and anti-angiogenic anticancer agents: Similarities and differences. Drug Resistance Updates. 2003;**6**(3):111-127

[148] Yu D-d et al. Exosomes from adriamycin-resistant breast cancer cells transmit drug resistance partly by delivering miR-222. Tumor Biology. 2016;**37**(3):3227-3235

[149] Mezencev R et al. Acquired resistance of pancreatic cancer cells to cisplatin is multifactorial with cell context-dependent involvement of resistance genes. Cancer Gene Therapy. 2016;**23**(12):446-453

[150] Gainor JF, Shaw AT. Emerging paradigms in the development of resistance to tyrosine kinase inhibitors in lung cancer. Journal of Clinical Oncology. 2013;**31**(31):3987

[151] Galluzzi L et al. Systems biology of cisplatin resistance: Past, present and future. Cell Death & Disease. 2014;**5**(5):e1257-e1257

[152] Rueff J, Rodrigues AS. Cancer drug resistance: A brief overview from a genetic viewpoint. Cancer Drug Resistance. 2016;**1395**:1-18

[153] Safaei R et al. Abnormal lysosomal trafficking and enhanced exosomal export of cisplatin in drug-resistant human ovarian carcinoma cells. Molecular Cancer Therapeutics. 2005;**4**(10):1595-1604

[154] Feng Q et al. A class of extracellular vesicles from breast cancer cells activates VEGF receptors and tumour angiogenesis. Nature Communications. 2017;**8**(1):1-17

[155] Hosseini-Beheshti E et al. Exosomes confer pro-survival signals to alter the phenotype of prostate cells in their surrounding environment. Oncotarget. 2016;**7**(12):14639

[156] Jaiswal R et al. Microparticles shed from multidrug resistant breast cancer cells provide a parallel survival pathway through immune evasion. BMC Cancer. 2017;**17**(1):1-12

[157] Fletcher JI et al. ABC transporters as mediators of drug resistance and contributors to cancer cell biology. Drug Resistance Updates. 2016;**26**:1-9

[158] Okabe M et al. Profiling SLCO and SLC22 genes in the NCI-60 cancer cell lines to identify drug uptake transporters. Molecular Cancer Therapeutics. 2008;**7**(9):3081-3091

[159] de Lima LT et al. Reduced ABCG2 and increased SLC22A1 mRNA expression are associated with imatinib response in chronic myeloid leukemia. Medical Oncology. 2014;**31**(3):851

[160] Kalayda GV, Wagner CH, Jaehde U. Relevance of copper transporter 1 for cisplatin resistance in human ovarian carcinoma cells. Journal of Inorganic Biochemistry. 2012;**116**:1-10

[161] Huang Y, Sadée W. Membrane transporters and channels in chemoresistance and-sensitivity of tumor cells. Cancer Letters. 2006;**239**(2):168-182

[162] Sousa D, Lima RT, Vasconcelos MH. Intercellular transfer of cancer drug resistance traits by extracellular vesicles. Trends in Molecular Medicine. 2015;**21**(10):595-608

[163] Simons M, Raposo G. Exosomes–vesicular carriers for intercellular communication. Current Opinion in Cell Biology. 2009;**21**(4):575-581

[164] Samuel P, Fabbri M, Carter DRF. Mechanisms of drug resistance in cancer: The role of extracellular vesicles. Proteomics. 2017;**17**(23-24):1600375

[165] Dong Y et al. Tumor endothelial expression of P-glycoprotein upon microvesicular transfer of TrpC5 derived from adriamycin-resistant breast cancer cells. Biochemical and Biophysical Research Communications. 2014;**446**(1):85-90

[166] Jaiswal R et al. Breast cancer-derived microparticles display tissue selectivity in the transfer of resistance proteins to cells. PLoS One. 2013;**8**(4): e61515

[167] de Souza PS et al. Microparticles induce multifactorial resistance through oncogenic pathways independently of cancer cell type. Cancer Science. 2015;**106**(1):60-68

[168] Calin GA, Croce CM. MicroRNA signatures in human cancers. Nature Reviews Cancer. 2006;**6**(11):857-866

[169] Samuel P et al. miRNAs and ovarian cancer: A miRiad of mechanisms to induce cisplatin drug resistance. Expert Review of Anticancer Therapy. 2016;**16**(1):57-70

[170] Aigner A. MicroRNAs (miRNAs) in cancer invasion and metastasis: Therapeutic approaches based on metastasis-related miRNAs. Journal of Molecular Medicine. 2011;**89**(5):445-457

[171] Ling H et al. MicroRNAs in testicular cancer diagnosis and prognosis. The Urologic Clinics of North America. 2016;**43**(1):127-134

[172] Si W et al. The role and mechanisms of action of microRNAs in cancer drug resistance. Clinical Epigenetics. 2019;**11**(1):1-24

[173] Binenbaum Y et al. Transfer of miRNA in macrophage-derived exosomes induces drug resistance in pancreatic adenocarcinoma. Cancer Research. 2018;**78**(18):5287-5299

[174] Kanlikilicer P, Recep B, Merve D, Mohammed HR, Cristina I, Burcu A, Rahul M, et al. "Exosomal miRNA confers chemo resistance via targeting Cav1/p-gp/M2-type macrophage axis in ovarian cancer." EBioMedicine. 2018;**38**: 100-112

[175] Fang Y et al. Exosomal miRNA-106b from cancer-associated fibroblast promotes gemcitabine resistance in pancreatic cancer. Experimental Cell Research. 2019;**383**(1):111543

[176] Wu H et al. Circulating exosomal microRNA-96 promotes cell proliferation, migration and drug resistance by targeting LMO7. Journal of Cellular and Molecular Medicine. 2017;**21**(6):1228-1236

[177] Mikamori M et al. MicroRNA-155 controls exosome synthesis and promotes gemcitabine resistance in pancreatic ductal adenocarcinoma. Scientific Reports. 2017;**7**(1):1-14

[178] Gu Y et al. Suppressing the secretion of exosomal miR-19b by gw4869 could regulate oxaliplatin sensitivity in colorectal cancer. Neoplasma. 2019;**66**(1):39-45

[179] Wei Y et al. Exosomal miR-221/222 enhances tamoxifen resistance in recipient ER-positive breast cancer cells. Breast Cancer Research and Treatment. 2014;**147**(2):423-431

Chapter 13

Extracellular Vesicles and Ovarian Cancer

Diego Aviles, David Warshal, Lauren Krill and Olga Ostrovsky

Abstract

Extracellular vesicles (EVs) are a varied group of cell-derived, microscopic, fluid-filled pouches released from cells into neighboring microenvironments that are quickly gaining recognition as a potentially powerful tool against epithelial ovarian cancer (EOC). Recent studies show that not only do EVs play an integral part in the development of cancer through intercellular communication, cell survival, and immune modulation but also may assist with early diagnosis and improved treatments. EOC currently has few effective screening options for early detection of this disease; and, therefore, it is detected at an advanced stage where it is more likely to recur, develop chemoresistance, and ultimately become fatal. Newer research has evaluated EVs as biomarkers for early screening and diagnosis and as novel targets for treatment of EOC. Moreover, EVs are possible targets for novel immunomodulatory therapies to directly target cancer cells or make cancer cells more susceptible to other treatment modalities. Therefore, EVs present an exciting, promising approach which may improve clinical outcome for EOC patients.

Keywords: extracellular vesicles, exosomes, epithelial ovarian cancer, ovarian cancer, diagnosis, prognosis, novel therapy, gynecologic oncology

1. Introduction

Extracellular vesicles (EVs) are a varied group of cell-derived, microscopic, fluid-filled pouches that cells release into the neighboring microenvironment. Recent studies show that not only do EVs play an integral part in the development of cancer through intercellular communication, cell survival, and immune modulation but also may assist with the early diagnosis and improved treatment of diseases such as epithelial ovarian cancer (EOC) [1]. EVs are quickly gaining recognition as an important enabler of EOC propagation and may potentially serve as a powerful tool in inhibiting and even reversing the progression of this disease. Historically, EOC has been a frustrating gynecologic malignancy characterized by its furtive early course that leads to an advanced presentation at initial diagnosis with subsequent poor outcomes [1]. Public health entities are ineffective at screening for early disease, leaving patients with few warnings to herald a lurking predator that affects 1–3% of women throughout their lifetime [2]. Once an EOC has manifested, the primary treatment options are surgery in combination with chemotherapy. While initially effective, these treatments are often fruitless at abating the malignancy due to the persistence of microscopic disease and the development of chemoresistance [3]. EOC patients desperately need new treatments, and EVs may provide an opportunity to gain an improved understanding about EOC proliferation and metastasis while hopefully providing novel, effective treatments.

1.1 Why is epithelial ovarian cancer so hard to treat, and how can EVs help?

Worldwide, ovarian cancer is the seventh most common malignancy among women; and over 280,000 cases were diagnosed in 2012 alone [1, 4]. In the United States ovarian cancer is the fifth deadliest cancer among women and is the deadliest cancer originating in the female reproductive system [2]. The most common type of ovarian cancer is EOC, making up more than 90% of cases [5]. EOC encompasses numerous histologic subtypes, including serous, mucinous, endometrioid, and clear cell types; additionally, EOC can proliferate rapidly, known as high-grade disease, or have a more insidious course, known as low-grade disease [6]. Interestingly, in the past decade researchers discovered that high-grade serous EOC originates in the fallopian tubes and then migrates to the ovary; so clinicians treat EOC and fallopian tube cancer as the same entity [7]. Additionally, high-grade serous fallopian tube, ovarian, and primary peritoneal cancer are all considered the same clinical entity based on common behaviors and treatments [8]. The chapter will primarily discuss high-grade epithelial ovarian carcinoma of the ovary, fallopian tubes, and peritoneum because it is the predominant subtype of EOC and because publications prioritize this subtype when studying EVs.

EOC is a difficult disease. When testing detects this malignancy at an early stage, 80% of these patients are free of cancer at 5 years [8]. However, the early signs of EOC are nonspecific and insidious, ranging from abdominal discomfort or pain to bloating and early satiety [1]. Unfortunately, these vague symptoms lead to a late diagnosis for most patients, with about 80% of patients diagnosed with advanced disease that is more challenging to cure [9]. While surgery and chemotherapy are initially effective in treating advanced EOC, most patients experience a relapse of the cancer that is chemoresistant, with a five-year survival under 30% [5]. Based on these grim outcomes, patients need new diagnostic and therapeutic tools to improve detection and treatment of EOC.

1.2 Why are EVs so exciting?

EVs are generating excitement within the field of gynecologic oncology because they not only help researchers to better understand how cancers grow and spread but also because they can assist with the diagnosis and management of EOC at every step of the disease course. EOC EVs carry a wide array of information including microRNA (miRNA), non-coding RNA, messenger RNA, DNA, lipids, glycans, and proteins that play a role in the proliferation and metastasis of this disease. In fact, patients with EOC are known to have an upregulation in EV secretion, transforming the microenvironment surrounding the cancer and causing normal cells to secrete tumorigenic factors [10]. Researchers will someday be able to detect EOC EVs readily in blood or urine for early detection of the cancer so that clinicians can treat it at an earlier stage, preventing metastasis and resistance to chemotherapy from ever occurring. By understanding the information carried inside of EVs, patients will have access to personalized treatment regimens specifically tailored to their cancer. By exploring different ideas, scientists will be able to unlock the potential for EVs to provide dramatic breakthroughs in the diagnosis and treatment of EOC.

2. Extracellular vesicles and their role in epithelial ovarian cancer

2.1 What is an extracellular vesicle?

For the past decade the classification of EVs has been based on size, ranging from exosomes that are 30–100 nm, microvesicles (MVs) that are 100–1000 nm, and apoptotic bodies that are 0.1–5 μm [1]. Exosomes are the smallest EV and appear to originate within the lumen of multivesicular bodies [1]. Oncosomes, a subtype of MVs, are released by budding from malignant cells [1]. As cells undergo apoptosis, they release apoptotic bodies [1]. One factor that limits this classification system is that some EVs that function as oncosomes are larger than the typical 100–1000 nm and can be as large as 1–10 μm [1]. When trying to isolate and study EVs, it became apparent that size did not adequately capture the breadth of heterogeneity among EVs with their varied functions and content. In 2019 the International Society of Extracellular Vesicles published a recommendation for the use of the term *extracellular vesicle* to encompass all types of EVs while still including a subclassification system that incorporated size [11]. Therefore, in this chapter the term *extracellular vesicle* will be used. Eventually, the optimal method of classification for EVs will be based on the specific phenotype and content of an EV that would better describe its origin and function. However, current testing methods and understanding of this topic need to be further studied.

2.2 How do extracellular vesicles impact epithelial ovarian cancer?

EVs provide a pertinent target for research because EOC cells exploit EVs for intercellular messaging. By hijacking the EV communication system, cancer cells distort key biological processes that enhance cancer survival, including angiogenesis, immunity, apoptosis, inflammation, migration, invasion, and even activation of secretion of tumorigenic factors by stem cells [1]. Malignant cells dramatically increase EV synthesis, manipulating crucial intercellular communication, exerting control over the tumor's surrounding environment, and transforming this microenvironment into a tumorigenic niche that facilitates chemoresistance and progression of disease [10]. Since EVs affect many aspects of EOC propagation and spread, they are suitable candidates for the development of new diagnostic and therapeutic management options.

3. Role of EVs in the diagnosis of epithelial ovarian cancer

3.1 How good are we at diagnosing epithelial ovarian cancer?

With the current tools available, clinicians are unable to reliably identify EOC early in its disease course, losing a valuable opportunity at early intervention and higher rates of cure. For patients who are diagnosed with EOC at stage I, disease that is confined to the ovaries, their five-year survival approaches 90% [2]. Survival drops precipitously for women with advanced stages of the disease, which is unfortunately the most common presentation. Attempts at establishing screening systems have certainly been investigated, with the United Kingdom famously conducting a randomized controlled study in which women were screened for EOC with a combination of serum markers and ultrasound [12]. When compared to women who underwent no screening, no impact was observed on overall survival from EOC [12]. Based on the results of this trial, no screening is currently recommended for the general population because modern diagnostic tests do not help patients that are diagnosed with EOC live longer and these same tests lead more women with benign ovarian diseases to have unnecessary procedures because the testing does not distinguish well between benign ovarian disease and cancer [12]. EVs present a promising new frontier for EOC screening because they are detectable in the serum and urine of patients, providing a potential novel method for diagnosing this cancer at an early stage when the patient can be cured more easily.

3.2 MicroRNA in EVs: how can they help to diagnose epithelial ovarian cancer?

One promising method for early cancer detection involves the analysis of EVs carrying microRNA, or miRNA, in the blood of patients. As strands of non-coding RNA that are 19–25 nucleotides in length, miRNAs are transcription products of DNA that regulate genes, a process that can activate or suppress the expression of different factors that can promote the growth and metastasis of EOC [1]. While most miRNA found in body fluids is cell-free and easily degradable, miRNA that is present in EVs is more stable, amplifying the role of this information in intercellular communication because it reaches cells more effectively [13]. When normal cells, such as stem cells, receive the miRNA from EVs, they produce tumorigenic factors that enhance the cancer's ability to survive and promote invasion and dissemination [1].

When compared to healthy individuals, patients with EOC have levels of certain circulating miRNAs carried in EVs that are often elevated [13]. Numerous specific miRNAs have already been linked to EOC. For example, miR-222-3p, which Ying et al. showed promotes the conversion of normal macrophages into tumor-supporting macrophages through the activation of the SOCS3/STAT3 pathway, is elevated in patients with EOC. Once normal macrophages are transformed, they exert immunosuppressive effects that assist EOC cells in evading identification while also secreting factors that promote migration and growth. Since EV miR-222-3p levels are increased in this cancer, its detection in serum can serve as a diagnostic biomarker for early detection [14].

In another study Cappellesso et al. identified elevated levels of EVs with miR-21, a known regulator of the tumor suppressor gene programmed cell death 4 (PDCD4), in patients with EOC compared to patients with benign ovarian disease [15]. The gene PDCD4 typically prevents cancer through the regulation of apoptosis. However, in EOC, the increased expression of miR-21 directly inhibits PDCD4, allowing the cancer cell to further mutate and to invade other tissues. Similarly to miR-222-3p, EVs with miR-21 can enhance diagnostic testing and clinical staging of EOC.

While looking at individual EV miRNAs can provide clues for early detection of EOC, their true value will come from evaluating the miRNAs in large groups as diagnostic panels that together will provide screening with high sensitivity and specificity. In their study Taylor and Gercel-Taylor reviewed a panel of 8 miRNAs found in EVs—miR-141, miR-214, miR-200a, miR-200b, miR-200c, miR-21, miR-205—that displayed distinct biological profiles between patients with benign ovarian disease and those with EOC [16]. By utilizing this panel of EV miRNAs and including other EV miRNAs, a simple blood sample may serve as a powerful test that can be employed by clinicians to apprehend EOC in asymptomatic populations before it lethally spreads.

Proteins are also transported in EVs and can potentially serve as biomarkers for early diagnosis of EOC. One example of these EV proteins is EpCAM, which is recognized for its role in tumorigenesis and tumor proliferation and is elevated in patients with EOC. However, the diagnostic utility of EpCAM and other proteins is limited because the proteins can be elevated in patients with benign ovarian disease, decreasing the specificity of these markers [16]. If such proteins are then implemented into screening protocols, patients may have false-positive test results and may subsequently undergo invasive procedures with their associated complications without any benefit.

However, some proteins carried by EVs appear to be specific to EOC. CD24, a known poor prognostic marker for EOC, can be detected within EVs in malignant ascites of EOC patients [17]. Additionally, about half of the blood samples from a cohort of EOC patients contained EV claudin-4, another protein that can potentially serve as a diagnostic marker [18]. With the development of new diagnostic panels that combine EV proteins and miRNAs, patients will 1 day obtain testing that identifies EOC early and gives them a better chance at a cure.

Even easier to obtain than blood, urine is another potentially rich source for EOC EVs. Studies have identified numerous EV miRNAs such as miR-92a and miR-30a-5p that are elevated in the urinary samples of patients with EOC when compared to healthy controls [19, 20]. Specifically, miR-30a-5p is elevated in EOC but decreased in other malignancies such as gastric and colon cancer, making it a potentially unique biomarker [20]. While EV miRNAs found in urine are a potentially exciting

Type of EV Content	Content
miRNA	miR-21
	miR-30a-5p
	miR-21
	miR-92a
	miR-141
	miR-155
	miR-181a
	miR-200a
	miR-200b
	miR-200c
	miR-205
	miR-214
	miR-222-3p
	miR-223
	miR-486
	miR-1908
Protein	CD24
	EpCam
	Claudin-4

Table 1.
Potential panel of EV biomarkers for the diagnosis of epithelial ovarian cancer [1, 16, 17, 19, 20, 22].

biomarker for diagnosing EOC, more research is required to further take advantage of this easily accessible opportunity.

While these many EV factors provide appealing options for future diagnostic applications, some barriers hinder the utilization of EVs in the clinical setting. Current methods for isolation and purification of EVs are still constrained, relying on identification of these vesicles by size, a non-specific criterion that does not distinguish EVs from large proteins and other types of vesicular structures. The purification process involves ultracentrifugation, a process that is inefficient and cumbersome, especially for serum samples [21]. Also, current methods of molecular identification are limited by the small size of EVs as well as by the difficulty in detecting the EV content [21]. Once scientists solve these issues and answer other questions regarding the viability and concentration of EVs in blood and urine samples, the detection of EOC EVs will bolster the strength of diagnostic tools (**Table 1**).

4. Role of EVs in the prognosis of epithelial ovarian cancer

EVs are positioned to provide valuable prognostic information for EOC because current prognostic tools struggle to accurately predict an individual's disease course and response to treatments. If there was a better understanding of how a patient's particular cancer would grow and which medicines would be effective against it, providers would better optimize treatment strategies that would extend a patient's life and even grant a better opportunity for cure. Currently, the prognosis for EOC is estimated based on generalized characteristics about this disease process within the context of the patient's health status and medical history [8]. Some factors include age, stage of the cancer at the time of diagnosis, and performance status [8]. In recent years genetic research has played a significant role in patient prognosis. BRCA mutations, a pathologic process that affects the repair of double-strand DNA breaks, place patients at increased lifetime risk for EOC but also confer an improved prognosis for EOC especially with new therapies that are targeted toward patients with these mutations. While these factors provide some helpful guidance regarding a patient's treatment outcomes, neither providers nor patients can accurately predict how an individual patient's EOC will respond to therapies. However, with EVs new factors are being identified that can help in better understanding which patients will respond to certain therapies and what personalized treatment regimens will best address the cancer.

An important part of caring for patients with EOC is selecting the best treatment for their specific tumor. When determining a patient's clinical management, the available prognostic information offers limited value in guiding clinicians about how to best care for their patients. However, novel therapeutic agents are demonstrating the need for refined prognostic tools that can identify a particular tumor's sensitivity or resistance to certain treatments. For example, the breakthrough use of PARP inhibitors for the treatment of EOC over the past decade served as an important demonstration of the necessity to discover new patient factors that facilitate targeted treatments [23]. In cancer cells with impaired repair of double-stranded DNA breaks, also known as homologous recombination deficiency (HRD), PARP inhibitors promote double-stranded breaks through the inhibition of secondary single-stranded DNA repair that triggers apoptosis [23]. When EOC with HRD is treated with a PARP inhibitor, these patients experience a significant improvement in the management of their cancer. Therefore, patients with EOC are now tested for homologous recombination deficiency [23]. While PARP inhibitors are clearly a success, patients need new biomarkers for individualized treatments; and EVs can be these new targets.

EOC quickly becomes resistant to front-line chemotherapy regimens, but it is currently not possible to predict which patients will develop chemoresistance [16]. Evidence from multiple studies suggest that EVs can predict which patients will have a tumor that is sensitive to chemotherapy [1]. For example, Yan studied a cohort of 50 patients and demonstrated an increase in serum EV annexin A3 levels, a protein involved in exocytosis and vesicle trafficking, among patients with resistance against primary chemotherapy drugs when compared to patients that are still sensitive to those chemotherapies [24]. In a second study protein RAB7A functioned as a potential mediator of chemoresistance [25]. Functioning as a key regulator of the influx of chemotherapy agents into cells, RAB7A is downregulated in chemoresistant cells, potentially affecting drug sequestration [25]. Finally, some groups are studying serum panels of EV miRNAs that are associated with chemoresistance and include miR-181a, miR-1908, miR-21, miR-486, and miR-223 [22]. Together, these different EV molecular targets may serve as prognostic biomarkers to identify chemoresistance in patients with EOC and help tailor the appropriate medication combination for each patient.

5. Can extracellular vesicles inspire novel cancer therapies?

Based on its role in tumor invasion, chemoresistance, angiogenesis, cancer metastasis, and immunologic suppression, EVs present a promising opportunity to target important regulators of EOC progression and to mobilize the immunologic response to combat the malignancy. EVs and associated miRNAs are generating excitement as novel therapeutic targets for drug development.

5.1 EVs as a drug delivery system

EVs can be employed as a drug delivery system that targets cancer cells directly. Harboring packages of chemotherapeutic agents, the EVs can be manufactured to express cell surface antigens and receptors that target it toward cancer cells and spare normal cells, maximizing cytotoxic effect while sparing healthy tissue. In recent work Tang developed a model in which they incubated tumor cells with chemotherapy; and the tumor cells subsequently packaged the chemotherapy into EVs [26]. Tang's group then took these EVs and demonstrated tumor-killing effect in mice with minimal side effects [26]. Therefore, if researchers translate this murine model into a therapy for EOC patients, EVs can be customized to carry the antitumor agents that are effective for a particular tumor. While chemotherapy affects tumor cells, it is also a treatment that damages healthy tissue, causing patients to experience a wide range of side effects. By having a treatment that can precisely target cancer cells while sparing normal tissues, clinicians could safely administer treatments to patients that could control or even cure EOC while avoiding harm.

5.2 Why are EVs so important in establishing the tumor microenvironment? Can the microenvironment become a target for new treatments?

In developing new therapeutic targets, one key area of focus is the immediate, small-scale environment surrounding cancer cells known as the microenvironment. The tumor microenvironment includes the extracellular matrix, neighboring blood

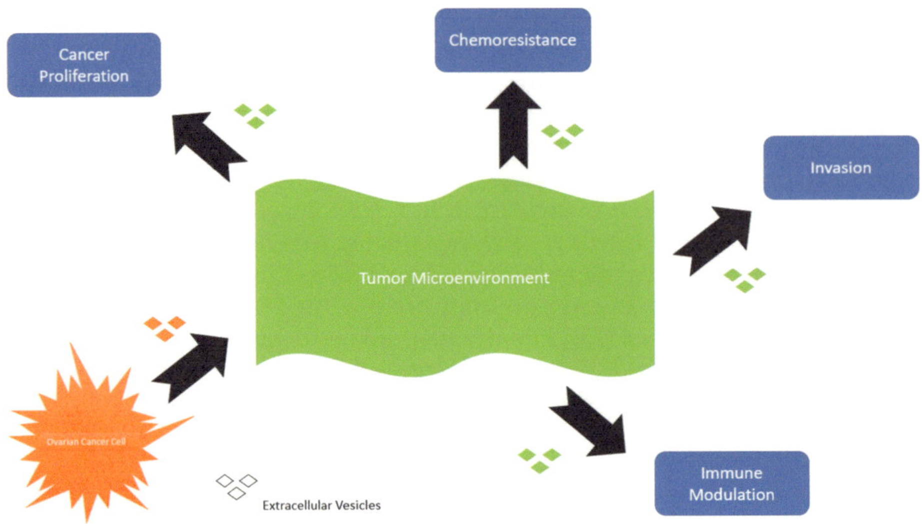

Figure 1.
The interaction between EOC cells and their microenvironment. Cancer cells release EVs that instruct normal cells to produce EVs that support cancer growth and invasion while also facilitating the development of chemoresistance and protection from the immune system [1, 2].

vessels, stromal cells such as macrophages and fibroblasts, stem cells, signaling molecules, and immune cells [22]. For EOC the microenvironment becomes an integral component for drug targeting because of the role that it plays in protecting the tumor from chemotherapy and the immune system as well as in promoting proliferation, invasion, and metastasis. By targeting EOC EVs that transform healthy tissue into this tumorigenic niche, researchers will enhance the effectiveness of current treatments by reversing chemoresistance as well as limit the deadly growth and spread of this disease (**Figure 1**).

EVs derived from EOC cells promote the transformation from a normal microenvironment into a tumorigenic one through intercellular communication that stimulates angiogenesis, immune suppression, and stromal invasion [27]. Specifically, altered expression of miRNA such as miR-214, miR-31, and miR-155 has been linked to the conversion of fibroblasts, a support cell within the connective tissue, into cancer-associated fibroblasts (CAFs)— cells that participate in cancer propagation, support of the tumorigenic microenvironment, alteration of the extracellular matrix, and metastasis [1]. The CAFs then produce EVs enriched with TGFβ1 that then trigger the invasive properties of the tumor. With almost deliberate malintent, the cancer cells drive their own invasive potential by directing the formation of CAFs that provide the necessary growth factors that allow the malignancy to spread. Following treatment with cisplatin, a frontline chemotherapy agent, EOC cells release EVs that promote tumorigenic activity of mesenchymal stem cells that eventually stimulate cancer progression [28]. By developing therapies targeted at these factors that transform the healthy tissue surrounding cancer into the tumorigenic microenvironment, scientists can inhibit the cancer's ability create its own protective environment that fuels its ability to grow and invade.

5.3 EVs and chemoresistance

Among the many challenges limiting the treatment of EOC, resistance to standard chemotherapy regimens exists as a frustrating inevitability in most patients with advanced disease; and EVs seem to play an integral role in this process. As the first-line regimen for EOC, platinum-based chemotherapy is the most effective treatment for EOC; yet 80% of patients with advanced EOC relapse, most within 2 years [3]. Following recurrence of the cancer, most people develop chemoresistance and succumb to the disease. An EOC that is *platinum-resistant* is defined as disease that recurs or progresses within 6 months of completion of the last treatment with a platinum-based regimen. Once a patient's cancer reaches this state, expectations for disease control change, with low response rates to subsequent chemotherapies and a median survival falling below 12 months [3].

While platinum resistance is a complicated, multifactorial process that still needs further elucidation, EVs may help to better understanding this transformation. EOC EVs function as an intercellular communication system. Interestingly and frighteningly, EVs excreted by platinum-resistant tumor cells are capable of inducing resistance in other tumor cells [29]. While this mechanism is not well understood, once the EVs that mediate this process are better defined, they can become targets for possible therapeutic intervention. Furthermore, by understanding the EV content that conveys chemoresistance between cancer cells, scientists can alter the EVs to send information directly to tumor cells that reverses this resistance, allowing first-line treatments to again become effective.

The cytotoxic effect of platinum-based drugs such as cisplatin relies on the uptake of the chemotherapy into cells followed by DNA binding, leading to the formation of DNA crosslinks and breaks that result in apoptosis [3]. In patients that develop platinum resistance, some of their cancer cells exhibit reduced uptake or increased efflux of platinum agents, a process that EVs may facilitate [1, 3]. Transport proteins that have been implicated in this mechanism of drug resistance such as the lysosomal proteins ATPase copper-transporting alpha and beta have been found in EOC EVs, allowing cancer cells to survive against chemotherapy [22]. Is it possible to negate the effect of these EVs through targeted therapies? By disrupting this EV communication system with antibodies or other novel therapies, researchers can provide hope to these patients by overcoming chemoresistance and making their chemotherapy more effective.

5.4 EVs and immunosuppression

One important technique that allows EOC to proliferate and spread is the ability to suppress the immune system. By further understanding the elaborate underlying mechanisms through which EOC EVs dampen immunity, researchers will be able to block immune escape by the cancer cells, producing new treatments. By preventing immune suppression within the tumorigenic microenvironment, ovarian cells that were previously protected within this nurturing space would be freshly susceptible to immune cells that could find and eliminate the cancer cells [1].

Given the significant promise for novel treatments for EOC that reactivate the suppressed immune system, studies are already underway that target EOC EVs. One therapy utilizes dendritic cells as a map that directs the immune system toward the cancer. In one study these dendritic cells, known for presenting specific foreign antigens to the immune system for identification and targeting, were exposed to EVs isolated from the ascites of EOC patients [1]. The dendritic cells then presented tumor-specific antigens from the cancer EVs to resting T cells that subsequently differentiated and then killed EOC cells [30]. Dendritic cells may be harvested from a patient with EOC, cultured with isolated EOC EVs, and then reintroduced to the patient as an autologous injection that then directs the patient's own T cells to eradicate the cancer. This concept elegantly demonstrates the potential for unleashing the immune system on cancer cells using EVs.

Another interesting avenue for treating EOC is through the utilization of immunoglobulins that directly target EVs. The serum of patients with EOC is more immunologically reactive when compared to the serum of healthy patients and patients with benign ovarian disease, indicating a robust immune response against the malignancy. As many studies have proven before, the natural immunoreactivity that the human body mounts against EOC is insufficient because the cancer employs tactics to evade the immune system, a process in which EVs play a significant role [22]. While immune evasion is a hallmark characteristic of EOC, the immune system may be mobilized against the cancer by findings ways to target EVs with immunoglobulins. Researchers can develop antibodies that specifically target EOC EVs, tagging them for the immune system so that they can be destroyed, effectively dismantling the vital EV communication system for the cancer cells and limiting the cancer's ability to grow and spread. While this novel use of EVs is exciting, more research is needed to use this method. Mainly, scientists need to better characterize EVs to develop targets for immunoglobulins. Also, it is difficult for antibodies to target the content within EVs because it is protected by the vesicular walls, so proteins on the vesicle wall may provide a unique target for the antibodies. As scientists better understand the unique protein signatures of EOC EVs, immunity-based therapeutics may provide promising new avenues for treating these patients.

5.5 EVs and angiogenesis

Angiogenesis, a vital component of cancer proliferation and progression, has become an important focus in the care of patients with EOC. Ovarian cancers have previously been recognized for their role in promoting angiogenesis; so, by targeting these specific EVs in combination with other antitumor treatments, more effective regimens may be developed for combating this cancer. In the study GOG 218, Burger et al. conducted a clinical trial in which they incorporated a vascular endothelial growth factor (VEGF) inhibitor into the standard primary chemotherapy regimen for advanced EOC [31]. While patients on the VEGF inhibitor experienced a longer period of progression-free survival, they did not live any longer when compared to those who did not receive the treatment. While the inhibition of VEGF, a family of proteins recognized for stimulating the formation of blood vessels, clearly has some effect on tumor growth, other factors appear to be at play that limit the effectiveness of this therapy. One explanation is that EVs play a role in angiogenesis that circumvents the use of VEGF. Ovarian cancer-derived EVs that contain proteins such as CD147, metastasis-associated protein 1, and activating transcription factor 2 appear to have a key effect on angiogenesis that promotes cancer proliferation [32, 33]. A treatment for EOC could include antibodies or some other novel therapy that targets cancer EVs that carry these proteins that stimulate angiogenesis and then eliminate the ability for the cancer to develop its own blood supply.

Another appealing area of active research is the study of common dietary supplements that may have antiangiogenic properties through the production of antiangiogenic EVs. A promising supplement, Amla extract, derived from the Indian Gooseberry tree, has long been suspected to have cancer preventative properties [34]. One recent study tested the supplement on EOC cells and noted increased expression of EV miR-375 which appears to block the proangiogenic proteins SNAIL1 and IGF1R [34]. With a better understanding of the mechanism of this supplement and many others, scientists may 1 day provide dietary recommendations that can enhance a patient's standard chemotherapy regimen or even derive a novel pharmacologic treatment that blocks blood vessel formation, helping to better destroy EOC cells [4].

6. Conclusion and clinical relevance

EOC remains a disease with a generally poor prognosis due to its asymptomatic early stages, ineffective screening mechanisms, and its predilection to develop chemoresistance with recurrence of disease. EVs are exciting within the field of EOC research because they provide the potential for many interventions that can save the lives of patients, ranging from diagnosing the cancer at earlier stages, identifying the optimal treatment for each individual patient, and even developing novel therapeutics that are more effective than the current regimens. With the ineffectiveness of screening tests, panels of EVs that can be detected in blood or urine provide hope for highly sensitive and specific tools that can give an accurate diagnosis of cancer in asymptomatic patients. Alternatively, by exploiting EVs to overcome chemoresistance, clinicians can redeploy existing treatments that typically become obsolete during a patient's disease course. The demand for new diagnostic tools and therapies for patients with EOC is high, and EVs can be the next frontier for seemingly miraculous advancements in cancer care.

Author details

Diego Aviles[1], David Warshal[1], Lauren Krill[1] and Olga Ostrovsky[2]*

1 Department of Gynecologic Oncology, MD Anderson Cancer Center at Cooper, Cooper University Health Care, Camden, NJ, USA

2 Department of Surgery, Division of Surgical Research, Cooper University Health Care, Camden, NJ, USA

*Address all correspondence to: ostrovsky-olga@cooperhealth.edu

References

[1] Lucidi A, Buca D, Ronsini C, Tinari S, Bologna G, Buca D, et al. Role of extracellular vesicles in epithelial ovarian cancer: A systematic review. International Journal of Molecular Sciences. 2020;**21**(22):8762. DOI: 10.3390/ijms21228762

[2] Islami F, Ward EM, Sung H, et al. Annual Report to the Nation on the Status of Cancer, Part 1: National Cancer Statistics. J Natl Cancer Inst. 2021;**113**(12):1648-1669. DOI: 10.1093/jnci/djab131. [Published online ahead of print, 2021 Jul 8]

[3] Pokhriyal R, Hariprasad R, Kumar L, Hariprasad G. Chemotherapy resistance in advanced ovarian cancer patients. Biomark Cancer. 2019;**11**: 1179299X19860815. Published 2019 July 5. DOI: 10.1177/1179299X19860815

[4] Zhang Y, Luo G, Li M, et al. Global patterns and trends in ovarian cancer incidence: age, period and birth cohort analysis. BMC Cancer. 2019;**19**(1):984. Published 2019 October 22. DOI: 10.1186/s12885-019-6139-6

[5] Li SS, Ma J, Wong AST. Chemo-resistance in ovarian cancer: Exploiting cancer stem cell metabolism. Journal of Gynecologic Oncology. 2018;**29**(2):e32. DOI: 10.3802/jgo.2018.29.e32

[6] Mahmood RD, Morgan RD, Edmondson RJ, Clamp AR, Jayson GC. First-line management of advanced high-grade serous ovarian cancer. Current Oncology Reports. 2020;**22**(6):64. Published 2020 Jun 4. DOI: 10.1007/s11912-020-00933-8

[7] Perets R, Wyant GA, Muto KW, et al. Transformation of the fallopian tube secretory epithelium leads to high-grade serous ovarian cancer in Brca;Tp53;Pten models. Cancer Cell. 2013;**24**(6): 751-765. DOI: 10.1016/j.ccr.2013.10.013

[8] Stewart C, Ralyea C, Lockwood S. Ovarian Cancer: An Integrated Review. Seminars in Oncology Nursing. 2019;**35**(2):151-156. DOI: 10.1016/j.soncn.2019.02.001. Epub 2019 March 11

[9] Torre LA, Trabert B, DeSantis CE, Miller KD, Samimi G, Runowicz CD, et al. Ovarian cancer statistics, 2018. CA: a Cancer Journal for Clinicians. 2018; **68**(4):284-296

[10] Nakamura K, Sawada K, Kobayashi M, et al. Role of the exosome in ovarian cancer progression and its potential as a therapeutic target. Cancers (Basel). 2019;**11**(8):1147. Published 2019 August 10. DOI: 10.3390/cancers11081147

[11] Thery et al. Minimal information for studies of extracellular vesicles 2018 (MISEV2018): A position statement of the International Society for Extracellular Vesicles and update of the MISEV2014 guidelines. Journal of Extracellular Vesicles. 2018;7:1. DOI: 10.1080/20013078.2018.1535750

[12] Jacobs IJ, Menon U, Ryan A, Gentry-Maharaj A, Burnell M, Kalsi JK, et al. Ovarian cancer screening and mortality in the UK Collaborative Trial of Ovarian Cancer Screening (UKCTOCS): A randomised controlled trial. Lancet. 2016;**387**(10022):945-956. DOI: 10.1016/S0140-6736(15)01224-6. Epub 2015 December 17. Erratum in: Lancet. 2016 March 5;387(10022):944. Erratum in: Lancet. 2016 March 5;387(10022):944

[13] Wang W, Yin Y, Shan X, Zhou X, Liu P, Cao Q, et al. The value of plasma based MicroRNAs as diagnostic

biomarkers for ovarian cancer. The American Journal of the Medical Sciences. 2019;**358**:256-267

[14] Xiang Y, Quanfeng W, Xiaoli W, Qinyi Z, Xinjing W, Lu J, et al. Epithelial ovarian cancer-secreted exosomal miR-222-3p induces polarization of tumor-associated macrophages. Oncotarget. 2016;**7**(28):43076-43087

[15] Cappellesso R, Tinazzi A, Giurici T, Simonato F, Guzzardo V, Ventura L, et al. Programmed cell death 4 and microRNA 21 inverse expression is maintained in cells and exosomes from ovarian serous carcinoma effusions. Cancer Cytopathology. 2014;**122**(9):685-693

[16] Taylor DD, Gercel-Taylor C. MicroRNA signatures of tumor-derived exosomes as diagnostic biomarkers of ovarian cancer. Gynecologic Oncology. 2008;**110**(1):13-14

[17] Runz S, Keller S, Rupp C, Stoeck A, Issa Y, Koensgen D, et al. Malignant ascites-derived exosomes of ovarian carcinoma patients contain CD24 and EpCAM. Gynecologic Oncology. 2007;**107**(3):563-571

[18] Li J, Sherman-Baust CA, Tsai-Turton M, Bristow RE, Roden RB, Morin PJ. Claudin-containing exosomes in the peripheral circulation of women with ovarian cancer. BMC Cancer. 2009;**9**:244

[19] Zavesky L, Jandakova E, Turyna R, Langmeierova L, Weinberger V, Minar L. Supernatant versus exosomal urinary microRNAs. Two fractions with different outcomes in gynaecological cancers. Neoplasma. 2016;**63**(1):121-132

[20] Zhou J, Gong G, Tan H, Dai F, Zhu X, Chen Y, et al. Urinary microRNA30a-5p is a potential biomarker for ovarian serous adenocarcinoma. Oncology Reports. 2015;**33**(6):2915-2923

[21] Zhao Z, Yang Y, Zeng Y, He M. A microfluidic ExoSearch chip for multiplexed exosome detection towards blood-based ovarian cancer diagnosis. Lab on a Chip. 2016;**16**(3):489-496

[22] Li X, Wang X. The emerging roles and therapeutic potential of exosomes in epithelial ovarian cancer. Molecular Cancer. 2017;**16**:92. DOI: 10.1186/s12943-017-0659-y

[23] Mirza MR, Coleman RL, González-Martín A, Moore KN, Colombo N, Ray-Coquard I, et al. The forefront of ovarian cancer therapy: update on PARP inhibitors. Annals of Oncology. 2020;**31**(9):1148-1159. DOI: 10.1016/j.annonc.2020.06.004. Epub 2020 Jun 20. Erratum in: Ann Oncol. 2021 Aug;32(8):1066-1067

[24] Yan XD, Yin J, Yao H, et al. Increased expression of annexin A3 is a mechanism of platinum resistance in ovarian cancer. Cancer Research. 2010;**70**:1616-1624

[25] Guerra F, Paiano A, Migoni D, Girolimetti G, Perrone AM, De Iaco P, et al. Modulation of RAB7A protein expression determines resistance to cisplatin through late endocytic pathway impairment and extracellular vesicular secretion. Cancers (Basel). 2019;**11**(1): 52. DOI: 10.3390/cancers11010052

[26] Tang K, Zhang Y, Zhang H, Xu P, Liu J, Ma J, et al. Delivery of chemotherapeutic drugs in tumour cell-derived microparticles. Nature Communications. 2012;**3**:1282

[27] Tang MK, Wong AS. Exosomes: Emerging biomarkers and targets for ovarian cancer. Cancer Letters. 2015;**367**(1):26-33

[28] Vera N, Acuña-Gallardo S, Grünenwald F, Caceres-Verschae A, Realini O, Acuña R, et al. Small

extracellular vesicles released from ovarian cancer spheroids in response to cisplatin promote the pro-tumorigenic activity of mesenchymal stem cells. International Journal of Molecular Sciences. 2019;**20**(20):4972. DOI: 10.3390/ijms20204972

[29] Pink RC, Samuel P, Massa D, Caley DP, Brooks SA, Carter DR. The passenger strand, miR-21-3p, plays a role in mediating cisplatin resistance in ovarian cancer cells. Gynecologic Oncology. 2015;**137**(1):143-151

[30] Li QL, Bu N, Yu YC, Hua W, Xin XY. Exvivo experiments of human ovarian cancer ascites-derived exosomes presented by dendritic cells derived from umbilical cord blood for immunotherapy treatment. Clin Med Oncol. 2008;**2**: 461-467. DOI: 10.4137/cmo.s776

[31] Burger RA, Brady MF, Bookman MA, Fleming GF, Monk BJ, Huang H, et al. Incorporation of bevacizumab in the primary treatment of ovarian cancer. The New England Journal of Medicine. 2011;**365**(26):2473-2483. DOI: 10.1056/ NEJMoa1104390

[32] Millimaggi D, Mari M, D'Ascenzo S, Carosa E, Jannini EA, Zucker S, et al. Tumor vesicle-associated CD147 modulates the angiogenic capability of endothelial cells. Neoplasia. 2007;**9**(4): 349-357

[33] Yi H, Ye J, Yang XM, Zhang LW, Zhang ZG, Chen YP. High-grade ovarian cancer secreting effective exosomes in tumor angiogenesis. Int J Clin Exp Pathol. 2015;8:5062-70 Shender VO, Pavlyukov MS, Ziganshin RH, Arapidi GP, Kovalchuk SI, Anikanov NA, Altukhov IA. Proteome-metabolome profiling of ovarian cancer ascites reveals novel components involved in intercellular communication. Molecular & Cellular Proteomics. 2014;**13**(12):3558-3571

[34] De A, Powers B, De A, et al. Emblica officinalis extract downregulates pro-angiogenic molecules via upregulation of cellular and exosomal miR-375 in human ovarian cancer cells. Oncotarget. 2016;**7**(21):31484-31500. DOI: 10.18632/ oncotarget.8966

Chapter 14

Engineering of Extracellular Vesicles as Nano Therapy for Breast Cancer

Sapna Jain and Manjari Singh

Abstract

Extracellular vesicles are membrane-derived nanoparticles that represent a novel mechanism of cell-to-cell communication. It is well reported that EVs play a central role in the tumor microenvironment by mediating intercellular signaling among cancer cells. This has resulted in the development of therapeutic strategies targeting various EV signaling pathways in cancer. However, because of their small size and endogenous origin, they have been extensively explored for cancer drug delivery. Hence, owing to their natural ability to mediate intercellular communication, high stability, and low immunogenicity, they have emerged as an attractive platform for cancer treatment. However, limited production and insufficient loading with therapeutic moieties are some of the issues constraining their clinical translation. In this chapter, recent research studies performed in an attempt to develop EVs as cancer biomarkers or drug delivery systems will be discussed. Further, it will also discuss various strategies such as direct and indirect cell surface modification, which can be employed to make EVs successful as cancer therapeutics. Furthermore, it will highlight the current and completed clinical trials using naturally derived EVs as cancer therapeutics.

Keywords: breast cancer, extracellular vesicles, engineering, drug delivery, biomarker

1. Introduction

In women, breast cancer is a prevalent cause of cancer worldwide [1]. It affected 2.3 million women globally in 2020, with 685,000 deaths. It has been diagnosed in 7.8 million women in the past 5 years, making it the most common type of cancer in the world [1]. Although breast cancer diagnostic methods and therapeutic procedures have improved in the past decade, the long-term survival of these patients remains low due to a high rate of postsurgical relapse. The efficacy of breast cancer treatment is limited by drug toxicity, multidrug resistance, and a lack of definitive prognostic biomarkers [2]. Thus, there is an urgent need to develop novel biomarkers and therapeutics to cure the disease.

In recent years, many studies have suggested that intercellular communication plays a key role in driving various cellular functions and homeostasis in physiological as well as pathological conditions such as cancer, cardiovascular diseases, and neurological disorders. Cancer development is mainly dependent on interactions between cancerous cells and their microenvironment components. Some of these interactions are mediated by extracellular vesicles, which alter the phenotype of recipient cells [3–5].

Extracellular vesicles (EVs) are spherical nanoparticles shed by all types of cells, including archaea, prokaryotes, eukaryotes, and fungi in the extracellular milieu [6]. These typically range from 30 nm to 5 μm in diameter based on their type and vary widely in composition [7]. In addition to being released during disease pathology, EVs allow various cells to send and receive messages to crosstalk with other cells, thus carrying out various biological functions [7]. These are mainly composed of different proteins, lipids, nucleic acids, and enzymes [8]. EVs circulate through many body fluids, such as blood, serum, and urine. Owing to their structural similarity to the parental source, they are considered potential biomarkers for diseases such as cancer [9]. To study the characteristics and functions of EVs, they are isolated using different techniques such as differential ultracentrifugation, size-exclusion, and ultrafiltration [10].

EVs are generally categorized into exosomes, microvesicles, and apoptotic bodies according to their release mechanism, size, and composition [3]. Exosomes are 30–150 nm in diameter and are formed by inward budding of the plasma membrane of the cell [9]. Microvesicles are formed by direct outward budding of the cell's plasma membrane and range in size from 100 to 1000 nm in diameter. Consequently, they are reported to contain mainly cytosolic and plasma membrane proteins, such as tetraspanins. Apoptotic bodies are shed during cell death into the extracellular space, ranging from 50 to 5000 nm in diameter. These generally contain intact organelles, glycosylated proteins, and chromatin, unlike the other two types of EVs. Among these, exosomes have been widely studied since their role in intercellular communication has been reported. This chapter will focus on exosomes and their potential applications as therapeutics for breast cancer.

2. Biogenesis, contents and functions of exosomes

Exosomes are generated by the endocytic pathway from late endosomes (LE) [11, 12]. LEs are formed by inward budding of the multivesicular body (MVB) membrane. LE membranes invaginate to form intraluminal vesicles (ILVs) within MVBs. During this process, some proteins are engulfed and packaged within the ILVs. ILVs then fuse with the cell's plasma membrane and release the vesicles into the extracellular space.

As reported previously, the formation of ILVs can occur either dependent or independent of the ESCRT complex. The ESCRT complex is a set of proteins that function together to facilitate the formation of MVBs, vesicle release, and protein cargo sorting [13–15]. ESCRT 0 has two subunits, HRS and STAM ½, which bind together and recognize specific ubiquitinated proteins in early endosomes. This leads to the recruitment of ESCRT 1 containing Tsg 101, Vps28, Vps37, and Mvb 12, which further recruits ESCRT II. ESCRT II is composed of four subunits, Vps22-EAP30, Vps36—EAP45 and Vps25—EAP20 which starts the invagination of endosomal membranes encapsulating different molecules/cargo such as proteins and nucleic acids.

The ESCRT II subunit Vsp25 then binds with Vsp20 to activate and recruit ESCRT III. It deubiquitinates proteins and allows complete membrane invagination, generating ILVs. Other adaptor proteins such as Vps4 interact with ESCRT III to finally start budding of the membrane, ECSRT subunit removal, and cargo delivery. Hence, the ESCRT complex regulates the whole process of vesicle budding and cargo sorting into exosomes [16, 17]. In cancer, an increased amount of exosomes is often observed in the bodily fluids of cancer patients as a result of deregulation of exosomal formation and secretion [18]. Specifically, in breast cancer, the amount of exosomes released by the human tumor cell line B42 clone 16 was much larger than that released by the parental normal mammary epithelial cells (HMEC B42), as shown by Azmi et al. [19].

Exosomes are composed of a heterogeneous set of cytosolic, nuclear, mitochondrial, ribosomal, and membrane-bound proteins derived from donor cells [20]. Some of these proteins are conserved irrespective of their origin; therefore, they are considered exosomal markers such as tetraspanins, ESCRT proteins, and major histocompatibility complex (MHC) molecules [21]. In addition, some proteins are related to the phenotype of producing cells, such as cancer-derived exosomes, which in turn determines their biological mechanisms. The lipid bilayer membrane of exosomes contains transmembrane proteins, transporter proteins, adhesion molecules, and lipid raft-associated proteins. Exosomes contain nucleic acids such as DNA (ssDNA, mtDNA, dsDNA, and RNA (mRNA, miRNA, and lncRNA) [22]. Exosomal miRNAs and mRNAs are transferred from donor cells to recipient cells, thus modulating the latter's phenotype. Although there are numerous reports indicating the presence of DNA within exosomes, the mechanisms leading to this phenomenon remain unclear. Exosomes also exhibit an exclusive set of lipids distributed in their bilayers, such as sphingolipids, arachidonic acid, cholesterol, phosphatidylserine, and

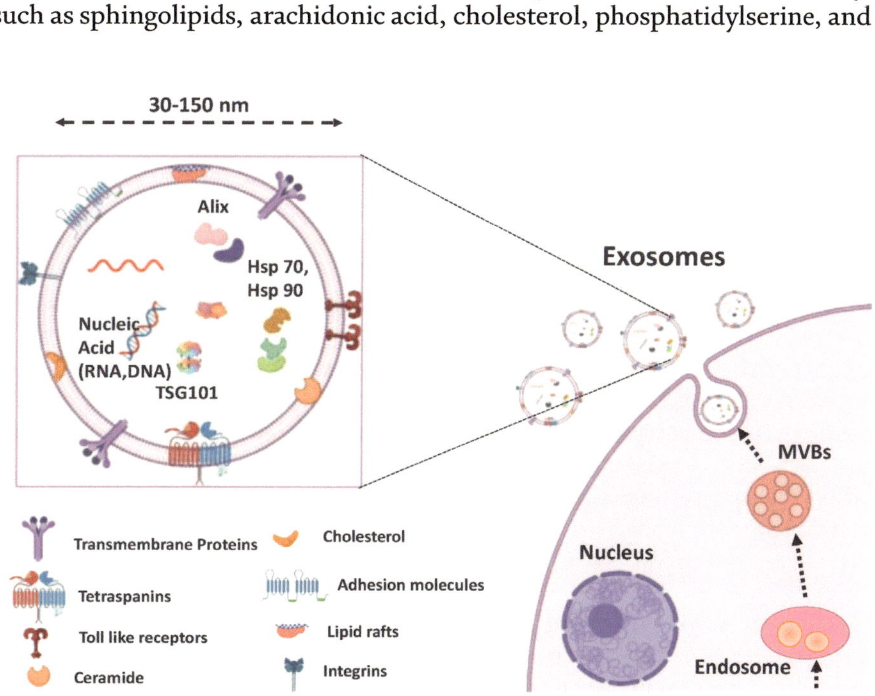

Figure 1.
Schematic representation of exosome biogenesis and composition. Exosomes originate from multivesicular bodies and shed into extracellular space packaging motley of proteins such as ESCRT associated protein, chaperones, along with ssDNA, RNA and dsDNA, miRNA and lncRNA.

ganglioside [23, 24]. Lipids such as lysobisphosphatidic acid are abundant in the inner membranes of multivesicular bodies and play a crucial role in exosome formation [25, 26]. ExoCarta is a database containing all the data on exosomal content, with over 47,000 protein, mRNA, and lipid entries. Furthermore, ExoCarta is an excellent source of information for exosome characterization (**Figure 1**) [27].

3. Functional role of exosomes in breast cancer development

Metastasis is the process by which primary tumor cells/tumor cells/cancer cells invade the surrounding tissues and colonize the blood vessels to proliferate and give rise to the tumor [28]. Controlling metastasis, which is mainly responsible for high patient mortality, is the main challenge in breast cancer therapy. Hence, several investigations are ongoing to understand the molecular mechanisms underlying metastasis in breast cancer. Recently, exosomes have attracted great attention as key players in regulating complex intracellular pathways from initiation to progression to metastasis in the development of breast cancer [29–31]. These mainly interact with the recipient cells in three ways: direct fusion with the cell membrane, interaction with the surface receptors, or internalization via endocytosis. Upon cellular uptake, exosomes deliver their cargo and initiate a cascade of events leading to various biological functions. Many breast cancer cell lines have been shown to release exosomes containing several proteins with signaling molecules, miRNAs, and long non-coding RNAs involved in migration, invasion, angiogenesis, and metastasis [32–34]. Proteomic profiling of exosomes secreted from breast cancer cell lines was shown to contain matrix metalloproteinases, which might be linked to the enhanced metastatic properties of breast cancer cells [32]. These findings suggest that exosomes act as key mediators in the tumor microenvironment by communicating various signaling molecules essential for breast cancer development [31].

Exosome-mediated transfer of genetic material from breast cancer cells has been shown to mediate resistance to chemotherapy and enhance tumor growth [35, 36]. Accumulating evidence suggests that exosomes may also play a role in the resistance of breast cancer radiotherapy and cancer immunotherapy [37, 38]. In breast cancer, drug-resistant cancer cells transmit resistance in drug-sensitive cells via the intercellular horizontal transfer of exosomal miRNAs [38]. Exosomes also transfer the drug efflux pump from docetaxel-resistant to sensitive ones in MCF-7 breast cancer cells [39]. Lv MM et al. showed that exosomes from drug-resistant cancer cells contain miRNAs that alter the phenotype of recipient breast cancer cells by altering their transcriptome [40]. Exosomes from stromal fibroblasts transmit non-coding RNA to breast cancer cells, thus contributing to treatment resistance by expanding therapy-resistant cells [41]. Thus, exosomes contribute to drug resistance in breast cancer.

4. Therapeutic implication of EVs in breast cancer

The role of exosomes in carcinogenesis has been extensively investigated in recent years. Cancer cells have been shown to use exosomes as a novel mechanism to transfer the malignant phenotype to normal healthy cells and establish a niche for tumor growth. Cancer cell-derived exosomes are reported to contain miRNAs, proteins, or long non-coding RNAs that mediate cancer development, growth, and progression [42, 43].

Exosomes derived from breast cancer cells contain a variety of proteins and RNAs that are transmitted among these cells as well as normal cells, thus altering the phenotype of healthy mammary epithelial cells. Wang J et al. showed that cancer exosomes were able to transform normal mammary epithelial cells into cancerous cells via transfer of microRNAs packaged within exosomes [44]. Similarly, Melo et al. showed enhanced expression of exosomal miR-10b in metastatic breast cells compared to non-metastatic or non-malignant breast cells [45]. Thus, it can be used as a therapeutic target for breast cancer therapy.

These characteristics make exosomes ideal biomarkers, and exosomal profiling in the absence of tissue holds great promise for early diagnosis. Owing to their crucial functional role in breast cancer, exosomes have been investigated for their potential development as breast cancer biomarkers and therapeutic targets. Singh R et al. have shown that psoralen reduces the formation and secretion of exosomes, thus reversing multidrug resistance in breast cancer cells [46]. The presence of diverse content within and on the surface of exosomes has led to their application as biomarkers, diagnostics, and drug delivery. A large number of exosomes circulate within bodily fluids of not only healthy individuals but also cancer patients, according to some studies. Since exosomes play various significant roles in breast cancer, exosomes can be developed as potential therapeutic agents in biomarkers, diagnostics, and drug delivery. Kumar et al. investigated the release of exosomes from breast cancer stem cells to characterize their constituent exosomal markers. They detected tetraspanin proteins, Alix, and tumor susceptibility gene-101 (TSG101) in breast cancer stem cell-derived exosomes. This study indicates that secreted exosomes can be utilized as biomarkers for breast cancer to understand their development, progression, and metastasis [47]. Kumar et al. showed that miRNAs 155 and 205 are expressed in serum exosomes derived from breast cancer cells and modulate the epithelial-to-mesenchymal transition (EMT), growth, and metastasis of cancer, suggesting their employability as breast cancer biomarkers [48]. Zhang et al. studied the role of long non-coding RNA MALAT 1 which is highly expressed in exosomes derived from breast cancer cells in tumor progression, representing a potential treatment strategy for breast cancer [49]. Dong et al. investigated the role of exosomal long non-coding RNA in the chemoresistance of HER2+ breast cancer cells. They found that exosomal lncRNA-SNHG14 was not only upregulated in trastuzumab-resistant cells but also transmitted the lncRNA into drug-sensitive cells, thus disseminating trastuzumab resistance. Furthermore, when compared to patients who responded to trastuzumab, the expression level of serum exosomal lncRNA-SNHG14 was higher in patients who were resistant. This suggests that lncRNA-SNHG14 is a promising therapeutic target for HER2+ breast cancer patients [50].

5. Engineering exosomes as therapeutics for breast cancer

Exosomes are emerging as promising therapeutic agents because of their role in tumor-related processes and their ability to deliver their cargo, such as proteins, lipids, and nucleic acids, into the tumor sites. However, their full clinical applicability has not yet been realized. This is because of many factors, including low yield and relatively low percentage loading to the therapeutic moiety. As such, new approaches for mass production and enhancement of the percent loading need to be explored. In general, these approaches are divided into two categories: passive and active loading, which are discussed in detail in the following sections.

Direct modification, also known as non-cell-based loading or exogenous loading, refers to the direct loading of therapeutic moieties such as siRNA, miRNA, drugs, and proteins after the isolation and purification of exosomes from the cells. This may encompass a series of procedures such as incubation, freeze-thaw cycles, sonication, and electroporation, and thus can further be categorized into passive and active loading. Passive loading includes loading of therapeutic moiety into exosomes by diffusion; on the other hand, active loading includes disrupting the exosomal membrane by electroporation, sonication, or freeze thawing, thus allowing the therapeutic moiety to enter into these vesicles. In passive drug loading, exosomes are incubated with drugs and allowed to diffuse into vesicles along a concentration gradient. Because exosomes consist of a lipid bilayer, the drug loading efficiency depends largely on the hydrophobicity of the drugs. Dong et al. loaded curcumin into milk exosomes by incubating at 4°C overnight and reported 70.46% drug loading using an incubation method [51]. Similarly, Sun et al. incorporated curcumin into exosomes derived from a mouse lymphoma cell line by incubating in PBS at room temperature (22°C) for 5 min and showed a binding capacity of 2.9 g curcumin to 1 g of exosomes [52]. Sun et al. packaged Cho-miR159 (cholesterol-modified miRNA 159) along with doxorubicin into exosomes derived from the human monocytic cell line THP-1 by incubating in PBS at 37°C to deliver to triple-negative breast cancer cells [53]. Linezolid was incorporated into exosomes derived from the mouse macrophage cell line RAW 264.7, by mixing both and incubating at 37°C for 1 h, resulting in ~5% drug loading. The exosomal formulation of linezolid was more effective against MRSA infections than the free drug [54]. Although several studies have reported the use of incubation with exosomes for drug or any therapeutic agent loading, it often suffers from issues of low percent drug loading, urging a requirement for improved methods for higher drug loading percent. Another method (less common) of passive loading includes incubating the exosome donor cells with the drugs/therapeutic agents. First, the donor cells are exposed to drugs or therapeutic agents, followed by isolation of released exosomes (supposedly) containing the loaded drugs or therapeutic agents. This method was used in a study by Pascucci et al., wherein they exposed bone marrow-derived mesenchymal stromal cells (MSCs) with a very high concentration of paclitaxel followed by incubation at 37°C for 24 h. After incubation, the cells were washed twice with PBS, trypsinized, and seeded in a fresh flask for 48 h. After 48 h, cell-conditioned medium was collected to isolate exosomes containing paclitaxel. They found that MSC-PTX-derived exosomes had a greater inhibitory effect on tumor cell proliferation (**Figure 2**) [55].

For active cargo loading, the exosomal membrane is temporarily disrupted using different methods and then restored once the drug/therapeutic agent was loaded. These methods may include sonication, extrusion, freeze-thawing, electroporation, use of membrane permeabilizers, conjugation using click chemistry, and antibodies against exosomal surface proteins. Electroporation uses an electric field to generate small pores in the exosomal membrane to disturb the phospholipid bilayer of exosomes. Drug/therapeutic agents can enter these vesicles via the generated pores. Once they entered, the pores were closed to recover the exosomal membrane integrity. This method has mostly been used to encapsulate siRNA or miRNA into exosomes and has been reported to enhance the percent loading compared to the simple diffusion method. Jia et al. loaded exosomes derived from RAW 264.7 cells with curcumin and superparamagnetic iron oxide nanoparticles (SPIONs) synchronously using optimal electroporation conditions of 400 V, 150 μF, and 1 ms discharge time. They observed that electroporation had no effect on the membrane integrity of exosomes and

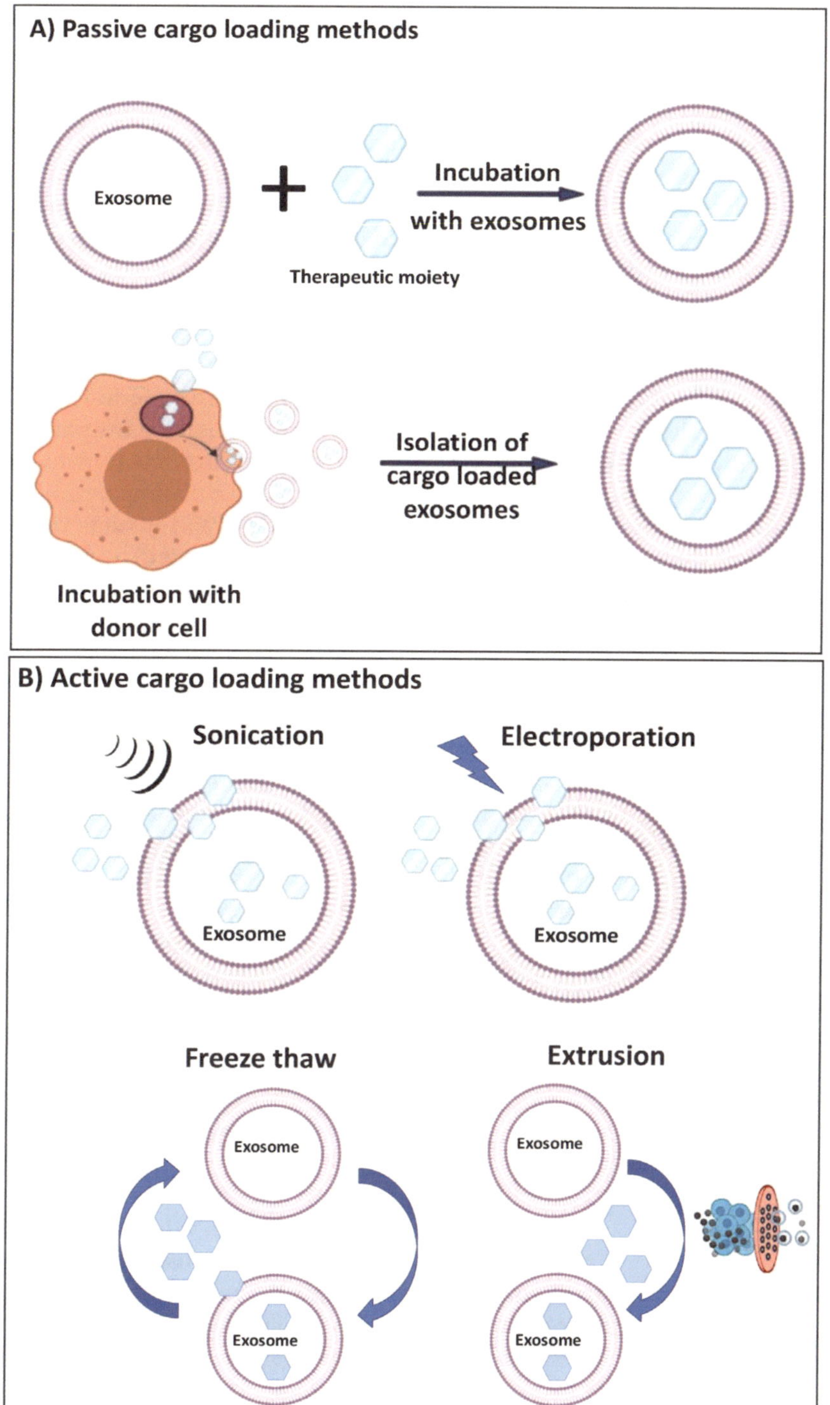

Figure 2.
Illustration representing different methods of cargo loading in exosomes. (A) Passive cargo loading is achieved by incubating the therapeutic moiety directly with isolated exosomes or by exposing to the exosome secreting donor cells followed by isolation of loaded exosomes. (B) Active cargo loading methods include use of physical treatments to disrupt the membrane integrity thus allowing entry of cargo in the interiors of exosomes. These treatments include sonication, electroporation, freeze thawing cycles and extrusion method.

efficiently encapsulated curcumin and SPIONs [56]. Similarly, Jia et al. incorporated doxorubicin into exosomes isolated from MDA-MB-231 and HCT-116 cell lines using an electroporation method, which resulted in ~1.5% drug loading [57]. According to published studies, although electroporation enhanced the percentage of drug loading in exosomes compared to the incubation method, it was still low. Therefore, scientists have employed sonication methods to load cargo more efficiently. The mechanical shear force of a sonicator/homogenizer probe is applied to disrupt the membrane integrity of exosomes, thus allowing the mixed drug/therapeutic agent to enter into the exosomes. In 2017, Kim et al. compared the incubation, electroporation, and sonication method of cargo loading in RAW264.7 cell derived exosomes to develop an exosomal formulation of paclitaxel (PTX). For the incubation method, the authors mixed and incubated PTX with exosomes at 37°C for 1 h. Using electroporation, exosomes and PTX were added to a pre-chilled electroporation cuvette and applied at 1000 kV for 5 ms followed by incubation at 37°C for half an hour to fully recover the exosome membrane. For sonication, the PTX-exosome mixture was sonicated at 20% amplitude, given 6 cycles of 30 s on/off for 3 min and a 2 min cooling period between each cycle. After sonication, the solution was incubated at 37°C for 1 h to fully recover the membrane of the exosomes. They showed the highest percent drug loading of ~28% using sonication followed by ~5% using electroporation and the lowest at ~1.4% with the incubation method [58].

In the extrusion method, exosomes mixed with the drug are passed through a syringe-based lipid extruder with a membrane ranging from to 10 to 400 nm pore size. In this process, the membrane of exosomes is disrupted by the extensive mechanical force of the extruder. In a study by Kim et al. when breast cancer cell-derived exosomes loaded with porphyrin were extruded, it altered the surface charge of blank exosomes, leading to cytotoxic effects [59]. On the other hand, in another study by Fuhrmann et al., loading cargo in exosomes using the extrusion method did not render them cytotoxic [60]. In the freeze-thaw method, the drug was first incubated with exosomes at ambient temperature and then frozen at −80°C. The mixture was then repeatedly thawed at room temperature to ensure drug loading into these vesicles. The main disadvantage is that this method often leads to particle aggregation, resulting in a wide size distribution. This method has also been reported to result in a lower percent drug loading than other methods, such as sonication.

6. Exosomes for drug delivery in breast cancer: progress and future promise

Although exosomes have been shown to mediate cancer development, they are an emerging platform for drug delivery to cancerous sites because of their excellent biocompatibility, low immunogenicity (since they are derived from the patient's own cells), good tolerance, and remarkable biodistribution. Owing to their small size, they can readily pass through different bodily barriers such as the blood-brain barrier [61]. Compared to synthetic nanoparticles, exosomes are relatively easy to manipulate through surface modification in order to enhance their targeting efficiency to cancer cells. Recently, the use of exosomes for drug delivery in breast cancer cells has been proven to be efficient. Alvarez-Erviti et al. delivered the chemotherapeutic drug doxorubicin to breast cancer tissues in a mouse model [62]. First, they engineered these cells by expressing Lamp2b, a lysosome-associated membrane glycoprotein 2b, on their surface and fused with a targeting peptide for integrins. They then isolated exosomes from immature dendritic cells (with low immunogenicity because of the absence of immunostimulatory markers on their surface) and used an electroporation technique to load doxorubicin within. They have shown that the exosomal formulation of doxorubicin has greater efficiency in targeting mouse tumors and hence, exhibits a novel propitious approach in breast cancer treatment in the clinical context. Li et al. loaded milk exosomes with doxorubicin to target CD44 overexpressed human breast cancer cell lines and found an exosomal formulation capable of delivering the drug into cancerous sites in a target-specific manner [63]. Vakshiteh et al. used dental pulp-derived mesenchymal stem cells to isolate exosomes and loaded them with miRNA, which was then targeted to breast cancer cells. They found that exosomes significantly decreased the proliferation of cancer cells and reduced the migratory and invasive properties of breast cancer cells *in vitro* [64, 65]. These studies indicate that exosomes are promising candidates for drug delivery in breast cancer therapy.

Currently, there are some hurdles in realizing the clinical potential of exosomes as drug delivery nanovehicles. These include low yield, long-term stability, and lack of understanding of their therapeutic effects. Hence, more research is required to develop techniques that can be used universally to enhance the yield in a time-efficient manner and increase the stability of exosomes.

Conflict of interest

The authors declare that they have no competing interests.

Author details

Sapna Jain[1] and Manjari Singh[2]*

1 Translational Health Science and Technology Institute (THSTI), Haryana, India

2 Assam University, Silchar, India

*Address all correspondence to: manjari.pharm@gmail.com; drmanjari.singh@aus.ac.in

References

[1] How common is breast cancer? [Internet]. Cancer.org. Available from: https://www.cancer.org/cancer/breast-cancer/about/how-common-is-breast-cancer.html [Accessed: September 24, 2021]

[2] Harbeck N, Gnant M. Breast cancer. Lancet. 2017;**389**(10074):1134-1150

[3] Raposo G, Stoorvogel W. Extracellular vesicles: Exosomes, microvesicles, and friends. The Journal of Cell Biology. 2013;**200**(4):373-383

[4] Yu D-D, Wu Y, Shen H-Y, Lv M-M, Chen W-X, Zhang X-H, et al. Exosomes in development, metastasis and drug resistance of breast cancer. Cancer Science. 2015;**106**(8):959-964

[5] Adem B, Vieira PF, Melo SA. Decoding the biology of exosomes in metastasis. Trends in Cancer. 2020;**6**(1):20-30

[6] Deatherage BL, Cookson BT. Membrane vesicle release in bacteria, eukaryotes, and archaea: A conserved yet underappreciated aspect of microbial life. Infection and Immunity. 2012;**80**(6):1948-1957

[7] Doyle LM, Wang MZ. Overview of extracellular vesicles, their origin, composition, purpose, and methods for exosome isolation and analysis. Cells. 2019;**8**(7):727

[8] Zaborowski MP, Balaj L, Breakefield XO, Lai CP. Extracellular vesicles: Composition, biological relevance, and methods of study. Bioscience. 2015;**65**(8):783-797

[9] Pang B, Zhu Y, Ni J, Thompson J, Malouf D, Bucci J, et al. Extracellular vesicles: The next generation of biomarkers for liquid biopsy-based prostate cancer diagnosis. Theranostics. 2020;**10**(5):2309-2326

[10] Konoshenko MY, Lekchnov EA, Vlassov AV, Laktionov PP. Isolation of extracellular vesicles: General methodologies and latest trends. BioMed Research International. 2018;**2018**: 8545347

[11] Willms E, Cabañas C, Mäger I, Wood MJA, Vader P. Extracellular vesicle heterogeneity: Subpopulations, isolation techniques, and diverse functions in cancer progression. Frontiers in Immunology. 2018;**9**:738. DOI: 10.3389/fimmu.2018.00738

[12] Hessvik NP, Llorente A. Current knowledge on exosome biogenesis and release. Cellular and Molecular Life Sciences. 2018;**75**(2):193-208

[13] Kalluri R, LeBleu VS. The biology, function, and biomedical applications of exosomes. Science. 2020;**367**(6478): eaau6977

[14] Williams RL, Urbé S. The emerging shape of the ESCRT machinery. Nature Reviews Molecular Cell Biology. 2007;**8**(5):355-368

[15] Henne WM, Buchkovich NJ, Emr SD. The ESCRT pathway. Developmental Cell. 2011;**21**(1):77-91

[16] Vietri M, Radulovic M, Stenmark H. The many functions of ESCRTs. Nature Reviews Molecular Cell Biology. 2020;**21**(1):25-42

[17] Hurley JH, Hanson PI. Membrane budding and scission by the ESCRT machinery: It's all in the neck. Nature Reviews Molecular Cell Biology. 2010;**11**(8):556-566

[18] Teng F, Fussenegger M. Shedding light on extracellular vesicle biogenesis and bioengineering. Advance Science (Weinh). 2020;**8**(1):2003505

[19] Azmi AS, Bao B, Sarkar FH. Exosomes in cancer development, metastasis, and drug resistance: A comprehensive review. Cancer Metastasis Reviews. 2013;**32**(3-4): 623-642

[20] Riches A, Campbell E, Borger E, Powis S. Regulation of exosome release from mammary epithelial and breast cancer cells - a new regulatory pathway. European Journal of Cancer. 2014;**50**(5): 1025-1034

[21] Mathivanan S, Ji H, Simpson RJ. Exosomes: Extracellular organelles important in intercellular communication. Journal of Proteomics. 2010;**73**(10):1907-1920

[22] van Niel G, Porto-Carreiro I, Simoes S, Raposo G. Exosomes: A common pathway for a specialized function. Journal of Biochemistry. 2006;**140**(1):13-21

[23] Poliakov A, Spilman M, Dokland T, Amling CL, Mobley JA. Structural heterogeneity and protein composition of exosome-like vesicles (prostasomes) in human semen. The Prostate. 2009;**69**(2): 159-167

[24] Vidal M, Sainte-Marie J, Philippot JR, Bienvenue A. Asymmetric distribution of phospholipids in the membrane of vesicles released during *in vitro* maturation of guinea pig reticulocytes: Evidence precluding a role for "aminophospholipid translocase.". Journal of Cellular Physiology. 1989;**140**(3):455-462

[25] Chu Z, Witte DP, Qi X. Saposin C-LBPA interaction in late-endosomes/ lysosomes. Experimental Cell Research. 2005;**303**(2):300-307

[26] Minciacchi VR, Freeman MR, Di Vizio D. Extracellular vesicles in cancer: Exosomes, microvesicles and the emerging role of large oncosomes. Seminars in Cell & Developmental Biology. 2015;**40**:41-51

[27] Bissig C, Lenoir M, Velluz M-C, Kufareva I, Abagyan R, Overduin M, et al. Viral infection controlled by a calcium-dependent lipid-binding module in ALIX. Developmental Cell. 2013;**25**(4):364-373

[28] Mathivanan S, Fahner CJ, Reid GE, Simpson RJ. ExoCarta 2012: Database of exosomal proteins RNA and lipids. Nucleic Acids Research. 2012;**40** (Database issue):D1241-D1244

[29] Valastyan S, Weinberg RA. Tumor metastasis: Molecular insights and evolving paradigms. Cell. 2011;**147**(2): 275-292

[30] Lowry MC, Gallagher WM, O'Driscoll L. The role of exosomes in breast cancer. Clinical Chemistry. 2015;**61**(12):1457-1465

[31] Liu T, Hooda J, Atkinson JM, Whiteside TL, Oesterreich S, Lee AV. Exosomes in breast cancer - mechanisms of action and clinical potential. Molecular Cancer Research. 2021;**19**(6):935-945

[32] Giordano C, La Camera G, Gelsomino L, Barone I, Bonofiglio D, Andò S, et al. The biology of exosomes in breast cancer progression: Dissemination, immune evasion and metastatic colonization. Cancers (Basel). 2020;**12**(8):2179

[33] Kruger S, Abd Elmageed ZY, Hawke DH, Wörner PM, Jansen DA, Abdel-Mageed AB, et al. Molecular characterization of exosome-like vesicles from breast cancer cells. BMC Cancer. 2014;**14**(1):44

[34] Gangoda L, Liem M, Ang C-S, Keerthikumar S, Adda CG, Parker BS, et al. Proteomic profiling of exosomes secreted by breast cancer cells with varying metastatic potential. Proteomics. 2017;**17**(23-24):1600370

[35] Palazzolo G, Albanese NN, DI Cara G, Gygax D, Vittorelli ML, Pucci-Minafra I. Proteomic analysis of exosome-like vesicles derived from breast cancer cells. Anticancer Research. 2012;**32**(3):847-860

[36] Ciravolo V, Huber V, Ghedini GC, Venturelli E, Bianchi F, Campiglio M, et al. Potential role of HER2-overexpressing exosomes in countering trastuzumab-based therapy. Journal of Cellular Physiology. 2012;**227**(2):658-667

[37] Shedden K, Xie XT, Chandaroy P, Chang YT, Rosania GR. Expulsion of small molecules in vesicles shed by cancer cells: Association with gene expression and chemosensitivity profiles. Cancer Research. 2003;**63**(15):4331-4337

[38] Hayes JD, Wolf CR. Molecular mechanisms of drug resistance. The Biochemical Journal. 1990;**272**(2): 281-295

[39] Chen W-X, Liu X-M, Lv M-M, Chen L, Zhao J-H, Zhong S-L, et al. Exosomes from drug-resistant breast cancer cells transmit chemoresistance by a horizontal transfer of microRNAs. PLoS ONE. 2014;**9**(4):e95240

[40] Lv M-M, Zhu X-Y, Chen W-X, Zhong S-L, Hu Q, Ma T-F, et al. Exosomes mediate drug resistance transfer in MCF-7 breast cancer cells and a probable mechanism is delivery of P-glycoprotein. Tumour Biology. 2014;**35**(11): 10773-10779

[41] Jaiswal R, Gong J, Sambasivam S, Combes V, Mathys J-M, Davey R, et al. Microparticle-associated nucleic acids mediate trait dominance in cancer. The FASEB Journal. 2012;**26**(1):420-429

[42] Boelens MC, Wu TJ, Nabet BY, Xu B, Qiu Y, Yoon T, et al. Exosome transfer from stromal to breast cancer cells regulates therapy resistance pathways. Cell. 2014;**159**(3):499-513

[43] Greening DW, Gopal SK, Mathias RA, Liu L, Sheng J, Zhu H-J, et al. Emerging roles of exosomes during epithelial-mesenchymal transition and cancer progression. Seminars in Cell & Developmental Biology. 2015;**40**:60-71

[44] Wang J, Hendrix A, Hernot S, Lemaire M, De Bruyne E, Van Valckenborgh E, et al. Bone marrow stromal cell-derived exosomes as communicators in drug resistance in multiple myeloma cells. Blood. 2014;**124**(4):555-566

[45] Melo SA, Sugimoto H, O'Connell JT, Kato N, Villanueva A, Vidal A, et al. Cancer exosomes perform cell-independent microRNA biogenesis and promote tumorigenesis. Cancer Cell. 2014;**26**(5):707-721

[46] Singh R, Pochampally R, Watabe K, Lu Z, Mo Y-Y. Exosome-mediated transfer of miR-10b promotes cell invasion in breast cancer. Molecular Cancer. 2014;**13**(1):256

[47] Wang X, Xu C, Hua Y, Sun L, Cheng K, Jia Z, et al. Exosomes play an important role in the process of psoralen reverse multidrug resistance of breast cancer. Journal of Experimental & Clinical Cancer Research. 2016;**35**(1):186

[48] Kumar D, Gupta D, Shankar S, Srivastava RK. Biomolecular characterization of exosomes released from cancer stem cells: Possible implications for biomarker and treatment

of cancer. Oncotarget. 2015;**6**(5):3280-3291

[49] Gorczynski RM, Zhu F, Chen Z, Kos O, Khatri I. A comparison of serum miRNAs influencing metastatic growth of EMT6 vs 4THM tumor cells in wild-type and CD200R1KO mice. Breast Cancer Research and Treatment. 2017;**162**(2):255-266

[50] Zhang P, Zhou H, Lu K, Lu Y, Wang Y, Feng T. Exosome-mediated delivery of MALAT1 induces cell proliferation in breast cancer. Oncotargets and Therapy. 2018;**11**:291-299

[51] Dong H, Wang W, Chen R, Zhang Y, Zou K, Ye M, et al. Exosome-mediated transfer of lncRNA-SNHG14 promotes trastuzumab chemoresistance in breast cancer. International Journal of Oncology. 2018;**53**(3):1013-1026

[52] Vashisht M, Rani P, Onteru SK, Singh D. Curcumin encapsulated in milk exosomes resists human digestion and possesses enhanced intestinal permeability *in vitro*. Applied Biochemistry and Biotechnology. 2017;**183**(3):993-1007

[53] Sun D, Zhuang X, Xiang X, Liu Y, Zhang S, Liu C, et al. A novel nanoparticle drug delivery system: The anti-inflammatory activity of curcumin is enhanced when encapsulated in exosomes. Molecular Therapy. 2010;**18**(9):1606-1614

[54] Gong C, Tian J, Wang Z, Gao Y, Wu X, Ding X, et al. Functional exosome-mediated co-delivery of doxorubicin and hydrophobically modified microRNA 159 for triple-negative breast cancer therapy. Journal of Nanobiotechnology. 2019;**17**(1):93

[55] Yang X, Shi G, Guo J, Wang C, He Y. Exosome-encapsulated antibiotic against intracellular infections of methicillin-resistant *Staphylococcus aureus*. International Journal of Nanomedicine. 2018;**13**:8095-8104

[56] Pascucci L, Coccè V, Bonomi A, Ami D, Ceccarelli P, Ciusani E, et al. Paclitaxel is incorporated by mesenchymal stromal cells and released in exosomes that inhibit *in vitro* tumor growth: A new approach for drug delivery. Journal of Controlled Release. 2014;**192**:262-270

[57] Jia G, Han Y, An Y, Ding Y, He C, Wang X, et al. NRP-1 targeted and cargo-loaded exosomes facilitate simultaneous imaging and therapy of glioma *in vitro and in vivo*. Biomaterials. 2018;**178**:302-316

[58] Toffoli G, Hadla M, Corona G, Caligiuri I, Palazzolo S, Semeraro S, et al. Exosomal doxorubicin reduces the cardiac toxicity of doxorubicin. Nanomedicine. 2015;**10**(19):2963-2971

[59] Kim MS, Haney MJ, Zhao Y, Mahajan V, Deygen I, Klyachko NL, et al. Development of exosome-encapsulated paclitaxel to overcome MDR in cancer cells. Nanomedicine. 2016;**12**(3):655-664

[60] Fuhrmann G, Serio A, Mazo M, Nair R, Stevens MM. Active loading into extracellular vesicles significantly improves the cellular uptake and photodynamic effect of porphyrins. Journal of Controlled Release. 2015;**205**:35-44

[61] Haney MJ, Klyachko NL, Zhao Y, Gupta R, Plotnikova EG, He Z, et al. Exosomes as drug delivery vehicles for Parkinson's disease therapy. Journal of Controlled Release. 2015;**207**:18-30

[62] Alvarez-Erviti L, Seow Y, Yin H, Betts C, Lakhal S, Wood MJA. Delivery

of siRNA to the mouse brain by systemic injection of targeted exosomes. Nature Biotechnology. 2011;**29**(4):341-345

[63] Tian Y, Li S, Song J, Ji T, Zhu M, Anderson GJ, et al. A doxorubicin delivery platform using engineered natural membrane vesicle exosomes for targeted tumor therapy. Biomaterials. 2014;**35**(7):2383-2390

[64] Li D, Yao S, Zhou Z, Shi J, Huang Z, Wu Z. Hyaluronan decoration of milk exosomes direct tumor-specific delivery of doxorubicin. Carbohydrate Research. 2020;**493**(108032):108032

[65] Vakhshiteh F, Rahmani S, Ostad SN, Madjd Z, Dinarvand R, Atyabi F. Exosomes derived from miR-34a-overexpressing mesenchymal stem cells inhibit *in vitro* tumor growth: A new approach for drug delivery. Life Sciences. 2021;**266**(118871):118871

Chapter 15

Roles of Extracellular Vesicles in Cancer Metastasis

Eman Helmy Thabet

Abstract

Extracellular vesicles (EVs) are biological active vesicles and carriers of information in intercellular communication. In cancer settings, EVs especially exosomes (Exo), play a focal role in modulating the tumor microenvironment mainly by increasing tumor proliferation, facilitating the crosstalk between tumor and tumor-neighboring cells, and influencing the host immune response. Amongst these functions in tumor growth, Exo modulate fundamental steps of tumor progression, such as growth, invasion, and immune modulation. On the endocrine level, Exo released from tumors were shown to mediate distant cell-cell communication processes via secretory factors and miRNAs, which result in the set-up of pro-tumorigenic microenvironments supportive of metastatic dissemination. This is achieved through processes such as fibroblast activation, extracellular matrix ECM production, angiogenesis, and immune modulation.

Keywords: tumor microenvironment, exosomes, tumor progression, metastasis, targeted therapy

1. Introduction

Extracellular vesicles (EVs) are traditionally classified into three types: exosomes (Exo), microvesicles (MVs), and apoptotic vesicles. Several theories exist on how tumor cells alter their neighboring cells and matrix ultimately changing their behavior into an invasive one. This typically would involve the transport of materials from tumor cells to their adjacent surroundings. These materials include a wide range of soluble cytokines, RNA species, enzymes, and proteins. Most of which are carried in nano-sized carriers such as EVs. EVs are classified according to their size and the mechanism of genesis. The first class of EVs known as MVs or when secreted from cancer cells, are called oncosomes [1]. MVs formation is originated by the outward budding of the cell surface at specific regions along the plasma membrane enriched with high concentrations of lipids, such as cholesterol and glycosphingolipids, and proteins such as Flotillin-1 and 2 [2]. Exo represent the second major class of EVs [3]. They are formed when multivesicular bodies (MVBs) in the endo-lysosomal pathway accumulate intraluminal vesicles (ILVs) that consist of proteins and nucleic acids. Exo are smaller in size and range from 30 to 50 nm.

EVs can function in an autocrine, paracrine, and even endocrine fashion, and were shown to impact various cancer cell phenotypes, increasing their cell growth and promoting metastasis [4]. This secretome is released into the microenvironment and acts as cell-cell communicators. Tumor derived Exo (TDE) has appeared as imperative facilitators in cancer initiation, progression, metastasis, host immune suppression, and drug resistance [5]. TDE typically consists of high sphingolipids and cholesterol contents that contain major histocompatibility complex (MHC) molecules, heat shock proteins, and tetraspanin (CD63, CD81, and CD9). Additionally, tumor antigens such as Mart1, gp100, TRP, and Her2-neu have been discovered in TDE [5]. TDE also contains surface and soluble proteins and RNA species such as mRNAs and miRNAs. mRNAs conveyed in EVs result in proteins synthesis in target cells, while miRNAs alter their gene expression [6]. The *protein cargo* of TDE includes extracellular matrix (ECM) proteins, cell adhesion proteins, cell-surface receptor tyrosine kinases, chaperones, cytosolic and nuclear signaling proteins, as well as DNA and RNA binding proteins. Several types of nucleic acids that have been identified in EVs include RNA transcripts, microRNAs, long non-coding RNAs (lncRNAs), and DNA [7].

2. The role of MVs/TDE regional preconditioning

Tumor development is a multistep process that starts by cellular reprogramming of cells to acquire the hallmarks of cancer cells to gain and maintain abnormal growth and invasive capacity [8]. The complex process of tumor formation and spreading additionally requires a rewiring of the surrounding stromal cells. This can be induced by intrinsic cell events such as genetic or epigenetic aberrations or by external factors from direct or indirect cell communication [9]. In cancer, EVs especially Exo, have been shown to be essential for various steps during tumor initiation and progression. EVs disrupt signaling and gene expression regulation in the recipient cell by horizontally transferring bioactive chemicals between cancer cells and the surrounding stroma. As a result, malignant cells can change the phenotype of surrounding benign cells to one that supports tumor growth and metastasis, creating a favorable environment for cancer progression and spread. EVs play several roles in priming the surrounding environment preparing it for metastasis and invasion. The role of EVs in promoting tumor progression has been elucidated in studies on mixed populations of EVs. The function of EVs largely depends on their bioactive cargo, in particular the shuttling of tumor-specific proteins to the surrounding cells. While researchers have mainly studied the RNA content of EVs, however, the focus is starting to shift towards the EVs proteome [10].

The protein content of MVs within mixed populations of EVs was discovered to be significantly diverse from that of the Exo proteome, and is supplemented in proteins involved in microtubule, actin, and cytoskeleton networks, ARF6, its effector phospholipase D2, and parts of the endosomal sorting complex required for transport family (ESCRT-I) [11]. By transporting these molecules, MVs can impact nearby tumor cells and stromal cells.

2.1 MVs mediated tumor-invasion

One example in which MVs shed by the cancer cells were shown to enhance tumor cell proliferation is in multiple myeloma. This effect was shown to be related to the amelioration of the Extracellular Matrix Metalloproteinase Inducer (EMMPRIN/CD147) on the tumor MVs. This protein is known to be overexpressed in solid tumors, some lymphomas, and leukemias [12]. Another study in breast cancer cells found that the highly glycosylated version of EMMPRIN exists in high quantities in breast cancer cell-derived MVs and enhances tumor invasion through activation of p38/MAPK signaling [11]. Interestingly, it was found that MVs from patient Blood with metastatic breast cancer had a similar high-EMMPRIN expression, along with the tumor marker Mucin-1 (MUC1/CA 15-3) [11]. Additionally, the truncated oncogenic form of the epidermal growth factor receptor (EGFR), EGFRvIII, commonly expressed in aggressive brain tumor cells, is associated with pro-tumorigenic tumor–tumor crosstalk via MVs. It was discovered that EGFRvIII was present in MVs released by U373 glioma cells, allowing them to transfer malignant features from highly aggressive tumor cells to the more benign tumor cells, EGFRvIII-negative, thereby facilitating their oncogenic transformation [11]. Hence, MVs are convenient communicators within the TME, as they can either mediate the horizontal transfer of oncogenic material or activate oncogenic signaling pathways in neighboring cancer cells, enhancing their survival, proliferative, and angiogenic potential and triggering their transformation into an aggressive phenotype.

2.2 Tumor-immune cells crosstalk

Alongside the tumor–tumor communication, MVs were proven to facilitate the crosstalk between the tumor and its surrounding stroma and immune cells which ultimately leads to cancer immune evasion. In breast cancer cells, the secretion of both tumor MV and TDE induced the expression of Wnt5a in tumor-associated macrophages. Macrophage Wnt5a promoted ß-catenin-independent Wnt signaling in breast cancer cells when delivered by macrophage-derived MVs and Exo, resulting in enhanced tumor invasion. This shows how EV-based cell-cell communication can drive tumor-associated immune cells to stimulate tumor growth [11]. MVs-enriched preparations induced the differentiation of monocytes producing anti-inflammatory cytokines such as IL-10. In line with this, early stimulation with tumor MV triggered macrophage polarization towards an anti-inflammatory phenotype with decreased anti-tumor cytotoxic potential. Additionally, as T cells represent the first line of the immune defense, tumor cells appear to suppress T cell activity and diminish antitumoral immune response via MVs-mediated cell-cell communication. For instance, leukemia-derived MVs deliver miRNAs to T cells, which alters T cell phenotype [13] (**Figure 1**). Moreover, MVs released by irradiated breast cancer cells were shown

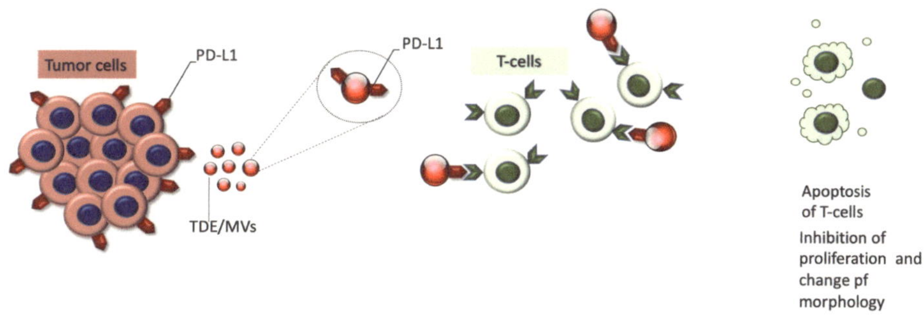

Figure 1.
Exosome PD-L1 (similar to tumor PD-L1) can bind to PD-1 on T cells, induce T cell apoptosis, and inhibit T cell activation and proliferation [14].

to carry abundant immune-suppressive proteins, such as programmed cell death ligand 1 (PD-L1) which inhibited cytotoxic T cell activity and enabled tumor growth (**Figure 1**) [15].

2.3 TDE-mediated epithethelial mesenchymal transition

TDEs, through their miRNAs proteins, DNAs, mRNAs lncRNAs, initiate the transformation of epithelial cells to mesenchymal cells. This transformation was due to the loss of epithelial E-cadherin expression, cell-cell adhesion and cell polarity, and gaining of vimentin expression [16].

3. Pre-metastatic niche formation

The complex and heterogeneous microenvironment of both primary or metastatic tumor is comprised of a network of cellular and acellular constituents. The cellular compartment consists of tumor cells and assorted non-transformed cells, such as cancer-associated fibroblasts (CAFs), macrophages, and endothelial cells. The non-cellular part is formed by secreted factors and components of the ECM. The tumor microenvironment modulates tumor progression by providing inhibitory or stimulatory growth signals [17]. Thus pre-metastatic niche refers to the microenvironment, that is primed to allow tumor cells to colonize in and disseminate to distant sites. The main machineries of the premetastatic niche formation include tumor-derived secreted factors (TDSFs), EVs bone marrow-derived cells (BMDCs), suppressive immune cells and host stromal cells [4], and inflammation. Chronic inflammation is a driving force for tumor development and metastasis. Thus, the local inflammatory microenvironment is one of the essential factors for the pre-metastatic niche formation and driving force for metastasis.

3.1 TDE in upregulation of inflammatory molecules and premalignant niche formation

Tumor development and metastasis are aided by chronic inflammation. As a result, one of the most important variables in the establishment of a pre-metastatic niche is the local inflammatory microenvironment. Tumor cells can be induced to create TDSFs such as vascular endothelial growth factor (VEGF), tumor necrosis factor alpha (TNF-α), transforming growth factor (TGF-β), and interleukin-2 by the local inflammatory microenvironment. These TDSFs then exert a paracrine effect on myeloid cells, initiating their migration to potential pre-metastatic niche formation sites [18]. Host stromal cells in the pre-metastatic niche may upregulate the expression of inflammatory factors in response to TDSF activation. The recruitment of BMDCs or immune cells to the pre-metastatic niche speeds up the release of inflammatory factors. Exo from tumors also transport inflammatory substances into the bloodstream, where they reach the pre-metastatic niche. In the pre-metastatic niche, an inflammatory milieu supportive to tumors is eventually generated [18].

In a study conducted by Hoshino, he showed that the proinflammatory cytokine s100 was upregulated up to four folds when Kupffer cells were treated with integrin intact Exo, as compared to those treated with integrin knocked out Exo. Hoshino speculated that the activation of Src, and its phosphorylation might be a causative pathway [19].

3.2 Cellular compartment of the pre-metastatic niche

3.2.1 Cancer-associated fibroblasts

TDE and MV were also shown to modify fibroblasts in the tumor stroma. When normal human fibroblasts were exposed to oral squamous carcinoma derived MV [20] the fibroblasts were altered into a cancer phenotype. This switch to CAFs was largely mediated via metabolic reprogramming of the fibroblasts to aerobic glycolysis, with an increase in glucose uptake and lactate secretion. Some TDEs contain surface TGF-β along with betaglycan, which could trigger SMAD-dependent signaling and regulate the differentiation of fibroblasts to myofibroblasts [21]. This was further proved by co-culturing the generated CAF with cancer cells which led to enhanced cancer cell invasion and migration, creating a bidirectional cross-talk that favors tumor promotion and spread. The MVs-induced fibroblast activation and spreading seem to occur in the matrix milieu in the tumor periphery [22]. In prostate cancer, TDE were shown to induce the expression of RANKL and Metalloproteinases in CAFs, through miR-100, -21, and -139, further promoting its metastasis [23]. Hypoxia seems to stimulate prostate cancer cells to release protein-rich Exo which further induces activation of CAFs [24], promotes epithethelial mesenchymal transition (EMT), stemness, and angiogenesis by prostate cancer cells.

Additionally, TDE were also described as regulators of metabolism in the tumor microenvironment, for example, breast cancer tumors could suppress glucose uptake by lung fibroblasts, via secretion of Exo containing miR-122, increasing glucose availability and facilitating metastasis [25]. The cell-to-cell communication mediated by Exo is also affected by the genetic profile of the recipient fibroblasts. For example, fibroblasts lacking the BRCA1, a tumor suppressor gene, internalize larger amounts of serum-derived Exo when compared to BRCA1 containing fibroblasts [26]. Furthermore, these cells were found to undergo a malignant transformation when exposed to Exo derived from sera of cancer patients, implying that oncosuppressor genes can prevent exosome information from tumor cells from being integrated and thus shelter these cells from their pro-oncogenic signals [26].

3.2.2 Macrophages and immune cells

Tumor MVs extravasate through the vessel wall in pancreatic cancer, reach the liver microcirculation and are picked up by perivascular macrophages to prime the liver metastatic niche in a CD36-dependent manner. Furthermore, tumor MVs produced from the B16F10 melanoma cell line was discovered to cause metastases in BALB/c mice, which are generally resistant to the B1610 tumor cell line [27]. TDEs also protect cancer cells from apoptosis by selectively effluxing apoptosis inducer proteins that are delivered by T cells or natural killer (NK) cells. TDEs also reduce the effects of therapy by preventing drug efflux or concealing the binding site of monoclonal antibodies, which could lead to the emergence of chemotherapy-resistant cell populations [28].

Exosome-derived programmed death receptor 1 (PD-1) and programmed death-ligand 1 have been linked to an immunological escape mechanism in recent years. PD-1 is mostly found on macrophages, activated T cells, and B cells, whereas PD-L1 is abundant in tumor tissues, antigen-presenting cells (APCs), and stromal cells [29]. T lymphocytes can recognize and destroy tumor cells in normal circumstances. When PD-1 attaches to PD-L1, however, it sends an inhibitory signal to T cells, causing them to die and inhibiting their activation and proliferation. As a result, blocking the PD-1/PD-L1 pathway may boost the immune response by increasing the killing effect of T cells [30]. T lymphocytes can recognize and destroy tumor cells in normal circumstances. When PD-1 attaches to PD-L1, however, it sends an inhibitory signal to T cells, causing them to die and inhibiting their activation and proliferation (**Figure 1**). As a result, blocking the PD-1/PD-L1 pathway may boost the immune response by increasing the killing effect of T cells. As a result, Exo containing PD-L1 suppress the immune system in the pre-metastatic milieu and promote the establishment of a pre-metastatic niche [31].

3.2.3 Endothelial cells and vascular barriers

Angiogenesis within the primary tumor is also influenced by tumor MVs and TDE. Normal endothelial cells (ECs) were shown to endocytose tumor EVs, which triggered PI3K/Akt signaling and increased EC motility and tube formation ability [32]. Tumor MVs and TDE also release VEGF, a pro-angiogenic substance that stimulates ECs [33]. Similarly, MVs produced from multiple myeloma cells have been demonstrated to transfer CD138, a myeloma cell marker, to ECs, promoting their proliferation, invasion, and production of the angiogenic mediators IL-6 and VEGF, resulting in tube formation [50] (**Figure 2**). MVs change the environment around the main tumor and create pre-metastatic niches from afar. This was originally attributed to their procoagulant activity, which encouraged the production of microthrombi and facilitated the extravasation of trapped circulating tumor cells. ECs are important components of the tumor microenvironment because they provide a pathway for nutrients and trophic substances [34].

3.2.3.1 Neovascularization

TDE enriched in vascular cell adhesion molecule (VCAM)-1 and intercellular adhesion molecule (ICAM)-1 has been demonstrated to regulate the process of

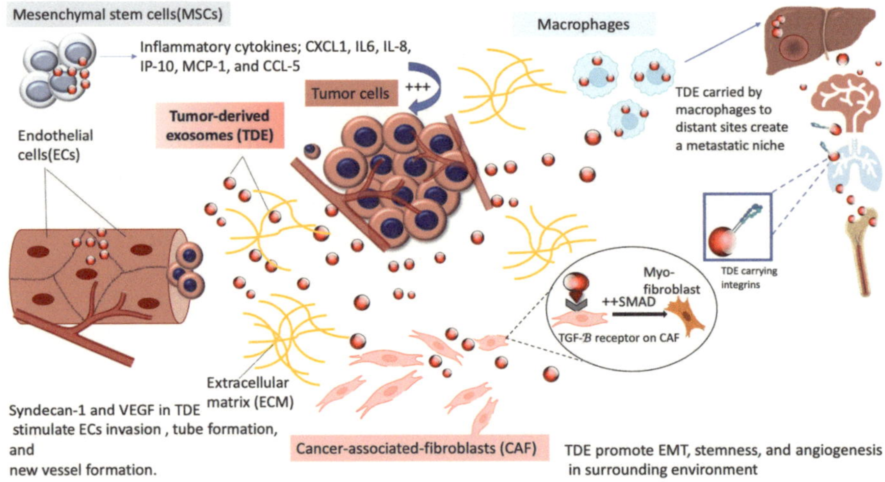

Figure 2.
Possible mechanisms of pre-metastatic niche formation. The figure delineates how TDEs can modulate its surroundings of ECM, cancer-CAFs, immune cells, ECs, and MSCs all in favor of tumor support and progression. TDE can carry integrins to distant sites and create a pre-metastatic niche.

neovascularization in myeloid leukemia [35]. Furthermore, enhanced vascularization has been linked to the packaging of miR-92a in Exo derived from leukemia[36] and CO-029/D6.1A Tetraspanin in Exo produced from pancreatic cancer [37]. Upregulation of Heparanase in tumor cells, such as myeloma and breast malignancies, has also been linked to increased exosome production and exosomal packing of Syndecan-1, VEGF, and hepatocyte growth factor, resulting in enhanced endothelial invasion through the ECM [38]. Exo produced from skin cancer can also enhance angiogenesis by transferring the EGFR [39] and miR-9 to ECs [26]. Furthermore, melanoma-derived Exo have been found to condition sentinel lymph nodes prior to the installation of melanoma cells and subsequent metastasis by upregulating Collagen 18 and Laminin 5, as well as producing angiogenic growth factors [26].

Another significant component in altering tumor-EC communication is hypoxia. Hypoxic glioblastoma cells, for example, release Exo that interact with ECs, promoting proliferation and angiogenesis both in vitro and in vivo [40], and also prompting tissue factor/Factor VIIa dependent activation of hypoxic ECs [26].

3.2.3.2 Vascular leakage

Exo from melanoma cause pulmonary vascular leakiness and upregulate tumor cell recruitment genes such Stabilin 1, Vitronectin, Integrins, and Ephrin receptor b4 in lymph nodes, forming pre-metastatic niches [41]. Furthermore, breast cancer-derived Exo enriched in miR-105 alter the expression of Claudin 5, Zonula Occludens protein 1, and Occludin, which promotes metastasis by disrupting vascular endothelial barriers [42]. Exo produced from brain tumors include miR-181c, which regulates EC actin dynamics and promotes the breakdown of the blood-brain barrier by three times. Protein Kinase-1 Degradation Requires Phosphoinositol [43]. Similarly, glioblastoma cells release Exo with high quantities of VEGF-A, which promote EC permeability and angiogenesis in vitro [44].

3.2.4 Tumor-stem/progenitor-non-transformed cell communication

TDE can promote pro-tumorigenic microenvironments via promoting tumor-stem/progenitor cell contact, in addition to its well-known actions in differentiated cells. Melanoma-derived Exo, for example, stimulate BMDCs by transferring the oncoprotein MET, resulting in the mobilization of vasculogenic and hematopoietic bone marrow progenitor cells to ensure vascular proliferation and immunosuppression at pre-metastatic niches [45]. Communication between tumor stem/progenitor cells is also critical in bone metastasis. Exo from bone metastatic prostate cancer PC3 cells were found to influence the process of bone metastasis by modulating both osteoclast genesis and osteoblast proliferation. Exo generated from osteoblasts, on the other hand, have been demonstrated to stimulate PC3 prostate cancer cell proliferation [46].

TDE was also demonstrated to influence the development of myeloid precursor cells into myeloid-derived suppressor cells (MDSCs), which are known to aid tumor progression by permitting immune escape [47]. Exo produced from breast carcinomas have been found to be taken up by bone marrow cells and to convert these cells' development pathways toward MDSCs via Prostaglandin E2 and TGF-β, boosting COX2, IL6, VEGF, and Arginase1 accumulation by MDSCs [48].

TDE can also cause alterations in mesenchymal stem cells (MSCs), which help to promote and maintain tumor-promoting inflammatory environments. For example, HSP70+ lung tumor-derived exosomes (TDEs) activate NF-kB and cause MSCs to secrete IL-6, IL-8, and MCP1 via TLR2-mediated signaling, causing MSCs to become more inflammatory and tumor supportive [49]. According to De Veirman et al. [50], myeloma-derived Exo transfer miR-146a to mesenchymal cells, stimulating them to secrete numerous cytokines and chemokines including CXCL1, IL6, IL-8, IP-10, MCP-1, and CCL-5 (**Figure 2**). Another example is Exo produced by KMBC cholangiocarcinoma cells, which cause MSCs to upregulate IL-6, and hence KMBC cell proliferation [51].

4. Mechanisms of TDE in tumor metastasis

4.1 Tumor cell proliferation and anti-apoptotic effect

One of the proposed mechanisms of tumorigenicity of TDE is the induction of tumor cell proliferation. Studies involving various cancer cells such as, chronic myeloid leukemia and in human gastric cancer, showed that this proliferative potential is via an autocrine induction through the phosphatidylinositol 3-kinase/protein kinase B (PI3K/AKT) and MAPK/ERK signaling pathways. Additionally, through the transference of lncRNAs (reviewed in [49]).

In addition, glioblastoma-derived Exo were shown to induce proliferation of the human glioma U87 cell line [40] in a mechanism dependent on the chloride intracellular channel protein 1 (CLIC1) [52]. In a more specific context linked to prostate cancer treatment, prostate cancer LNCaP cells grown in the presence of androgens generate Exo high in CD9, which enhance the growth of androgen-depleted LNCaP cells. Another example involves the promotion of in vivo growth of murine melanomas by systemic treatment of mice with melanoma-derived Exo, which accelerated growth and inhibited apoptosis of melanoma tumors in vivo [26].

4.2 Invasiveness and motility

TDE can alter the migratory behavior of recipient malignant cells. Exo produced from nasopharyngeal cancer-bearing EMT-inducing signals such as TGF-β and hypoxia-inducible factor 1 alpha (HIF1a) [53], matrix metalloproteinases (MMPs) Notch1, LMP1 Casein Kinase II and Annexin A2, were shown to enhance the migratory capacity of the tumor recipient cells. Another example involves Exo derived from hypoxic prostate cancer cells, which prompted invasiveness and motility of naïve human prostate cancer cells (reviewed in [26]) through the neighboring stroma and to nearby cells.

4.3 Chemoresistance

Exo have been found to have a role in tumor-tumor communication by transferring chemoresistance. Exo have been linked to the transfer of Docetaxel resistance in prostate cancer since Corcoran and colleagues first discovered it [54]. The transfer of cisplatin resistance in lung cancer is achieved by donor resistant cells producing Exo with low levels of miR-100-5p, which leads to enhanced expression of the mammalian target of rapamycin (mTOR) protein and chemoresistance in recipient cells [55].

MiRNA packed in Exo from drug-resistant cells can modulate the expression of specific target genes in breast cancer, such as miR-23a targeting Sprouty2, miR-222 targeting PTEN, miR-452 targeting APC4, and miR-24 targeting p27, thereby modulating chemoresistance in recipient cells that integrate these Exo. In fact, exosomal miR-222 plays a key role in this process, as the silencing of miR-221/222 prevents the transmission of resistance [56].

In addition to miRNAs, the transfer of exosomal mRNAs that encode drug-resistant proteins may result in chemoresistance in the receiving cell. GSTP1 exosomal mRNA from Adriamycin-resistant breast cancer cells, for example, confers resistance to previously susceptible cells. The presence of GSTP1 in circulating Exo from patients' peripheral blood was linked to a worse outcome in breast cancer patients receiving Adriamycin [57]. A supporting stroma is required for an optimum metabolic and physiological environment for tumor growth. Fibroblasts are the most abundant cells in most solid tissues, participating in environmental cue responses and being a common target of tumor-derived signals [58].

4.4 Integrins in metastasis

Integrins are a wide family of cell adhesion receptor proteins such as alpha3beta1, alpha6beta1, alpha6beta4, and alpha7beta1. Their roles have been implicated in tumor metastasis and mesenchymal transformation. TDE carry these integrins from primary tumor sites to distant sites such as lung, lymph nodes, brain, and bone creating pre-metastatic niches (**Figure 2**) [59].

5. Clinical applications of TDE

TDEs are involved in the advancement of several forms of cancer. Because of their abundance, TDEs may serve as noninvasive diagnostic and prognostic tools for various cancers. Additionally, blocking exosome secretion can slow the growth of some malignancies. Hence, Exo have been a popular target for developing cancer treatment techniques because of this property. Decreasing the expression of the exosomal proteins, Rab27a and Rab27b, inhibit exosome secretion without matching changes in soluble proteins secretions [60]. Several drugs used in the pharmaceutical industry such as Ketoconazole (an anti-fungal) sphingomyelinase (a hydrolase enzyme that is responsible for degrading sphingomyelin) [61], are additionally Rab27a inhibitors. These drugs can be re-directed as cancer modulators for their possible effects on attenuating TDE tumor progressive effects.

Furthermore, TDE owing to its small size, cancer-homing, and nontoxic nature, TDE can be re-directed to serve as a drug delivery system. Exo have been proven in several investigations to act as drug delivery vehicles, transporting anti-cancer chemicals to target cells [62]. For example, adriamycin and paclitaxel, target cancer cells via exosomal encapsulation and have low toxicity and immunogenicity [63].

6. Conclusions

EVs modulate the environment that favors tumor growth and progression. EVs provide a method of cell-cell communication, and through their rich cargo of ECM proteins, cell adhesion proteins, tyrosine kinases, chaperones, signaling proteins, DNA and RNA binding proteins, they create a pre-metastatic niche. By priming nearby and distant cells into becoming cancerous, they promote tumor metastasis. Several mechanisms have been discovered for their actions including, promotion of migratory behavior, chemoresistance, anti-apoptosis, vascular leakage, and immune modulation. Understanding how TDE and MVs create a pre-metastatic niche and how halting the trafficking of such vesicles can produce a revolutionizing new era in the field of cancer therapeutics. By preventing TDE-promoted metastasis and tumor progression, coupled with conventional radio and chemotherapy, the survival rates of cancer patients can significantly improve.

Conflict of interest

The author declares no conflicts of interest.

Author details

Eman Helmy Thabet
Center of Excellence in Regenerative Medicine and Its Research Application (CERRMA), Alexandria University, Egypt

*Address all correspondence to: e_thabet00@alexmed.edu.eg

References

[1] Raposo G, Stoorvogel W. Extracellular vesicles: Exosomes, microvesicles, and friends. Journal of Cell Biology. 2013;**200**(4):373-383

[2] Sezgin E, Levental I, Mayor S, Eggeling C. The mystery of membrane organization: Composition, regulation and roles of lipid rafts. Nature Reviews Molecular Cell Biology. 2017;**18**(6): 361-374

[3] Desrochers LM, Antonyak MA, Cerione RA. Extracellular vesicles: Satellites of information transfer in cancer and stem cell biology. Developmental Cell. 2016;**37**(4):301-309

[4] Liu Y, Cao X. Characteristics and significance of the pre-metastatic niche. Cancer Cell. 2016;**30**(5):668-681

[5] Taghikhani A, Farzaneh F, Sharifzad F, Mardpour S, Ebrahimi M, Hassan ZM. Engineered tumor-derived extracellular vesicles: Potentials in cancer immunotherapy. Frontiers in Immunology [Internet]. 2020;**11**:221 Available from: https://www.frontiersin.org/article/10.3389/fimmu.2020.00221

[6] Valadi H, Ekström K, Bossios A, Sjöstrand M, Lee JJ, Lötvall JO. Exosome-mediated transfer of mRNAs and microRNAs is a novel mechanism of genetic exchange between cells. Nature Cell Biology. 2007;**9**(6):654-659

[7] Latifkar A, Cerione RA, Antonyak MA. Probing the mechanisms of extracellular vesicle biogenesis and function in cancer. Biochemical Society Transactions. 2018;**46**(5):1137-1146

[8] Hanahan D, Weinberg RA. Hallmarks of cancer: The next generation. Cell. 2011;**144**(5):646-674

[9] Gupta GP, Massagué J. Cancer metastasis: Building a framework. Cell. 2006;**127**(4):679-695

[10] Sheehan C, DSouza-Schorey C. Tumor-derived extracellular vesicles: Molecular parcels that enable regulation of the immune response in cancer. Journal of Cell Science. 2019;**132**(20): jcs235085

[11] Menck K, Sivaloganathan S, Bleckmann A, Binder C. Microvesicles in cancer: Small size, large potential. International Journal of Molecular Sciences. 2020;**21**(15):5373

[12] Arendt BK, Walters DK, Wu X, Tschumper RC, Huddleston PM, Henderson KJ, et al. Increased expression of extracellular matrix metalloproteinase inducer (CD147) in multiple myeloma: Role in regulation of myeloma cell proliferation. Leukemia. 2012;**26**(10): 2286-2296

[13] Arkhypov I, Lasser S, Petrova V, Weber R, Groth C, Utikal J, et al. Myeloid cell modulation by tumor-derived extracellular vesicles. International Journal of Molecular Sciences. 2020;**21**(17):6319

[14] Guo Y, Ji X, Liu J, Fan D, Zhou Q, Chen C, et al. Effects of exosomes on pre-metastatic niche formation in tumors. Molecular Cancer. 2019;**18**(1): 1-11

[15] Zhao X, Shao C. Radiotherapy-mediated immunomodulation and anti-tumor abscopal effect combining immune checkpoint blockade. Cancers. 2020;**12**(10):2762

[16] Kalluri R, Weinberg RA. The basics of epithelial-mesenchymal transition.

The Journal of Clinical Investigation. 2009;**119**(6):1420-1428

[17] Correia AL, Bissell MJ. The tumor microenvironment is a dominant force in multidrug resistance. Drug Resistance Updates. 2012;**15**(1-2):39-49

[18] Guo Y, Ji X, Liu J, Fan D, Zhou Q, Chen C, et al. Effects of exosomes on pre-metastatic niche formation in tumors. Molecular Cancer [Internet]. 2019;**18**(1):39. DOI: 10.1186/s12943-019-0995-1

[19] Hoshino A, Costa-Silva B, Shen T-L, Rodrigues G, Hashimoto A, Tesic Mark M, et al. Tumour exosome integrins determine organotropic metastasis. Nature. 2015;**527**(7578):329-335

[20] Dourado MR, Korvala J, Åström P, De Oliveira CE, Cervigne NK, Mofatto LS, et al. Extracellular vesicles derived from cancer-associated fibroblasts induce the migration and invasion of oral squamous cell carcinoma. Journal of Extracellular Vesicles. 2019;**8**(1):1578525

[21] Webber J, Steadman R, Mason MD, Tabi Z, Clayton A. Cancer exosomes trigger fibroblast to myofibroblast differentiation. Cancer research. 2010;**70**(23):9621-9630

[22] Liao Z, Tan ZW, Zhu P, Tan NS. Cancer-associated fibroblasts in tumor microenvironment - Accomplices in tumor malignancy. Cellular immunology. 2019;343. Article ID: 103729. https://doi.org/10.1016/j.cellimm.2017.12.003

[23] Sánchez CA, Andahur EI, Valenzuela R, Castellón EA, Fullá JA, Ramos CG, et al. Exosomes from bulk and stem cells from human prostate cancer have a differential microRNA content that contributes cooperatively over local and pre-metastatic niche. Oncotarget. 2016;**7**(4):3993

[24] Taddei ML, Cavallini L, Ramazzotti M, Comito G, Pietrovito L, Morandi A, et al. Stromal-induced downregulation of miR-1247 promotes prostate cancer malignancy. Journal of Cellular Physiology. 2019;**234**(6):8274-8285

[25] Fong MY, Zhou W, Liu L, Alontaga AY, Chandra M, Ashby J, et al. Breast-cancer-secreted miR-122 reprograms glucose metabolism in premetastatic niche to promote metastasis. Nature Cell Biology. 2015;**17**(2):183-194

[26] Maia J, Caja S, Strano Moraes MC, Couto N, Costa-Silva B. Exosome-based cell-cell communication in the tumor microenvironment. Frontiers in Cell and Developmental Biology. 2018;**6**:18

[27] Costa-Silva B, Aiello NM, Ocean AJ, Singh S, Zhang H, Thakur BK, et al. Pancreatic cancer exosomes initiate pre-metastatic niche formation in the liver. Nature Cell Biology. 2015;**17**(6):816-826

[28] Yang C, Robbins PD. The roles of tumor-derived exosomes in cancer pathogenesis. Clinical and Developmental Immunology. 2011;**2011**:1-11. Article ID: 842849

[29] Jiang X, Wang J, Deng X, Xiong F, Ge J, Xiang B, et al. Role of the tumor microenvironment in PD-L1/PD-1-mediated tumor immune escape. Molecular Cancer. 2019;**18**(1):1-17

[30] Jiang X, Wang J, Deng X, Xiong F, Ge J, Xiang B, et al. Role of the tumor microenvironment in PD-L1/PD-1-mediated tumor immune escape. Molecular Cancer [Internet]. 2019;**18**(1):10. DOI: 10.1186/s12943-018-0928-4

[31] Yin Z, Yu M, Ma T, Zhang C, Huang S, Karimzadeh MR, et al.

Mechanisms underlying low-clinical responses to PD-1/PD-L1 blocking antibodies in immunotherapy of cancer: A key role of exosomal PD-L1. Journal for Immunotherapy of Cancer. 2021;**9**(1):1-9, e001698

[32] DeRita RM. The Signaling Axis of β1 Integrins in Prostate Cancer Extracellular Vesicle Biology. Philadelphia, Pennsylvania: Thomas Jefferson University; 2019

[33] Melincovici CS, Boşca AB, Şuşman S, Mărginean M, Mihu C, Istrate M, et al. Vascular endothelial growth factor (VEGF) - key factor in normal and pathological angiogenesis. Romanian Journal of Morphology and Embryology. 2018;**59**(2):455-467

[34] Bian X, Xiao Y-T, Wu T, Yao M, Du L, Ren S, et al. Microvesicles and chemokines in tumor microenvironment: Mediators of intercellular communications in tumor progression. Molecular Cancer. 2019;**18**(1):1-13

[35] Taverna S, Flugy A, Saieva L, Kohn EC, Santoro A, Meraviglia S, et al. Role of exosomes released by chronic myelogenous leukemia cells in angiogenesis. International Journal of Cancer. 2012;**130**(9):2033-2043

[36] Umezu T, Ohyashiki K, Kuroda M, Ohyashiki JH. Leukemia cell to endothelial cell communication via exosomal miRNAs. Oncogene. 2013;**32**(22):2747-2755

[37] Gesierich S, Berezovskiy I, Ryschich E, Zöller M. Systemic induction of the angiogenesis switch by the tetraspanin D6. 1A/CO-029. Cancer Research. 2006;**66**(14):7083-7094

[38] Bandari SK, Tripathi K, Rangarajan S, Sanderson RD. Therapy-induced chemoexosomes: Sinister small extracellular vesicles that support tumor survival and progression. Cancer Letters. 2020;**493**:113-119

[39] Gluszko A, Mirza SM, Piszczatowska K, Kantor I, Struga M, Szczepanski MJ. The role of tumor-derived exosomes in tumor angiogenesis and tumor progression. Current Issues in Pharmacy and Medical Sciences. 2019;**32**(4):193-202

[40] Skog J, Würdinger T, Van Rijn S, Meijer DH, Gainche L, Curry WT, et al. Glioblastoma microvesicles transport RNA and proteins that promote tumour growth and provide diagnostic biomarkers. Nature Cell Biology. 2008;**10**(12):1470-1476

[41] Hood JL, San RS, Wickline SA. exosomes released by melanoma cells prepare sentinel lymph nodes for tumor metastasis [Internet]. Cancer Research. 2011;**71**:3792-3801 Available from: https://cancerres.aacrjournals.org/content/canres/71/11/3792.full.pdf

[42] Orso F, Quirico L, Dettori D, Coppo R, Virga F, Ferreira LC, et al. Role of miRNAs in tumor and endothelial cell interactions during tumor progression. Seminars in Cancer Biology [Internet]. 2020;**60**:214-224 Available from: https://www.sciencedirect.com/science/article/pii/S1044579X19301695

[43] Tominaga N, Kosaka N, Ono M, Katsuda T, Yoshioka Y, Tamura K, et al. Brain metastatic cancer cells release microRNA-181c-containing extracellular vesicles capable of destructing blood–brain barrier. Nature Communications [Internet]. 2015;**6**(1):6716. DOI: 10.1038/ncomms7716

[44] Treps L, Perret R, Edmond S, Ricard D, Gavard J. Glioblastoma stem-like cells secrete the pro-angiogenic

VEGF-A factor in extracellular vesicles. Journal of Extracellular Vesicles [Internet]. 2017;**6**(1):1359479. DOI: 10.1080/20013078.2017.1359479

[45] Peinado H, Alečković M, Lavotshkin S, Matei I, Costa-Silva B, Moreno-Bueno G, et al. Melanoma exosomes educate bone marrow progenitor cells toward a pro-metastatic phenotype through MET. Nature Medicine [Internet]. 2012;**18**(6):883-891. DOI: 10.1038/nm.2753

[46] Ma Q, Liang M, Wu Y, Dou C, Xu J, Dong S, et al. Small extracellular vesicles deliver osteolytic effectors and mediate cancer-induced osteolysis in bone metastatic niche. Journal of Extracellular Vesicles [Internet]. 2021;**10**(4):e12068-e12068 Available from: https://pubmed.ncbi.nlm.nih.gov/33659051

[47] Nagaraj S, Gabrilovich DI. Myeloid-Derived Suppressor Cells. New York, NY: Springer New York; 2007. pp. 213-223 (Immune-Mediated Diseases)

[48] Harris DA, Patel SH, Gucek M, Hendrix A, Westbroek W, Taraska JW. Exosomes released from breast cancer carcinomas stimulate cell movement. Plos One [Internet]. 2015;**10**(3): e0117495. DOI: 1371/journal.pone.0117495

[49] Han C, Sun X, Liu L, Jiang H, Shen Y, Xu X, et al. Exosomes and their therapeutic potentials of stem cells. Stem Cells International [Internet]. 2015;**2016**:7653489. DOI: 10.1155/2016/7653489

[50] De Veirman K, Wang J, Xu S, Leleu X, Himpe E, Maes K, et al. Induction of miR-146a by multiple myeloma cells in mesenchymal stromal cells stimulates their pro-tumoral activity. Cancer Letters [Internet]. 2016;**377**(1):17-24 Available from: https://www.sciencedirect.com/science/article/pii/S0304383516302592

[51] Bai M, Fu W, Su G, Cao J, Gao L, Huang C, et al. The role of extracellular vesicles in cholangiocarcinoma. Cancer Cell International [Internet]. 2020;**20**:435 Available from: https://www.ncbi.nlm.nih.gov/pmc/articles/PMC7709354/

[52] Setti M, Osti D, Richichi C, Ortensi B, Del Bene M, Fornasari L, et al. Extracellular vesicle-mediated transfer of CLIC1 protein is a novel mechanism for the regulation of glioblastoma growth. Oncotarget [Internet]. 2015;**6**(31):31413-31427 Available from: https://pubmed.ncbi.nlm.nih.gov/26429879

[53] Aga M, Bentz GL, Raffa S, Torrisi MR, Kondo S, Wakisaka N, et al. Exosomal HIF1α supports invasive potential of nasopharyngeal carcinoma-associated LMP1-positive exosomes. Oncogene [Internet]. 2014;**33**(37):4613-4622 Available from: https://pubmed.ncbi.nlm.nih.gov/24662828

[54] Corcoran RB, Ebi H, Turke AB, Coffee EM, Nishino M, Cogdill AP, et al. EGFR-mediated reactivation of MAPK signaling contributes to insensitivity of BRAF-mutant colorectal cancers to RAF inhibition with vemurafenib [Internet]. Cancer Discovery. 2012;**2**:227-235 Available from: https://cancerdiscovery.aacrjournals.org/content/candisc/2/3/227.full.pdf

[55] Qin X, Yu S, Zhou L, Shi M, Hu Y, Xu X, et al. Cisplatin-resistant lung cancer cell-derived exosomes increase cisplatin resistance of recipient cells in exosomal miR-100-5p-dependent manner. International Journal of Nanomedicine [Internet]. 2017;**12**:3721-3733 Available from: https://pubmed.ncbi.nlm.nih.gov/28553110

[56] Li XJ, Ren ZJ, Tang JH, Yu Q. Exosomal microRNA MiR-1246 promotes cell proliferation, invasion and drug resistance by targeting CCNG2 in breast cancer. Cellular Physiology and Biochemistry. 2017;**44**(5):1741-1748

[57] Namee NM, O'Driscoll L. Extracellular vesicles and anti-cancer drug resistance. Biochimica et Biophysica Acta (BBA). Reviews on Cancer [Internet]. 2018;**1870**(2):123-136 Available from: https://www.sciencedirect.com/science/article/pii/S0304419X18300969

[58] Tlsty TD, Coussens LM. Tumor stroma and regulation of cancer development. Annual Review of Pathology: Mechanisms of Disease. 2006;**1**(1):119-150. DOI: 10.1146/annurev.pathol.1.110304.100224

[59] Bai S, Wei Y, Liu R, Xu R, Xiang L, Du J. Role of tumour-derived exosomes in metastasis. Biomedicine & Pharmacotherapy [Internet]. 2022;**147**:112657 Available from: https://www.sciencedirect.com/science/article/pii/S0753332222000452

[60] Zhang H, Lu J, Liu J, Zhang G, Lu A. Advances in the discovery of exosome inhibitors in cancer. Journal of Enzyme Inhibition and Medicinal Chemistry. 2020;**35**(1):1322-1330

[61] Airola MV, Shanbhogue P, Shamseddine AA, Guja KE, Senkal CE, Maini R, et al. Structure of human nSMase2 reveals an interdomain allosteric activation mechanism for ceramide generation. Proceedings of the National Academy of Sciences. 2017;**114**(28):E5549-E5558

[62] Alvarez-Erviti L, Seow Y, Yin H, Betts C, Lakhal S, Wood MJ. Delivery of siRNA to the mouse brain by systemic injection of targeted exosomes. Nature Biotechnology. 2011;**29**(4):341-345

[63] Yang T, Martin P, Fogarty B, Brown A, Schurman K, Phipps R, et al. Exosome delivered anticancer drugs across the blood-brain barrier for brain cancer therapy in Danio rerio. Pharmaceutical Research. 2015;**32**(6):2003-2014

Chapter 16

Tumor-Derived Exosome and Immune Modulation

Deepak S. Chauhan, Priyanka Mudaliar, Soumya Basu, Jyotirmoi Aich and Manash K. Paul

Abstract

Tumor cells, like most other cells, release exosomes called tumor-derived exosomes (TEX) and are vital for intercellular communication. TEX are membrane-bound extracellular vesicles (EVs), containing unique cargo reminiscent of the parent tumor cells and possess immunomodulatory functions. TEX carries factors that directly promote immunosuppression in the tumor microenvironment and indirectly attract immunosuppressive T-regulatory (Treg) cells. The tumor-secreted exosomes can transfer their cargo by multiple mechanisms like fusion, phagocytosis, and receptor-mediated endocytosis, activating the recipient cells. TEX directly engages and releases cytokines, inactivating natural killer (NK) cells and T-cells and activating apoptosis. Tumor-derived exosomes also release soluble factors to suppress dendritic cell (DC) maturation while activating the expansion of immune-suppressive cells like Myeloid-derived suppressor cells (MDSCs) and Regulatory T (Treg) cells. Several studies have shown the relevance of TEX containing tumor-associated antigens (TAA) in reducing the efficacy of cancer immunotherapy and adoptive cell therapy. Hence understanding the basic biology and mechanism of TEX-mediated immunosuppression is critical in discovering cancer biomarkers and finding better immunotherapy and cell therapy approaches. In this chapter, we have discussed TEX biogenesis, TEX's structural and molecular features, TEX-mediated immunosuppression, and its relation to immunotherapy.

Keywords: extracellular vesicles (EVs), tumor-derived exosomes (TEX), immune modulation, immunotherapy, TEX Cargo

1. Introduction

James E. Rothman, Randy W. Schekman, and Thomas C. Südhof pioneered and discovered the molecular principles regulating cellular cargo trafficking via extracellular vesicles and were jointly awarded the 2013 Nobel Prize in Physiology or Medicine. Since then extracellular vesicles (EVs)-mediated horizontally transport of cargo across donor to recipient cells, followed by phenotypic alterations in the latter, has aroused significant scientific attention. EVs are lipid bilayer-delimited particles spontaneously secreted practically from all kinds of cells. The EVs contain cargo, including proteins, nucleic acids, lipids, metabolites, and even organelles, representing the parent cell's physiological state [1–3]. The terminology and classification of EVs are still emerging. Exosomes are a subgroup of EVs with a size ranging from 30 to 150 nm, produced via the parent cell's endocytic pathway and engaged in intracellular communication. Exosomes are released from cells upon fusion of an intermediate endocytic compartment, multi-vesicular body (MVB), and plasma membrane. This process delivers intraluminal vesicles (ILVs) into the extracellular milieu and in circulation (**Figure 1**) [1, 2].

The conventional exosome secretion process involves a few key steps: ILVs formation and exosome biosynthesis within MVBs, MVB trafficking, and fusion with the parent cell's plasma membrane followed by released via exocytosis (**Figure 1**). Once the exosomes reach a recipient cell, they either engage with the recipient cell's surface molecules to promote juxtracrine downstream signaling, undergo fusion with the recipient cell's membrane to deposit their payloads into the cytosol, or are taken by the recipient cells via processes like phagocytosis, macropinocytosis, and receptor-mediated endocytosis [1, 2, 4]. The fate of internalized EVs is still poorly understood and may be determined by exosomal heterogeneity and mode of cellular uptake. Internalized exosomes go to the early endocytic pathway after being endocytosed. Early endosomal membrane fusion may deliver the soluble cargo into the cytoplasm; in contrast, EV-associated membrane proteins undergo retrograde transport to the trans-Golgi network. Endosomal recycling may deliver them to the plasma membrane or be degraded in the lysosomes [5–7].

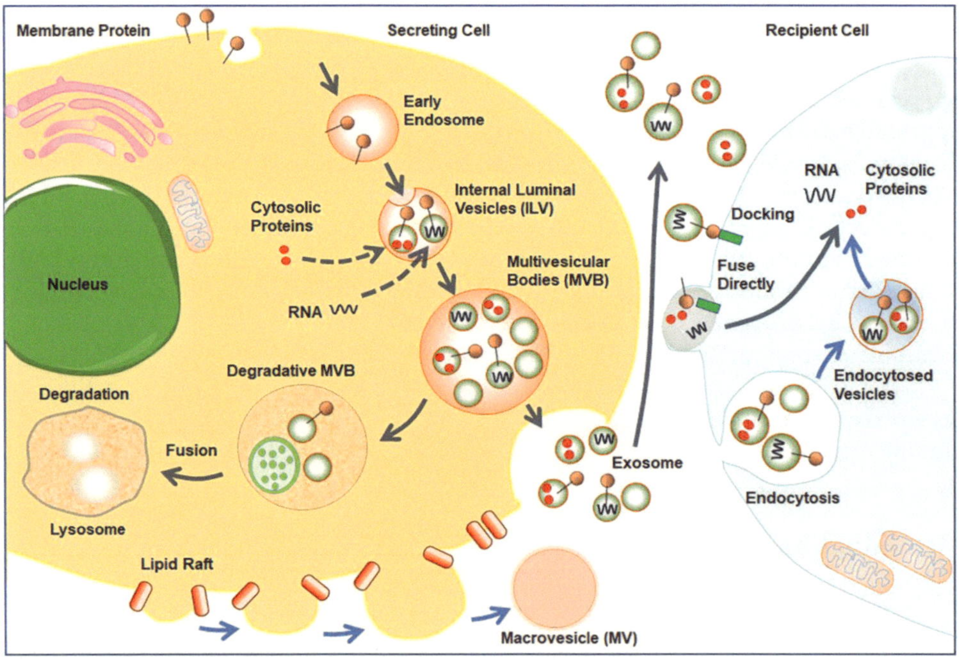

Figure 1.
Exosome biogenesis. The cytoplasmic outer layer protrudes to make up an initial secretory endosome (aka early endosome); intraluminal vesicles (ILVs) grow inwardly into the endosomal lumen constituting the multi-vesicular body (MVB). This process is known as the MVB biogenesis; these ILVs are secreted as exosomes when the MVB merges with the plasma membrane; but a few selected merges with the lysosome for degradation. The cargo of exosomes comprises of lipids, mRNA, miRNA, tRNA, lncRNA, DNA, proteins, adhesion molecules, receptors, and other functional compounds.

The endosomal sorting complex required for transports (ESCRTs) machinery is pivotal for the biogenesis of MVBs and ILVs [2, 8]. Exosomes contain distinctive cargos, including DNA, messenger RNAs (mRNA), micro RNAs (miRNA), transfer RNAs (tRNA), long non-coding RNA (lncRNA), proteins, lipids, and metabolites, among other biologically active molecules. These payloads are carefully processed and packed into the exosomes. The contents vary with each type of cell and are influenced by different cellular phenotypes and metabolic states, thereby imparting differential biological functionality [7, 9]. Exosome protein composition analysis has indicated that certain proteins are exclusive to the cell and tissue of origin, while others are found in all exosomes. Among some of the reported, 9769 exosomal proteins (exocarta.org) are conserved, and some are cell type-specific, like the major histocompatibility complex (MHC) class-I and class-II, other cell surface receptors, and proteases. Exosomes include proteins associated with membrane transport and fusion (e.g., annexin, nuclear-related protein Rab family GTPase (Rab-GTPase), SNAREs, and heat shock proteins (HSPs). While exosomes also have membrane-associated proteins (Tetraspanins, ICAM, etc.), MVB-related proteins (ALIX and TSG101), and other proteins such as actin, myosin, and adhesion molecules such as integrins. Specific proteins are widely used as exosomal markers, including the tetraspanins (CD9, CD63, CD81, CD82, Tspan8, CD151), Alix, and Tsg101 [5, 8].

Exosomes also include cell-specific or conserved lipid content (like cholesterol, sphingomyelin, phosphatidylserine, and saturated fatty acids). Lipids are involved in exosome biosynthesis as well as maintaining homeostasis in recipient cells, in addition to safeguarding exosome structure. Additionally, exosomes contain a variety of RNAs that are active and can influence the transcriptome of recipient cells [10]. Exosomes contribute to maintaining cellular homeostasis and cell-to-cell communications and are secreted by cells in normal physiological and pathological settings [5, 11].

There is an unprecedented need to study the role of exosomes and TEX to understand tumor progression that can aid in cancer diagnosis, prognosis, and therapeutic interventions. Tumor cells are reported to secrete more exosomes than healthy cells, thereby inhibition of exosome production, release, and reduction of circulating level may be an effective cancer therapy approach [12, 13]. Understanding the interaction of cancer cells with the body's immune system is key to cancer immunology and immunotherapy success. Immune suppressive features exist in the tumor microenvironment, limiting responses to immune-regulated assaults on the tumor [14]. Several immune cell types get functionally specialized and activated to fight and neutralize tumor cells and tumorigenesis. On the contrary, the tumor cells either evade immune identification, or induce an immunosuppressive tumor microenvironment (TME) to thwart the immunological onslaught. The tumor cells strategically use TEX-driven immunosuppression in the TME and within the tumor [15].

Recent studies show that tumor-derived exosomes are widely generated and contain a wide range of immunosuppressive chemicals. Here the role of TEX is extremely significant in intercellular communication and TME remodeling. This remodeling has the potential to aid cancer cells in avoiding detection by the immune system. According to several studies, TEX released by tumor cells modulates tumorigenesis, metastasis, and angiogenesis and facilitates drug resistance [14–16]. For example, when tumor cells are subjected to hypoxia, they release exosomes with increased angiogenic and metastatic potential, supporting the concept that tumor cells respond to a hypoxic milieu by releasing exosomes to promote angiogenesis or allow progression of the tumor cells to a more suitable habitat [16, 17]. TEXs are implicated in regulating the bioactivities of their target cells via the transmission of their oncogenic cargo. EVs/exosomes have also been termed "oncosomes" in these circumstances, and they transport active proteins, lipids, and nucleic acids to recipient target cells and control gene expression, therefore regulating their function. These exosomes may potentially aid in the establishment of metastatic niches. The molecular basis driving immune avoidance and the development of metastatic niches is still ambiguous [9, 15, 18]. In the tumor microenvironment, TEXs are copiously secreted and transport a range of immunosuppressive molecules. Associations between TEXs and immune cells can suppress immune cell function and the anti-tumor immune system, both directly and indirectly. Thus, TEXs are currently being investigated as prospective candidates for cancer immunotherapy due to these features and the discovery that large quantities of these TEXs in cancer patients correspond with tumor load and progression of the disease [15, 18].

2. Structural and molecular features of TEX

Electron microscopic (EM) evaluation suggests that the structural features of tumor-derived exosomes are comparable to that of most other exosomes [10]. Like many other cells, tumor cells release TEX, which are spherical membrane-bound vesicles that frequently have a diameter ranging from 30 to 150 nm, forming an aggregation of varying sizes. Considering that exosomes carry cell-type-specific molecules, it has been proposed that TEXs vary from normal healthy exosomes in regards to their structural, molecular, and biomechanical features [19]. TEX have lipid bilayer membrane structures that include transmembrane proteins and receptors. Numerous proteins, nucleic acids, and a diverse range of compounds can be discovered within the exosomal cavity, specific to the cancer cell type. The nucleic acid constituents have a role in intracellular transmission, chemotherapy resilience, micro-angiogenesis, tumor microenvironment alteration, immune response modulation, and tumor invasion and metastatic stimulation [16].

According to several proteomics studies, TEX contain membrane proteins, Rab family proteins, annexin, proteins associated with the Endosomal Sorting Complexes Required for Transport (ESCRT) complex like Alix TSG101, MHC molecules, heat shock proteins, and tetraspanins (CD63, etc.), all of which are endosomal pathway proteins. TEXs have also been reported to include tumor antigens such as Mart1, gp100, TRP, and Her2-neu, TGF-β, FasL, TRAIL (TNF-related apoptosis-inducing ligand), and beta-glycan [10, 19, 20]. TEX carry molecular cargo (**Table 1**) that comprises of specific lipids, MHC components, tumor-associated antigens, and other proteins derived in part from the surface of parent tumor cells [10, 15, 16, 19, 20]. These suggest that they can either activate or repress the immune system, although several studies reveal that TEXs enhance immunosuppression in the tumor microenvironment. Exosomes secreted by primary tumor cells can be transmitted to distant metastatic organs before tumor cells arrive at their final destination, according to growing data. This procedure allows for the formation of an accommodating pre-metastatic niche that promotes the proliferation of disseminated tumors. TEXs have immunosuppressive and immunostimulatory effects that are not limited to the tumor microenvironment [15, 16].

Ubiquitination is a feature of normal exosomal/TEX proteins that permits them to be identified by ESCRT-0, whereas deubiquitination is critical in sorting them into ILVs. Membrane transport and fusion proteins such as annexin, Rab-GTPase, and HSPs such as Hsp60, Hsp70, and Hsp90; Tetraspanins such as CD9, CD63, CD81, CD82,

TEX Cargo	Types of molecules	Functional activity
Protein	Heat shock proteins Membrane transport and fusion related proteins Enzymes Tetraspanins ICAM MVB related proteins Cell adhesion Proteins Receptor proteins	Regulation of proliferation, growth, metastasis, migration, angiogenesis, adhesion, immunological suppression, and a variety of other physiological processes related to the tumor cells; also modulate immunotherapy effects
Lipids	Phosphatidylserine Lactosylceramide	Can be potentially used as biomarkers in many cancers
Nucleic acids	DNA RNA miRNA mRNA lncRNA circRNA	Enhancing proliferation and growth of tumors, promoting angiogenesis, inhibiting immunological activities, and increasing metastasis all contribute to cancer and influence immune response stimulation

Table 1.
Cargo of TEX with their molecular features and functions.

CD106, Tspan8, intercellular adhesion molecules (ICAM); MVBs associated protein such as ALIX and TSG101; and certain other proteins such as integrins, cytoskeletal construction proteins like actin and myosin are content of TEX [6, 15, 20]. These proteins are critical for exosome functioning. Since tumor cells are always under stressful conditions such as hypoxia, acidosis, nutrient shortage, etc., Hsp90 expression is high in numerous cancer cells. Hsp90 is linked to poor tumor prognosis and tumor development in breast cancer, pancreatic cancer, and leukemia. A study suggested that certain TEX's expressed surface TGF-β and beta-glycan, which can stimulate the SMAD signaling and govern fibroblast development into myofibroblasts [19, 21].

Exosome-bound miRNAs may also aid tumor growth in a number of ways.

By decreasing the expression of E-cadherin in normal fibroblasts, miR-9 in exosomes originating from triple-negative breast cancer cells (TNBC) might increase tumor cell migration and enhance the transition of fibroblasts into cancer-associated fibroblasts. Exosomal miRNAs of mesenchymal stromal cells have also been demonstrated to be transported directly to tumor cells, promoting cancer growth and inducing treatment resistance in multiple myeloma, colorectal, and gastric cancer cells [6, 21]. lncRNA is a newly discovered regulatory RNA that can be packed into exosomes and functions as a messenger in intercellular communication to control tumor development and other related processes while also reshaping the tumor microenvironment. lncRNAs-ATB, for example, a new cancer-associated lncRNA that was anomalously exhibited in many cancers, is known for enhancing tumor progression and growth primarily by competitively anchoring miRNA to stimulate epithelial-mesenchymal transition (EMT) [21].

Although prevailing data indicates that TEX may have different immune-triggering/immunosuppressive activities based on the cargo they transmit and the functional capacity of immune cells in the tumor microenvironment, reconciling these two opposing factors of exosome features has proven to be challenging. TEX can potentially manipulate a wide variety of functions in target cells simultaneously, and the consequence is determined by the content of their cargo and the target cell's potential to accept or decline the conveyed signals.

3. TEX and immune modulation

The TEX are responsible for the tumor proliferation, metastasis, and antitumor response through immune and non-immune pathways. On the one hand, TEX interacts with immune cells and conveys negative signals, thereby interfering with their antitumor functions, while on the other hand, TEX promotes tumor development and facilitates tumor escape by reducing immune effector cell activity [20]. TEX transport immunoinhibitory, immunosuppressive mediators, and costimulatory molecules (like cytokines, MHC I and II, etc.) that directly or indirectly regulate immune cell formation, maturation, and antitumor activity [15, 20]. These molecules with unique cell surface motifs assist TEX in modulating the immune response, which works by the coordinated action and networking of different components. Human cancer cell exosomes may induce inter-and intracellular signals to the tumor microenvironment modulating immune cell infiltration (**Figure 2**). Recent studies show that it can suppress the immune response both in antigen-specific and non-antigen-specific fashion. For example, TEX can induce apoptosis by the transfer of FasL and TRAIL to activated T cells [22].

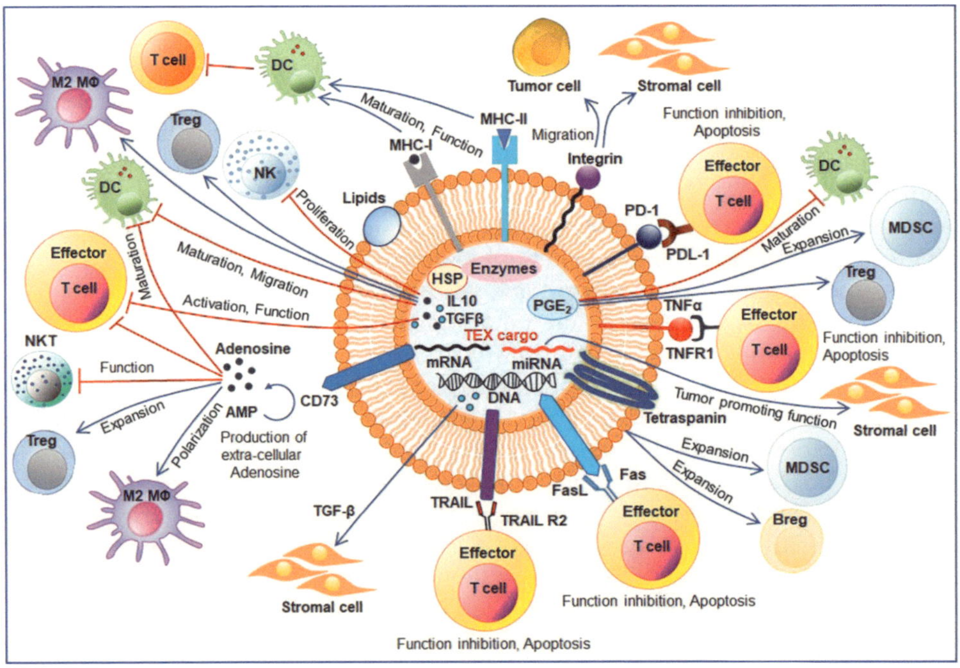

Figure 2.

Role of TEX in cancer progression. TEX are responsible for transferring oncogenic proteins and nucleic acid. It can promote angiogenesis and thrombosis by activating the endothelial cells. Also, it helps in the conversion of MSCs and fibroblast to myofibroblast to promote angiogenesis and metastasis. Further, it impairs the immune response by directing apoptosis in T cells and NK cells, promoting Treg cell activity, expanding MDSC, and inhibiting DC differentiation and function. Also, it assists in developing drug resistance by delivering multidrug resistance proteins and miRNA. Moreover, it helps in neutralizing antibodies and expelling the anticancer drugs. DC: dendritic cell; MDSC: myeloid-derived suppressor cell; Treg: regulatory T cell; Breg: regulatory B cells; M2 Mφ: M2 macrophage; NKT: natural killer T cells; NK: natural killer cells; MHC: major histocompatibility complex; TEX: tumor-derived exosomes; TGF-β: transforming growth factor-beta; IL-10: interleukin 10; TNFα: tumor necrosis factor-alpha; TNFR1: tumor necrosis factor receptor 1; TRAIL: TNF-related apoptosis-inducing ligand; TRAIL R2: TRAIL receptor 2; PGE_2: prostaglandin E2; HSP: heat shock protein.

TEX include membrane-bound NKG2D ligands including MICA, MICB, or ULBP1-6, which may directly suppress NK and CD8+ T cells [23]. Also, TEX is known to suppress the expression of CD3- ζ chains in T cells to prevent their activation, and NKG2D inhibition in natural killer (NK) cells prevents NK cell-dependent toxicity [24, 25]. Tumor-derived exosomes may also suppress the anti-tumor immune response by producing prostaglandin E2 (PGE_2). In the presence of TGF-β, PGE_2 promotes the growth of myeloid-derived suppressor cells (MDSCs) and their suppressive function. PGE_2 also inhibits NK cell cytolysis and IFN-γ production, as well as T cell IL-2 production and responsiveness [26]. TEX can also modulate the antigen-presenting cells; for example, TEX miRNA may bind to TLRs, triggering an inflammatory response. For example, miR21 and -29a secreted from exosomes of lung cancer cells bind to the human and murine TLRs and stimulate the secretion of pro-inflammatory cytokines like IL-6 and TNF-α [27]. TEX can also disrupt the differentiation of peripheral blood monocytes into functional dendritic cells. For example, TEX-released by colorectal and melanoma cells, for example, was shown to impede CD14+ monocyte differentiation into dendritic cells instead of causing them to highly immunosuppressive MDSCs (**Figure 2**) [25]. Tumor-derived exosomes have emerged as an important factor in the loss of antigen presenting cell function and decreased anti-tumor immune responses in patients with cancer [28].

The miRNA, HSP 70, prostaglandin E2, and TGF-β are found in the TEX and play an essential role in the differentiation of the monocytes [29]. It has been reported that the above factors can also be transported distantly by the TEX towards altering the function and differentiation of myeloid cells for favoring the MDSCs at the metastatic sites [30]. After that, MDSC induces the regulatory T cells (Treg), which play a crucial role in the tumor-suppressive microenvironment. The CD4+ T cells are directed towards the Th2 and Treg due to the expression of cytokines, TGF-β, MMPs, and growth factors in MDSCs [31]. TEX has been shown to convert the CD4+FoxP3+ T cells into Tregs via IL-10 and TGF, which are very suppressive and resistant to apoptosis [32]. Also, it has been reported that CD11b+ TEX in the tumor-bearing mice can suppress the specific response to tumor antigens via the MHC class-I independent and MHC class II-dependent pathways [33]. It suggests that TEX first stimulates the antigen-presenting cells containing CD11 in the tumor microenvironment, which then secretes the CD11b and MHC class II immunosuppressive vesicles in the circulation. Adenosine synthesis in T cells was reported to be increased by Treg coincubated with TEX, which have CD39 and CD73 ectonucleotidases [34, 35]. TEX-mediated adenosine production is implicated in suppressing activated B cells and may in-turn activate B cells into regulatory B cells (**Figure 2**).

Exosomes may promote innate and adaptive immunity, as seen in infected macrophages, which produce exosomes containing bacterial cell wall components that activate uninfected macrophages [36]. Among the adjuvants found in tumor exosomes is heat shock protein 70 (hsp70), which may stimulate anti-cancer immune responses. Researchers discovered that Hsp70/Bag-4-positive human pancreatic and colon cancer cells secrete exosomes that promote the migration and cytolytic activity of NK cells [37]. They also showed that Hsp70-positive exosomes operate in macrophage activation, as measured by TNF production [38]. Biomolecular cargo found in exosomes from DCs might facilitate the development of cell-free DC-based cancer vaccines [36, 39]. Exosomes from other cellular sources may also activate an immune response. When human alveolar epithelial cells were treated with TNF-α+ mature DC exosomes, they in turn produced inflammatory mediators such as IL-8, MCP-1, MIP-1, RANTES, and TNF-α as a result [39]. Advanced stage NSCLC or metastatic melanoma patients who received DC exosomes in phase I clinical trials showed enhanced NK cells activity [40, 41]. Injection of DC exosomes restored NKG2D levels in patients with metastatic melanoma, and tumor regression in mice was encouraged [42]. Though exosomes and TEX might induce immune activation or immunosuppression in the tumor microenvironment, but most reports for TEX suggest an immunosuppressive mode of action.

4. Mechanism of TEX-mediated immune suppression

Immunostimulatory TEX from tumors may spread to distant tissues and organs, impairing systemic anti-tumor immune responses. Signaling molecules in the tumor microenvironment aid in tumor development and inhibits the immune response, with T cells being highly vulnerable to TEX-driven negative messages [7, 43]. Unlike other leukocytes which engulf other cells, T cells interact with the TEX using the ligands and surface receptors. It leads to the signal-driven influx of Ca^{2+} in the cells and activation and suppression of pathways along with downstream modified responses. Two major receptors on the T cells that are negatively regulated by the TEX are the interleukin 2 receptor (IL-2R) and T cell receptor (TCR) [44]. The TCR zeta chain is downregulated consistently on the incubation of TEX with the T cell. In addition, JAK expression and phosphorylation are diminished, which is responsible for the production of IL-2, IL-7, and IL-15, all of which are deleterious to T cell proliferation. Further, TEX downregulates the proliferation of CD4+ T cells but upregulates the expansion of CD8+ T cells [45]. However, in the case of normal cell-derived exosomes, the proliferation of all T cells is experienced [46]. Also, TEX regulates the STAT5 function in the T cells, as STAT5 phosphorylation in the case of CD4+ T cells increases while the phosphorylation of STAT5 in CD8+ T cells decreases [47]. The level of CD69 on the surface of CD4+ T cells is also reduced due to TEX immunosuppressive function.

T cell inhibitory and apoptotic receptors are directly engaged by tumor-derived exosomes (TEX). To enhance T cell death, TEX expresses Fas ligand (FasL), TNF-alpha, TRAIL, and Galectin-9 that interacts with counterpart T cell receptors like FAS, TNF R1, TRAIL R2, and Tim3 [48]. The TEX is also found to inhibit the antigen-specific T cell, as in the case of human melanoma, wherein specific T cells are generated via the melanoma-specific pulsing of the DC cells. Also, evidence suggests the presence of membrane-bound FasL and PD-L1 on TEX [49]. TEX-mediated apoptosis of CD8+ T cells is associated with canonical features like the caspase 3 cleavage, annexin V binding, loss of mitochondria membrane potential, DNA fragmentation, and cytochrome C release [45]. It suggests the involvement of extrinsic and an intrinsic mechanism for the cascade of apoptosis of the CD8+ T cells. Also, the PI3K/AKT pathways are the main target for TEX-driven apoptosis in CD8+ T cells: AKT dephosphorylation in a time-dependent manner decreases the expression of BCL-xL, BCL-2, and MCL-1 along with the increase in BAX was observed when TEX was incubated with CD8+ T cells [50]. All these data indicate that TEX may not be internalized by the T cells; instead, the negative signaling by the surface receptors modulates the function of T cell response. Also, TEX-driven transcription changes are regulated by the type of recipient cell, activation status, and presence or absence of exosomes. TEX is also responsible for the apoptosis of CD8+ T cells with the help of death ligands and interaction via the Fas/FasL pathway [51]. Also, it has been shown in the lymphoma animal model that TEX protects cancer cells from complement-dependent cytolysis by binding to the complements [52].

Cancer cells upregulate PD-L1 to avoid immune identification by causing anergy in PD-1+ T cells. Although immune checkpoint inhibitors have shown extraordinary effectiveness, most patients do not react to PD-1/PD-L1 inhibition. Paracrine immunosuppression may now include recently discovered exosomal-PD-L1 as well as cell-to-cell interaction [53]. By both direct and indirect means, exosomes seem to be capable of immunomodulating PD-L1 (**Figure 3**). Recent research shows that PD-L1 is active on exosome membranes and may promote tumor development by inhibiting CD8+ T cell proliferation and cytotoxicity [53]. Exosomes from human breast cancer cells contain PD-L1, while PD-L1 knockout (KO) cells do not [54]. In a unique experiment, two groups of mice were treated with exosome-expressing PD-L1^{FLAG} and exosome-expressing no PD-L1 (PD-L1-KO). Animals treated with exosome-expressing PD-L1^{FLAG} showed considerably larger tumor volumes than mice treated with exosome-expressing no PD-L1, demonstrating that exosomal PD-L1 stimulates breast cancer tumor development. Exosomes expressing PD-L1 dramatically reduced T cell death, demonstrating that PD-L1 signaling may limit T cell killing of cancer cells [55]. Exosomes isolated from the WM9 and WM164 human melanoma cell lines express PD-L1, and may function as an anti-PD-L1 antibody sink [34, 38]. Chen et al., also reported exosomal PD-L1 in human non–small cell lung cancer (NSCLC) and breast cancer cell lines. Recent findings suggest that PD-L1 exosome's immunosuppressive effects are not limited to the tumor microenvironment, and that exosomes might cause systemic alterations in adaptive immune components (**Figure 3**) [49]. Functional modifiers that stimulate PD-L1 expression in target cells might be delivered via tumor-derived exosomes. T cell dysfunction may be caused directly by PD-L1+ monocytes or macrophages as a result of a PD-L1-PD-1 interaction [56].

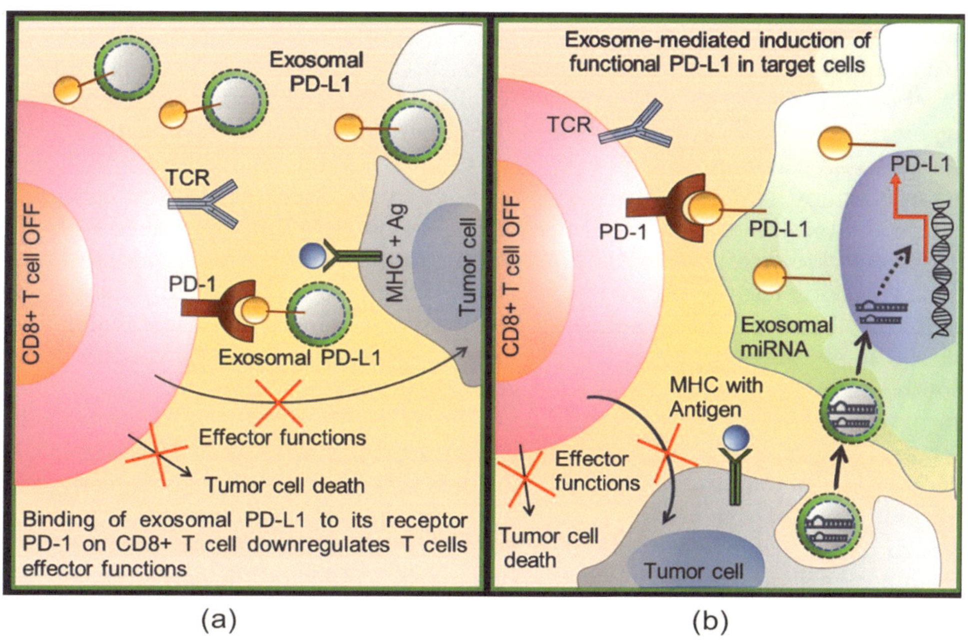

Figure 3.
TEX and PD-L1: By activating the PD-L1/PD1 axis, cancer cells adapt and exploit the immune system to elude immune monitoring. a. Exosomal-PD-L1 interacts with T-cells through paracrine signaling, inactivating T cell effector activity. b. Induction of functional PD-L1 in target cells by exosome cargo (miRNA). The PD-L1 on the surface then interacts with T cells, inactivating effector activity.

Further investigation is needed regarding exosomal PD-L1 and exosome-induced PD-L1 in immune modulation and cancer. Exosomal-PD-L1-based paracrine immune regulation may help create novel therapeutic options.

Another interesting aspect of TEX-mediated immunomodulation involves the anti-phagocytic surface proteins called 'do not eat me' signals, such as the cluster of differentiation (CD) 47 help healthy and normal cells evade macrophage-mediated phagocytosis, while a loss of 'do not eat me' signals in apoptotic or senescent cells leads to their systemic clearance [57]. Cancer cells use a similar technique to avoid macrophage-mediated clearance by overexpressing 'do not eat me' signals on their surface. TEX overexpressing CD47 renders phagocytic inactivation by interacting with SIRPα of immune cells and thereby enjoys prolonged circulation [58]. Downregulation of CD47 expression or inhibition of CD47-SIRPα can be an interesting approach to activate cancer cell/TEX phagocytosis [59].

5. TEX and cancer immunotherapy

There is a controversy regarding the role of exosomes in immunotherapy due to their immunostimulatory and immunoinhibitory action. However, it is imperative that TEX is responsible for the impairment of immune response and take part in the active progression of cancer as it contains several immunoinhibitory, tumor antigens, and invasive molecules to directly or indirectly suppress the immune system. It suppresses the proliferation and differentiation of the immune cells, including the remodeling of genetic materials. For example, tumor-associated antigens (TAA) bind to the antibodies produced against the cancer cells, thus reducing its efficacy in reaching the primary cancer tissue [18]. The efficacy of trastuzumab is severely affected by the TAA cargo of TEX [3]. Also, TEX has been found to play a role in inhibiting antibody-dependent cell-mediated toxicity (ADCC), which is majorly responsible for cancer prevention and the primary activity of humanized antibodies [60]. Evidence suggests that TEX also neutralizes the beneficial effect of immunotherapies. For example, TEX carrying HER2 or other TAA lessen the potential of antibody-driven immunotherapy. The TEX are present in all body fluids, which can neutralize the therapeutic antibodies, thus blocking access to the tumor. Also, the engineered T and NK cells are susceptive to the TEX carrying immunoinhibitory ligands like FasL, which is majorly responsible for the apoptosis of the adapted T cells [22]. The relapsed case of myelogenous leukemia showed the enriched presence of exosomes with immunosuppressive cargo in the plasma. These patients also responded poorly to the adoptive NK92 therapy [61]. It was speculated that a negative immunosuppressive exosome-based signaling was behind the failure of natural killer-based adoptive cell therapy. This was later confirmed by co incubating the patient-derived exosomes with NK-92 cells, wherein function and antileukemia activity of NK-92 cells were severely affected, while the vice-versa was experienced on the blocking of immunosuppressive exosomes in ex-vivo studies. Thus, it was established that TEX interferes with the immune cells and limits the therapy's therapeutic efficacy. The new target of checkpoint inhibitors is the TEX PD-L1 to mitigate the resistance faced by the current antibody approaches [62]. It could also be used as a predictor for anti-PDL1 therapy.

Also, the Tim-3 and Galectin-9 is the upcoming checkpoint inhibitor that negatively regulates the antitumor immune response [63]. The TEX are also responsible for the variations in the PD-1/PD-L1 treatment in different patients. The low surface expressivity of PD-L1 may be the reason for the low response or resistance to immunotherapy [64]. Also, there is less information on whether the exosomal PD-L1 function varies with cancer type or not. The role of exosomal PD-L1 is to be elucidated further, which may help understand the early diagnosis and better therapeutic outcomes. Exosomes usually confer resistance by manipulating the cell-cell communication in the tumor microenvironment [10]. For example, the exosomes emanating from the macrophages are responsible for the drug resistance in ovarian cancer treatment in hypoxic conditions.

Further, HER2 overexpressing exosomes were found to neutralize the effect of trastuzumab [3]. In gastric cancer cells, the miRNA-21 containing exosomes secreted by the M2 macrophages were found to inhibit the action of cisplatin [65]. The miRNAs containing exosomes are responsible for converting the monocytes to MDSCs. It has also been shown in the breast cancer model of animals that exosomes release is responsible for developing the premetastatic niche in the lungs under the influence of chemotherapy [66, 67]. Thus, the role of TEX in the development of protumor immunity and progression of metastases is involved. Also, the immunosuppressive molecules affect the antitumor activities and maturation of immune cells. Although the dual role of TEX is studied like one is the development of metastatic niche and increasing the invasiveness of cancer cells while on the other side is helping in inducing the tumor-specific immune response for the cell lysis. Thus, the role of TEX in controlling the effectiveness of immunotherapy is an interesting area to be explored.

6. TEX and cancer vaccine

The use of exosomes in cancer treatment is made possible by the fact that exosomes can serve as delivery vehicles for genes and biological therapeutics. Tumor cells must be targeted explicitly while limiting adverse effects on healthy tissues to treat cancer effectively. Due to their ability to deliver their contents inside cancer cells, exosomes play a significant role in improving the therapeutic index of cancer treatment [68]. Exosomes exhibit these abilities because of the multivalent display of their surface fractions derived from cells, and it is impossible to recreate this complexity in synthetic nanoparticles [68]. Researchers have researched the development of therapeutic cancer vaccines, also known as active specific immunotherapy, and discovered that exosomes offer a great deal of promise for cancer immunotherapy and therapeutic cancer vaccines [69]. Exosomes from different cell types, such as immune cells, cancer cells, and normal cells, are compared for their effectiveness of exosome-based cancer immunotherapy. B cells secrete exosomes harboring MHC class II peptides that aid in antigen presentation to CD4+ T lymphocytes. Dendritic cell-derived exosomes (DEXs) play a critical role in anti-tumor immune response and may trigger a particular Cytotoxic T lymphocytes (CTLs) response and activate a T cell-dependent anti-tumor response [13].

TEXs, are modified for cancer vaccine development as they are a natural source of tumor antigens and can activate APCs to display them effectively [68]. Several studies have reported positive outcomes with TEX-based vaccination and observed activation of T-cell mediated antitumor immune reactions and tumor reduction [70, 71]. Researchers found that TEX immunization not only protected against tumor development and stimulated Th1 immune responses in melanoma animal models but also might limit lung metastasis [72]. In another study, immunization of syngeneic mice with exosomes generated by L1210 leukemia cells reduced tumor growth and provided resistance against subsequent tumor challenges [73]. In another study, both T lymphocyte proliferation and specific CTL activity were all stimulated when treated with HeLa cell-derived TEX [74]. Though many studies support the use of TEX-based vaccination, it is unlikely that TEXs alone can initiate acceptable levels of anti-tumor immunity due to their role in immunosuppression and the low immunogenicity of their components. Multiple strategies are devised to maximize antigen immunogenicity for effective tumor vaccination. TEXs might be modified by the following approaches (a) incorporating electroporated siRNA, (b) engineered to express tumor-associated and pathogenic antigens concomitantly, (c) tagging known immune boosters like CpG DNA and TLR ligands, (d) Direct fusion of TEXs with antigens, and (e) using external stimulus to increase TEX release [68, 75]. Innovative approaches are needed to make TEX-based cancer vaccines a reality.

7. Conclusion

Exosomes play a crucial role in facilitating the cell-cell environment under normal and pathological conditions. TEXs are emerging as the key immunoregulatory players for cancer cell-to-tumor microenvironment communication, influencing tumor development and metastasis. It's exciting and intriguing that the TEX can train immune effector cells to suppress or stimulate the immune system. TEX-driven interactions may be direct or indirect, and the presence or absence of immune recipient cells in the TME may alter the outcome. Oncotransducers, which generate juxtracrine or paracrine signals, may alter suppressive pathways formed in immune recipient cells and resulting in faster tumor growth. Immunotherapies are unlikely to function at its full potential due to TEX -based antibody sequestration. Reports also suggest that TEX-induced immunostimulatory signals can modify the TME to enhance immune activation rather than tumor progression. Hence more investigations are warranted. TEX-engineering to activate the anti-tumor potentials may lead to future therapeutics. Presently, the key challenge of the tumor biomarker identification and validation procedure is dependent on decoding the messages in TEX cargo and correlating them to clinical data. The practical usage of TEX may revolutionize cancer detection, diagnosis and therapy.

Acknowledgements

M.K.P. acknowledges S. Dubinett, B. Gomperts, and V. Hartenstein for providing constant support and mentoring.

Conflict of interest

The authors declare no conflict of interest.

Author details

Deepak S. Chauhan[1†], Priyanka Mudaliar[2†], Soumya Basu[3], Jyotirmoi Aich[2] and Manash K. Paul[4*]

1 Faculté de Pharmacie, Université de Montréal, Montréal, QC, Canada

2 School of Biotechnology and Bioinformatics, DY Patil Deemed to Be University, Navi Mumbai, Maharashtra, India

3 Cancer and Translational Research Laboratory, Dr. D.Y. Patil Biotechnology and Bioinformatics Institute, Dr. D.Y. Patil Vidyapeeth, Pune, Maharashtra, India

4 Department of Pulmonary and Critical Care Medicine, David Geffen School of Medicine, University of California Los Angeles (UCLA), Los Angeles, CA, USA

*Address all correspondence to: paul_cancerbiotech@yahoo.co.in

† Equal contribution.

References

[1] Madamsetty VS, Mukherjee A, Paul MK. Bioinspired nanoparticles-based drug delivery systems for cancer theranostics. In: Patra C, Ahmad I, Ayaz M, Khalil AT, Mukherjee S, Ovai M, editors. Biogenic Nanoparticles for Cancer Theranostics. Elsevier; 2021. pp. 189-228. ISBN 9780128214671. DOI: 10.1016/B978-0-12-821467-1.00008-2

[2] Yáñez-Mó M, Siljander PRM, Andreu Z, Bedina Zavec A, Borràs FE, Buzas EI, et al. Biological properties of extracellular vesicles and their physiological functions. Journal of Extracellular Vesicles. 2015;**4**(1):27066

[3] Ciravolo V, Huber V, Ghedini GC, Venturelli E, Bianchi F, Campiglio M, et al. Potential role of HER2-overexpressing exosomes in countering trastuzumab-based therapy. Journal of Cellular Physiology. 2012;**227**(2): 658-667

[4] Sun Y-Z, Ruan J-S, Jiang Z-S, Wang L, Wang S-M. Extracellular vesicles: A new perspective in tumor therapy. BioMed Research International. 2018;**2018**:1-7

[5] Gurung S, Perocheau D, Touramanidou L, Baruteau J. The exosome journey: From biogenesis to uptake and intracellular signalling. Cell Communication and Signaling. 2021;**19**(1):47

[6] Dai J, Su Y, Zhong S, Cong L, Liu B, Yang J, et al. Exosomes: Key players in cancer and potential therapeutic strategy. Signal Transduction and Targeted Therapy. 2020;**5**(1):145

[7] Seo N, Akiyoshi K, Shiku H. Exosome-mediated regulation of tumor immunology. Cancer Science. 2018; **109**(10):2998-3004

[8] Zhang L, Yu D. Exosomes in cancer development, metastasis, and immunity. Biochimica et Biophysica Acta (BBA): Reviews on Cancer. 2019;**1871**(2): 455-468

[9] Liu J, Wu S, Zheng X, Zheng P, Fu Y, Wu C, et al. Immune suppressed tumor microenvironment by exosomes derived from gastric cancer cells via modulating immune functions. Scientific Reports. 2020;**10**(1):14749

[10] Mashouri L, Yousefi H, Aref AR, Ahadi A, Molaei F, Alahari SK. Exosomes: Composition, biogenesis, and mechanisms in cancer metastasis and drug resistance. Molecular Cancer. 2019;**18**(1):75

[11] Gurunathan S, Kang M-H, Jeyaraj M, Qasim M, Kim J-H. Review of the isolation, characterization, biological function, and multifarious therapeutic approaches of exosomes. Cells. 2019;**8**(4):307

[12] Sun W, Luo J-d, Jiang H, Duan DD. Tumor exosomes: A double-edged sword in cancer therapy. Acta Pharmacologica Sinica. 2018;**39**(4):534-541

[13] Thakur A, Parra DC, Motallebnejad P, Brocchi M, Chen HJ. Exosomes: Small vesicles with big roles in cancer, vaccine development, and therapeutics. Bioactive Materials. 2022;**10**:281-294

[14] Yang C, Robbins PD. The roles of tumor-derived exosomes in cancer pathogenesis. Clinical and Developmental Immunology. 2011;**2011**:1-11

[15] Whiteside TL, Diergaarde B, Hong C-S. Tumor-derived exosomes (TEX) and their role in immuno-oncology.

International Journal of Molecular Sciences. 2021;**22**(12):6234

[16] Zhang X, Zhou C. The function of tumor-derived exosomes. Journal of BUON. 2019;**24**(3):897-904

[17] Park JE, Tan HS, Datta A, Lai RC, Zhang H, Meng W, et al. Hypoxic tumor cell modulates its microenvironment to enhance angiogenic and metastatic potential by secretion of proteins and exosomes. Molecular & Cellular Proteomics. 2010;**9**(6):1085-1099

[18] Whiteside TL. Tumor-derived exosomes and their role in cancer progression. Advances in Clinical Chemistry. 2016;**74**:103-141

[19] Huang T, Deng C-X. Current progresses of exosomes as cancer diagnostic and prognostic biomarkers. International Journal of Biological Sciences. 2019;**15**(1):1-11

[20] Olejarz W, Dominiak A, Żołnierzak A, Kubiak-Tomaszewska G, Lorenc T. Tumor-derived exosomes in immunosuppression and immunotherapy. Journal of Immunology Research. 2020;**2020**:1-11

[21] Xu Z, Zeng S, Gong Z, Yan Y. Exosome-based immunotherapy: A promising approach for cancer treatment. Molecular Cancer. 2020;**19**(1):160

[22] Andreola G, Rivoltini L, Castelli C, Huber V, Perego P, Deho P, et al. Induction of lymphocyte apoptosis by tumor cell secretion of fasl-bearing microvesicles. Journal of Experimental Medicine. 2002;**195**(10):1303-1316

[23] Duan S, Guo W, Xu Z, He Y, Liang C, Mo Y, et al. Natural killer group 2D receptor and its ligands in cancer immune escape. Molecular Cancer. 2019;**18**(1):29

[24] Taylor DD, Gercel-Taylor C, Lyons KS, Stanson J, Whiteside TL. T-cell apoptosis and suppression of T-cell receptor/CD3-zeta by Fas ligand-containing membrane vesicles shed from ovarian tumors. Clinical Cancer Research. 2003;**9**(14):5113-5119

[25] Valenti R, Huber V, Filipazzi P, Pilla L, Sovena G, Villa A, et al. Human tumor-released microvesicles promote the differentiation of myeloid cells with transforming growth factor-β–mediated suppressive activity on T lymphocytes. Cancer Research. 2006;**66**(18):9290-9298

[26] Finetti F, Travelli C, Ercoli J, Colombo G, Buoso E, Trabalzini L. Prostaglandin E2 and cancer: Insight into tumor progression and immunity. Biology. 2020;**9**(12):434

[27] Fabbri M, Paone A, Calore F, Galli R, Gaudio E, Santhanam R, et al. MicroRNAs bind to Toll-like receptors to induce prometastatic inflammatory response. Proceedings of the National Academy of Sciences. 2012;**109**(31):E2110-E21E6

[28] Hosseini R, Asef-Kabiri L, Yousefi H, Sarvnaz H, Salehi M, Akbari ME, et al. The roles of tumor-derived exosomes in altered differentiation, maturation and function of dendritic cells. Molecular Cancer. 2021;**20**(1):83

[29] Xiang X, Liu Y, Zhuang X, Zhang S, Michalek S, Taylor DD, et al. TLR2-mediated expansion of MDSCs is dependent on the source of tumor exosomes. The American Journal of Pathology. 2010;**177**(4):1606-1610

[30] Gabrilovich DI, Ostrand-Rosenberg S, Bronte V. Coordinated regulation of myeloid cells by tumours. Nature Reviews Immunology. 2012;**12**(4):253-268

[31] Condamine T, Gabrilovich DI. Molecular mechanisms regulating myeloid-derived suppressor cell differentiation and function. Trends in Immunology. 2011;**32**(1):19-25

[32] Wieckowski EU, Visus C, Szajnik M, Szczepanski MJ, Storkus WJ, Whiteside TL. Tumor-derived microvesicles promote regulatory T cell expansion and induce apoptosis in tumor-reactive activated CD8+T lymphocytes. The Journal of Immunology. 2009;**183**(6):3720-3730

[33] Yang C, Ruffner MA, Kim S-H, Robbins PD. Plasma-derived MHC class II+ exosomes from tumor-bearing mice suppress tumor antigen-specific immune responses. European Journal of Immunology. 2012;**42**(7):1778-1784

[34] Schuler PJ, Saze Z, Hong CS, Muller L, Gillespie DG, Cheng D, et al. Human CD4+CD39+regulatory T cells produce adenosine upon co-expression of surface CD73 or contact with CD73+exosomes or CD73+cells. Clinical & Experimental Immunology. 2014;**177**(2):531-543

[35] Whiteside TL. The effect of tumor-derived exosomes on immune regulation and cancer immunotherapy. Future Oncology. 2017;**13**(28):2583-2592

[36] Barros FM, Carneiro F, Machado JC, Melo SA. Exosomes and immune response in cancer: Friends or foes? Frontiers in Immunology. 2018;**9**:730

[37] Linder M, Pogge von Strandmann E. The role of extracellular HSP70 in the function of tumor-associated immune cells. Cancers. 2021;**13**(18):4721

[38] Vega VL, Rodríguez-Silva M, Frey T, Gehrmann M, Diaz JC, Steinem C, et al. Hsp70 translocates into the plasma membrane after stress and is released into the extracellular environment in a membrane-associated form that activates macrophages. The Journal of Immunology. 2008;**180**(6):4299-4307

[39] Obregon C, Rothen-Rutishauser B, Gerber P, Gehr P, Nicod LP. Active uptake of dendritic cell-derived exovesicles by epithelial cells induces the release of inflammatory mediators through a TNF-α-mediated pathway. The American Journal of Pathology. 2009;**175**(2): 696-705

[40] Morse MA, Garst J, Osada T, Khan S, Hobeika A, Clay TM, et al. A phase I study of dexosome immunotherapy in patients with advanced non-small cell lung cancer. Journal of Translational Medicine. 2005;**3**(1):9

[41] Escudier B, Dorval T, Chaput N, André F, Caby M-P, Novault S, et al. Vaccination of metastatic melanoma patients with autologous dendritic cell (DC) derived-exosomes: Results of the first phase I clinical trial. Journal of Translational Medicine. 2005;**3**:10

[42] Pockley G, Viaud S, Terme M, Flament C, Taieb J, André F, et al. Dendritic cell-derived exosomes promote natural killer cell activation and proliferation: A role for NKG2D ligands and IL-15Rα. PLoS One. 2009;**4**(3):e4942

[43] Lim AR, Rathmell WK, Rathmell JC. The tumor microenvironment as a metabolic barrier to effector T cells and immunotherapy. eLife. 2020;**9**:e55185

[44] Fu W, Lei C, Liu S, Cui Y, Wang C, Qian K, et al. CAR exosomes derived from effector CAR-T cells have potent antitumour effects and low toxicity. Nature Communications. 2019;**10**(1):4355

[45] Zhou J, Yang Y, Wang W, Zhang Y, Chen Z, Hao C, et al. Melanoma-released exosomes directly activate the mitochondrial apoptotic pathway of CD4+ T cells through their microRNA

cargo. Experimental Cell Research. 2018;**371**(2):364-371

[46] Shao Q, Deng L, Liu H, Liu Z, Chen J, Jiang F, et al. Involvement of MM cell-derived exosomes in T lymphocytes immune responses. Oncology Letters. 2020;**20**(4):31

[47] Ye S-b, Li Z-L, Luo D-h, Huang B-j, Chen Y-S, Zhang X-s, et al. Tumor-derived exosomes promote tumor progression and T-cell dysfunction through the regulation of enriched exosomal microRNAs in human nasopharyngeal carcinoma. Oncotarget. 2014;**5**(14):5439-5452

[48] Kugeratski FG, Kalluri R. Exosomes as mediators of immune regulation and immunotherapy in cancer. The FEBS Journal. 2020;**288**(1):10-35

[49] Chen G, Huang AC, Zhang W, Zhang G, Wu M, Xu W, et al. Exosomal PD-L1 contributes to immunosuppression and is associated with anti-PD-1 response. Nature. 2018;**560**(7718):382-386

[50] Qu J-L, Qu X-J, Qu J-L, Qu X-J, Zhao M-F, Teng Y-E, et al. The role of cbl family of ubiquitin ligases in gastric cancer exosome-induced apoptosis of Jurkat T cells. Acta Oncologica. 2009;**48**(8):1173-1180

[51] Cai Z, Yang F, Yu L, Yu Z, Jiang L, Wang Q, et al. Activated T cell exosomes promote tumor invasion via FAS signaling pathway. The Journal of Immunology. 2012;**188**(12):5954-5961

[52] Boyiadzis M, Whiteside TL. The emerging roles of tumor-derived exosomes in hematological malignancies. Leukemia. 2017;**31**(6):1259-1268

[53] Kim DH, Kim H, Choi YJ, Kim SY, Lee J-E, Sung KJ, et al. Exosomal PD-L1 promotes tumor growth through immune escape in non-small cell lung cancer. Experimental & Molecular Medicine. 2019;**51**(8):1-13

[54] Yang Y, Li C-W, Chan L-C, Wei Y, Hsu J-M, Xia W, et al. Exosomal PD-L1 harbors active defense function to suppress T cell killing of breast cancer cells and promote tumor growth. Cell Research. 2018;**28**(8):862-864

[55] Daassi D, Mahoney KM, Freeman GJ. The importance of exosomal PDL1 in tumour immune evasion. Nature Reviews Immunology. 2020;**20**(4):209-215

[56] Haderk F, Schulz R, Iskar M, Cid LL, Worst T, Willmund KV, et al. Tumor-derived exosomes modulate PD-L1 expression in monocytes. Science Immunology. 2017;**28**;2(13):eaah5509

[57] Ma F, Vayalil J, Lee G, Wang Y, Peng G. Emerging role of tumor-derived extracellular vesicles in T cell suppression and dysfunction in the tumor microenvironment. Journal for ImmunoTherapy of Cancer. 2021;**9**(10):e003217

[58] Willingham SB, Volkmer JP, Gentles AJ, Sahoo D, Dalerba P, Mitra SS, et al. The CD47-signal regulatory protein alpha (SIRPa) interaction is a therapeutic target for human solid tumors. Proceedings of the National Academy of Sciences. 2012;**109**(17):6662-6667

[59] Koh E, Lee EJ, Nam G-H, Hong Y, Cho E, Yang Y, et al. Exosome-SIRPα, a CD47 blockade increases cancer cell phagocytosis. Biomaterials. 2017;**121**:121-129

[60] Battke C, Ruiss R, Welsch U, Wimberger P, Lang S, Jochum S, et al. Tumour exosomes inhibit binding of tumour-reactive antibodies to tumour cells and reduce ADCC. Cancer Immunology, Immunotherapy. 2011;**60**(5):639-648

[61] Hong C-S, Sharma P, Yerneni SS, Simms P, Jackson EK, Whiteside TL, et al. Circulating exosomes carrying an immunosuppressive cargo interfere with cellular immunotherapy in acute myeloid leukemia. Scientific Reports. 2017;**7**(1): 14684

[62] Cordonnier M, Nardin C, Chanteloup G, Derangere V, Algros MP, Arnould L, et al. Tracking the evolution of circulating exosomal-PD-L1 to monitor melanoma patients. Journal of Extracellular Vesicles. 2020;**9**(1): 1710899

[63] Gao J, Qiu X, Li X, Fan H, Zhang F, Lv T, et al. Expression profiles and clinical value of plasma exosomal Tim-3 and Galectin-9 in non-small cell lung cancer. Biochemical and Biophysical Research Communications. 2018;**498**(3):409-415

[64] Xie F, Xu M, Lu J, Mao L, Wang S. The role of exosomal PD-L1 in tumor progression and immunotherapy. Molecular Cancer. 2019;**18**(1):146

[65] Huber V, Vallacchi V, Fleming V, Hu X, Cova A, Dugo M, et al. Tumor-derived microRNAs induce myeloid suppressor cells and predict immunotherapy resistance in melanoma. Journal of Clinical Investigation. 2018;**128**(12):5505-5516

[66] Tian W, Liu S, Li B. Potential role of exosomes in cancer metastasis. BioMed Research International. 2019;**2019**:1-12

[67] Alzahrani FA, El-Magd MA, Abdelfattah-Hassan A, Saleh AA, Saadeldin IM, El-Shetry ES, et al. Potential effect of exosomes derived from cancer stem cells and MSCs on progression of DEN-induced HCC in rats. Stem Cells International. 2018;**2018**:1-17

[68] Naseri M, Bozorgmehr M, Zöller M, Ranaei Pirmardan E, Madjd Z. Tumor-derived exosomes: The next generation of promising cell-free vaccines in cancer immunotherapy. OncoImmunology. 2020;**9**(1):1779991

[69] Thakur A, Ke X, Chen Y-W, Motallebnejad P, Zhang K, Lian Q, et al. The mini player with diverse functions: Extracellular vesicles in cell biology, disease, and therapeutics. Protein & Cell. 2021. DOI: 10.1007/s13238-021-00863-6

[70] Ren G, Wang Y, Yuan S, Wang B. Dendritic cells loaded with HeLa-derived exosomes simulate an antitumor immune response. Oncology Letters. 2018;**15**:6636-6640

[71] Multhoff G, Yao Y, Wang C, Wei W, Shen C, Deng X, et al. Dendritic cells pulsed with leukemia cell-derived exosomes more efficiently induce antileukemic immunities. PLoS One. 2014;**9**(3):e91463

[72] Pizzo SV, Lee E-Y, Park K-S, Yoon YJ, Lee J, Moon H-G, et al. Therapeutic effects of autologous tumor-derived nanovesicles on melanoma growth and metastasis. PLoS One. 2012;**7**(3):e33330

[73] Bu N, Li Q-L, Feng Q, Sun B-Z. Immune protection effect of exosomes against attack of L1210 tumor cells. Leukemia & Lymphoma. 2009;**47**(5): 913-918

[74] Xiang X, Poliakov A, Liu C, Liu Y, Deng Z-b, Wang J, et al. Induction of myeloid-derived suppressor cells by tumor exosomes. International Journal of Cancer. 2009;**124**(11):2621-2633

[75] Gu X, Erb U, Büchler MW, Zöller M. Improved vaccine efficacy of tumor exosome compared to tumor lysate loaded dendritic cells in mice. International Journal of Cancer. 2015;**136**(4):E74-E84

Chapter 17

Extracellular Vesicles for Cancer Immunotherapy: Biomarkers and Beyond

Baranya Murugan and Suresh Sagadevan

Abstract

Extracellular vesicles (EVs), like exosomes and microvesicles, are membrane-bound vesicles released by most cell types in response to cellular stress as well as normal physiologic conditions. EV plays a vital part in cell communication and tumor immunology. Tumor-derived EVs carry a wide range of tumor neoantigens and have a distinct molecular signature that reflects the tumor's genomic complexities. These tumor-derived EVs provide a glance into the immunological tumor microenvironment and have a perspective to be a novel, minimally invasive cancer immunotherapy biomarker. Antibodies against immune checkpoint inhibitors like anti-programmed death-1 (PD-1) and its ligand (PD-L1) have changed the treatment of broad diversity of solid tumors such as non-small cell lung cancer, head, and neck squamous cell carcinoma, urothelial carcinoma, melanoma, etc. Invasive tissue biopsy is necessary for both histologic diagnosis and next-generation sequencing efforts. The latter has become increasingly widespread in today's healthcare. There is an unmet need for non-invasive or minimally invasive (e.g., plasma-based) biomarkers in both diagnosis and therapy monitoring. The selected investigation of EV in biospecimens, including plasma and saliva, can achieve this goal by potentially avoiding the need for tissue samples. In this chapter, we discuss the present challenges of biomarkers in cancer immunotherapy and the mechanistic role of tumor-derived EV in regulating the anti-tumor immune response.

Keywords: extracellular vesicles, cancer, immune response, biomarkers, immunotherapy

1. Introduction

Exosomes were initially discovered in 1981 as exfoliating vesicles from different normal and neoplastic cell lines. Exosomes correspond to the family of extracellular vesicles (EVs), composed of microvesicles, apoptotic bodies, and exosomes [1–3]. Exosomes arise from the endo-lysosomal pathway and originate from the endosomal compartment known as the multivesicular bodies [4, 5]. The microvesicles are developed by sprouting from the plasma membrane. The microvesicles size is around 100–1000 nm and originates from sprout and fusion of plasma membrane into extracellular space, and sharing out different models with the parental cells, involving membrane lipids, receptors, and different types of nucleic acids and proteins [6, 7]. Depending on their size and shape, exosomes can be divided into nine different subpopulations or categories, signifying that exosome that arises from a single cell line is different from morphologically functionally [8]. EVs play a potential role in many factors such as cellular homeostasis, physiology, and pathobiology [9, 10]. In the tumor milieu, EVs are isolated from cancer cells, immune cells, and non-immune host cells delivering as a critical component of the tumor microenvironment. Isolation of EV from various roots exhibits distinct roles in tumor immune cells, leading to tumor proliferation, metastasis, and drug resistance [11]. Examinations from the past years have shown a curious spike of exosomes initiating to fuse and communicate their vital functions. EV are linked with different cell types which also have required macromolecules comprising DNA, micro-RNA, messenger RNA, proteins, and lipids [12].

Immune checkpoint blockade therapy has modified an environment for the treatment of cancer [13]. Therefore, deeper knowledge on deciding the success and failure of this therapy is required. Tumor-derived EVs are linked in immunological crosstalk are likely to be an emerging biomarker for cancer immunotherapy [14, 15]. Tumor-derived EVs are notably interesting targets as they are discharged from cancer cells as to normal cells and it has been segregated from different biospecimens such as blood, urine, cerebrospinal fluid, and saliva [16, 17]. Therefore, this chapter focuses on the current availability of biomarkers for cancer immunotherapy.

2. Exosomes contents

The contents of the exosomes not only reflect the composition of the donor cell but also consider the controlled sorting mechanism. Composites of different proteins involving receptors, transcription factors, enzymes, extracellular matrix proteins, lipids, nucleic acids (DNA, mRNA, and miRNA) interior and present on the surface of the exosomes [18]. As compared to the other cell membranes, Exosomes manifested a superior expression of sphingomyelin, cholesterol, phosphatidylserine, and saturated fatty acids as compared to whole cell membranes, according to earlier reports [19]. The proteins present in the exosomes consist of the endosome, plasma, and nuclear proteins. An exosome of various cell types consisting of TSG101, Alix, Rab GTPases, heat shock proteins (HSP70, HSP90), integrins, tetraspanins (CD9, CD63, CD81), and MHC class II proteins. Furthermore, exosomes also carry genetic material like mRNA, long non-coding RNA, micro-RNA, and double-stranded DNA [20]. Furthermore, the contents of exosomes may differ from the cells of their parent cells due to the precise distribution of cargo into exosomes. As a result, there is not enough information on their back-and-forth mechanisms.

3. Extracellular vesicles as biomarkers

During the 1990s National Institute of Health (NIH) has explained biomarker as a biological molecule or gene that is found in body fluids estimated as a marker of a biological, pharmacological, pathogenic process for therapeutic intervention [21]. EVs are systematically involved in intercellular communication and exhibited as qualitatively and quantitatively in diseases like cancer, it also plays a significant role in drug resistance. EVs are unique and can serve as a valuable tool to aid in cancer detection, prognosis, and tracking therapeutic efficacy. EVs circulate in bodily fluids, making them an ideal candidate for non-invasive testing [22]. In addition, the lipid bilayer shields the biomacromolecules like RNA, proteins, and enzymatic activity, EVs constitute a safe vehicle for evaluating genetic sequences. Liquid biopsy is an emerging technique, that requires implying the circulating tumor cells (CTCs), cell-free DNA (cfDNA), and EVs, which has brought new understanding and aspects to the field of cancer therapeutic interventions [23]. CTCs are cancer cells that are slough off from the primary or metastatic sites through the circulation. According to a few recent studies, molecular profiling of CTCs can be used to monitor patients who are receiving therapies [24].

The foremost difficulty in examining CTCs such as rarity of the cells in circulation (one CTC is found in per billion blood cells). Up to date, cell search, targeting epithelial cell adhesion molecule, EpCAM to procure cell CTCs is the most prominent cell CTC enumeration assay. There are other biomarkers such as cfDNA known as short fragments of nucleic acids which are seen in body fluids, like blood or urine. It is demonstrated to be produced by the apoptotic degeneration of cellular DNA [25], and a fragment of cfDNA is procured from the tumor cells and is explained as circulating tumor DNA, which is known to imitate the genetic and epigenetic alterations of the initial tumor and its possible to be used as a diagnostic and prognostic biomarker for cancers. Also, NGS is the most common method for examination of the genetic information of the cfDNA [26]. As compared to NGS, digital PCR can only do the screening for the known variants and has the capability for the limited sample in one reaction. But NGS is a high cost and relatively time-consuming technique, which also necessitates bioinformatics skills for data analysis and interpretation.

EVs have great attention as like another kind of biopsy. EVs are mostly found in different kinds of body fluids such as blood, saliva, urine, bronchoalveolar fluid, breast milk, and semen. They reproduce the disease type by taking the molecules from benign cells, like miRNAs, proteins, long noncoding RNAs, and lipids. Out of

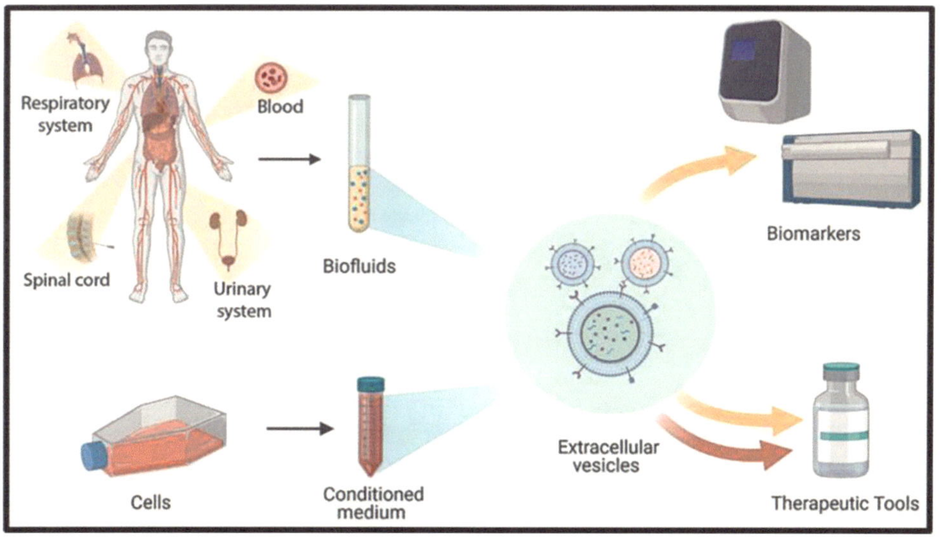

Figure 1.
Extracellular vesicles - body fluids as biomarkers and therapeutic clinical applications [28].

the components of EVs, the associated proteins and RNA of EVs are demonstrated to be used as a tumor biomarker for detecting cancer and observing the progression of cancer [27]. **Figure 1** and **Table 1** show the development of novel exosome-based biomarkers that could benefit cancer patients in a variety of ways.

Exosomal cargos	Cancer types	Clinical value	Biofluids	Refs
Glypican-1	Pancreatic cancer	Patients with early- and late-stage pancreatic cancer had higher levels of Glypican-1 positive exosomes as compared to healthy controls.	Serum	[29]
miRNA-375, 1290	Prostate cancer	Patients with castration-resistant prostate cancer who have high levels of both exosomal miRNA-375 and miRNA-1290 may have a worse prognosis.	Serum	[30]
miRNA-21	HCC	HCC cancer patients have increased levels of exosomal microRNA-21 than healthy normal	Serum	[31]
EML4-ALK fusion	NSCLC	Exosomal RNA of NSCLC patients found to include EML4-ALK fusion transcripts	Plasma	[32]
miRNA-19-3p, 21-5p, 221-3p	Lung adenocarcinoma	Lung adenocarcinoma patients have elevated levels of the miRNAs than healthy controls	Plasma	[33]
Phosphatidylserine	Ovarian cancer	Ovarian cancer has elevated levels of phosphatidylserine positive exosomes patients than normal controls	Plasma	[34]
LncRNA-p21	Prostate cancer	Prostate cancer patients were found to have elevated levels of exosomal lncRNA-p21 than healthy controls	Plasma	[35]
miRNA-21, 375	Prostate cancer	Prostate cancer patients were found to have increased levels of urinary exosomal miRNA-21 and miRNA-375 as compared with normal controls	Urine	[36]
Mesenchymal Stromal Cells	Prostate cancer	Patients with SARS-Cov-2 infection	Urine	[37]
Dendritic cells	Non-Small Cell Lung Cancer	Immunotherapy	Blood	[38]

Table 1.
Exosomes from different types of cancer patients—body fluids as biomarkers.

4. Cancer immunotherapy biomarkers and challenges

Tumor-derived extracellular vesicles, which transfer immunosuppressive chemicals like PD-L1, TGF1, FasL, TRAIL, and NKG2D ligands, are the major transporters of tumor immune evasion and potential therapeutic targets. Anti-programmed death-1 (PD-1) and its ligand (PD-L1) antibodies are immune checkpoint inhibitors (ICI), stimulating an antitumor immune response over blocking inhibitory immune signaling. Immune checkpoint inhibitors therapy has become potential in determining effectiveness in different cancers such as non-small cell lung cancer, head, and neck squamous cell carcinoma, melanoma, etc. [39]. Un happily only a few patients have responded to the checkpoint inhibitors and there is a critical need to find out the reason for an adaptive immune response [40].

PD-1 and PD-L1 are located on the exterior of the tumor cells, and it acted as a second signal when it is attached to the PD-1 receptors on T-cells. There are a few blocking antibodies that target PD-1 and PD-L1, like pembrolizumab (Merck) and nivolumab (Bristol Meyer-Squibb) enhance the anti-tumor immune response by opposing this inhibitory signal [41]. The response rate percent is depending on the type of tumor and is a distinctive group of patients gained a very less response. PD-L1 as a biomarker and its expression levels on tumor cells were studied using immunohistochemistry has shown to act as a prognostic marker as well prognostic to anti-PD-1 therapy [42]. The tissue-based testing method needs a decent tissue biopsy. A tissue biopsy test method also has a few advantages like bleeding, infection, and other procedural complication like causing pneumothorax in the case of parenchymal lung biopsy. Tissue can be isolated either from the primary tumor or a metastatic tumor lesion upon the different factors that are considered. Hence the expression levels may differ based on the tumor tissues isolated. Researchers have also seen [43]. These results of PD-L1 expression levels by immunohistochemistry have influenced the practitioners in a way to treat patients. There are also limitations for immunohistochemistry staining, distinct antibodies have unique sensitizers. Additionally, the threshold value of PD-L1 staining is still a discussion between pathologists and oncologists. Hence there is a great challenge for researchers, clinicians, and patients, antibody clones have influenced the distinctive epitope of the PD-L1 molecule with unique scoring systems based on the assays [44].

In inclusion to PD-L1, a few other immunosuppressive molecules such as TGFB1 and NKG2D ligands were also augmented in TD-EVs and were capable to prompt T-cell suppression [45]. NKG2D is a receptor that is activated by NK cells and a few subgroups of T cells and acts as a prime recognition receptor for the detection and elimination of cancer cells. These are stress-influenced self-proteins, which are released as soluble molecules through protease-mediated cleavage. An excretion of NKG2D ligands is regarded to maintain their expression levels related to the immune evasion mechanism occupied by tumor cells to avoid NKG2D-mediated immune observation [46]. It has been observed that TD-EVs from ovarian cancer and melanoma expressed NKG2D ligands and intercept activation of cytotoxic NK cells [47]. FASL and TRAIL expression on TD-EVs persuades apoptosis in dendritic cells (DCs) leading to immunosuppression and stimulating the progression of tumors [48]. TD-EVs on FASL terminated an antigen-specific effector T cells. There are few other immunosuppressive proteins like COX2, CD39/CD73, PDL1, FASL, TGFβ, CTLA4, TRAIL, etc., are appeared to relate to the TD-EVs [49].

For a better immunotherapy effect, the immunosuppressive part of TD-EVs would be impeded, to initiate the immune system. For this case, DCs are the foremost step in the immunity cycle for terminating the cancer cells via T-cell activation [50]. The surface membranes are easily attracted with the immune cells, in which the dendritic cells procured extracellular vesicles can likely be manipulated to behave as anti-cancer vaccines thereby leading to novel immunotherapy to fight against cancer. The surface molecules of the TD-EVs would be guided to deliver cancer treatment. Hence, the tumor-associated antigens, immunogenic peptides, and heat shock proteins would bring about cancer treatment shortly [51].

5. Extracellular vesicles towards cancer immunotherapy

An immunosuppressive molecule like PD-L1, TGF β1, FasL, TRAIL, and NKG2D ligands are carried by TD-EVs, these ligands are the most crucial mediators of tumor immune evasion and are also the feasible targets for immunotherapy [52]. PD-L1 is more expressed when the metastatic melanoma-derived exosomes are stimulated by interferon-γ, on EVs and impede the antitumor responses. When PD-L1 is expressed on the tumor cell surface, it enhances evasion of immune surveillance through interacting with the ligand, by curbing the T-cell function. EV released from metastatic melanoma brings a PD-L1 which terminates the cytotoxic function of CD8+ T cells. PD-L1 indicates EVs were released from human blood, which is not a soluble form of PD-L1 and is also connected with the head and neck cancer progression. Similarly, PD-L1 isolated from EVs from the supernatant of murine or human HNSCC cell lines inhibits the infiltration of $CD4^+$ T and $CD8^+$ T cells into the tumor microenvironment, therefore enhancing tumor progression [53]. Along with the PD-L1 ligand, the other immunosuppressive molecules like TGFβ1 and NKG2D were also augmented in TD-EVs thereby bringing out the suppression of T-cells. Another activating receptor NKG2D is activated by NK cells and by T-cells, thereby presenting a crucial receptor for recognizing and terminating the cancer cells [54]. The NKG2D ligands are stress-induced self-proteins that can be divided by proteases and released as soluble molecules. The release of NKG2D ligands in the extracellular environment is to fine-tune the surface expression levels, constituting the immune evasion mechanism employed by cancer cells to neglect the NKG2D-ligand interfered immune response [55].

Few reports demonstrated that ovarian and melanoma TD-EVs activate NKG2D ligands and inhibit the activation of cytotoxic NK cells [56]. Activation of FASL and TRAIL on TD-EVs causes apoptosis in DCs and peripheral blood mononuclear cells (PBMCs) thereby promoting tumor progression [57]. The other inhibitory immunosuppressive proteins like COX2, CD39/CD73, PDL1, CTLA4, FASL are related to TD-EVs. Apart from the immune-suppressive character of TD-EVs, it would obstruct for greater immunotherapy effect and trigger the immune system. For instance, DCs are the foremost process of the immunity cycle, it activates the T-cell and thereby eliminate tumor cells. The surface membrane components that communicate with other immune cells, DCs derived EVs likely to be used as cell-free antitumor vaccines thereby delivering a novel and potential immunotherapy towards cancer [58]. Therefore, with the above observation, TD-EVs showed a major part in tumor immune evasion and growth are presented in **Table 2**. Hence, peptides, antigens, and other small molecules such as heat shock proteins could be designed shortly for cancer therapy.

Treatment molecules	Cells used	Effect of treatment	Research	Results	Reference
Small molecules	Murine macrophages (RAW 264.7 cell line)	Paclitaxel is loaded into exosomes through sonication	The efficacy of paclitaxel for the treatment of multiple drug-resistant cancers was assessed when delivered via exosomes	Paclitaxel with exosomes entered the tumour cells and inhibited the growth of pulmonary metastases	[58]
Antigens and antibodies	HLA-DR15-positive human B cells	Exosomes tagged with Hsp65 antigen or antigenic peptide.	Evaluated the B cell-derived exosomes can activate T cells through MHC-mediated presentation of Hsp65 antigen	Exosomes tagged with Hsp65 could activate the T-Cells through MHC mediated presentation of Hsp65 antigen	[59]
siRNA and RNAi	Plasma cells of human	siRNA Electroporation	siRNA is delivered towards MAPK1 to monocytes and T cells	siRNA delivered and downregulated the MAPK-1 transcription	[60]
miRNA	Panc-1	DNA plasmids transfected for miR-155 and miR-125b2 into Panc-1 cells	Upon treatment with exosomes, murine macrophages overexpress miR-155 and miR-125b2	After the treatment, transfected Panc-1 cells were reprogrammed from M2 to M1 macrophage phenotype	[61]
Blocking exosomes biogenesis and secretion-GW4869	All types of cells	Exosome's biogenesis was inhibited	Combination therapy of gemcitabine and an inhibitor of exosomes	On the release of exosomes by cancer-associated fibroblasts on exposure to gemcitabine, GW4869 prevented chemoresistance	[62]

Table 2.
Exosomes—cancer therapeutic targets.

6. Conclusions

Exosomes are endogenous tiny vesicles that enable communication between neighbor or far cells [63]. Exosomes originating from various origins can be initiated in circulation and target specific cells inside distant tissues [64]. Because of its stability and capacity to bypass natural barriers. Exosomes are a suitable and potential nanocarrier vehicle for cancer immunotherapy and chemotherapy because of their characteristics. However, exosomes are demonstrated to be involved in the triggering of immune responses towards cancer cells and stable of an immunosuppressive milieu. The qualitative and quantitative data of EVs molecular signatures are mandatory in extra cellular-based tumor diagnosis, monitoring, and therapeutic delivery to distinct cancer subtypes. As previously stated, researchers are most interested in EVs and therapeutic targets. However, much more research on these EVs for cancer immunotherapy and chemotherapy is needed to gain a better understanding. In the view of EVs have received great attention and would be a potential and promising biomarker and a better candidate for the treatment of deadly disease, cancer.

Acknowledgements

The authors would like to acknowledge the financial support provided by a Research University grant from the University of Malaya (RU003-2021).

Conflict of interest

The authors declare no potential conflict of interest regarding the publication of this work.

Author details

Baranya Murugan* and Suresh Sagadevan
Nanotechnology and Catalysis Research Centre, University of Malaya, Kuala Lumpur, Malaysia

*Address all correspondence to: baranyaphd@gmail.com

References

[1] Zhang Y, Liu Y, Liu H, Tang WH. Exosomes: Biogenesis, biologic function, and clinical potential. Cell & Bioscience. 2019;**9**:19. DOI: 10.1186/s13578-019-0282-2

[2] Willms E, Cabañas C, Mäger I, Wood MJA, Vader P. Extracellular vesicle heterogeneity: Subpopulations, isolation techniques, and diverse functions in cancer progression. Frontiers in Immunology. 2018;**9**:738. DOI: 10.3389/fimmu.2018.00738

[3] Kang H, Kim J, Park J. Methods to isolate extracellular vesicles for diagnosis. Micro and Nano Systems Letters. 2017;**5**:15. DOI: 10.1186/s40486-017-0049-7

[4] Doyle LM, Wang MZ. Overview of extracellular vesicles, their origin, composition, purpose, and methods for exosome isolation and analysis. Cell. 2019;**8**:727. DOI: 10.3390/cells8070727

[5] Taylor D, Gercel-Taylor C. The origin, function, and diagnostic potential of RNA within extracellular vesicles present in human biological fluids. Frontiers in Genetics. 2013;**4**:142. DOI: 10.3389/fgene.2013.00142

[6] Yáñez-Mó M, Siljander PR-M, Andreu Z, Zavec AB, Borràs FE, Buzas EI, et al. Biological properties of extracellular vesicles and their physiological functions. Journal of Extracellular Vesicles. 2015;**4**:27066. DOI: 10.3402/jev.v4.27066

[7] Tang T-T, Lv L-L, Lan H-Y, Liu B-C. Extracellular vesicles: Opportunities and challenges for the treatment of renal diseases. Frontiers in Physiology. 2019;**10**:226. DOI: 10.3389/fphys.2019.00226

[8] He C, Zheng S, Luo Y, Wang B. Exosome theranostics: Biology and translational medicine. Theranostics. 2018;**8**:237-255. DOI: 10.7150/thno.21945

[9] Maia J, Caja S, Strano Moraes MC, Couto N, Costa-Silva B. Exosome-based cell-cell communication in the tumor microenvironment. Frontiers in Cell and Developmental Biology. 2018;**6**:18. DOI: 10.3389/fcell.2018.00018

[10] Darkwah S, Park EJ, Myint PK, Ito A, Appiah MG, Obeng G, et al. Potential roles of muscle-derived extracellular vesicles in remodeling cellular microenvironment: Proposed implications of the exercise-induced myokine, irisin. Frontiers in Cell and Developmental Biology. 2021;**9**:634853. DOI: 10.3389/fcell.2021.634853

[11] Becker A, Thakur BK, Weiss JM, Kim HS, Peinado H, Lyden D. Extracellular vesicles in cancer: Cell-to-cell mediators of metastasis. Cancer Cell. 2016;**30**:836-848. DOI: 10.1016/j.ccell.2016.10.009

[12] O'Brien K, Breyne K, Ughetto S, Laurent LC, Breakefield XO. RNA delivery by extracellular vesicles in mammalian cells and its applications. Nature Reviews Molecular Cell Biology. 2020;**21**:585-606. DOI: 10.1038/s41580-020-0251-y

[13] Darvin P, Toor SM, Sasidharan Nair V, Elkord E. Immune checkpoint inhibitors: Recent progress and potential biomarkers. Experimental & Molecular Medicine. 2018;**50**:1-11. DOI: 10.1038/s12276-018-0191-1

[14] Taghikhani A, Farzaneh F, Sharifzad F, Mardpour S, Ebrahimi M,

Hassan ZM. Engineered tumor-derived extracellular vesicles: Potentials in cancer immunotherapy. Frontiers in Immunology. 2020;**11**:221

[15] Ma F, Vayalil J, Lee G, Wang Y, Peng G. Emerging role of tumor-derived extracellular vesicles in T cell suppression and dysfunction in the tumor microenvironment. Journal for ImmunoTherapy of Cancer. 2021;**9**:1-15. DOI: 10.1136/jitc-2021-003217

[16] Tai Y-L, Chu P-Y, Lee B-H, Chen K-C, Yang C-Y, Kuo W-H, et al. Basics and applications of tumor-derived extracellular vesicles. Journal of Biomedical Science. 2019;**26**:35. DOI: 10.1186/s12929-019-0533-x

[17] Michela B. Liquid biopsy: A family of possible diagnostic tools. Diagnostics. 2021;**11**(8):1391. DOI: 10.3390/diagnostics11081391

[18] Kalluri R, LeBleu VS. The biology, function, and biomedical applications of exosomes. Science (New York, N.Y.). 2020;**367**:eaau6977. DOI: 10.1126/science.aau6977

[19] Huda MN, Nafiujjaman M, Deaguero IG, Okonkwo J, Hill ML, Kim T, et al. Potential use of exosomes as diagnostic biomarkers and in targeted drug delivery: Progress in clinical and preclinical applications. ACS Biomaterials Science & Engineering. 2021;**7**:2106-2149. DOI: 10.1021/acsbiomaterials.1c00217

[20] Silva M, Melo SA. Non-coding RNAs in exosomes: New players in cancer biology. Current Genomics. 2015;**16**: 295-303. DOI: 10.2174/1389202916666150707154719

[21] Lu H, Zhu X-C, Jiang T, Yu J-T, Tan L. Body fluid biomarkers in Alzheimer's disease. Annals of Translational Medicine. 2015;**3**:70. DOI: 10.3978/j.issn.2305-5839.2015.02.13

[22] Verma M, Lam TK, Hebert E, Divi RL. Extracellular vesicles: Potential applications in cancer diagnosis, prognosis, and epidemiology. BMC Clinical Pathology. 2015;**15**:6. DOI: 10.1186/s12907-015-0005-5

[23] Palmirotta R, Lovero D, Cafforio P, Felici C, Mannavola F, Pellè E, et al. Liquid biopsy of cancer: A multimodal diagnostic tool in clinical oncology. Therapeutic Advances in Medical Oncology. 2018;**10**:1758835918794630. DOI: 10.1177/1758835918794630

[24] Lin SY, Chang S-C, Lam S, Irene Ramos R, Tran K, Ohe S, et al. Prospective molecular profiling of circulating tumor cells from patients with melanoma receiving combinatorial immunotherapy. Clinical Chemistry. 2020;**66**:169-177. DOI: 10.1373/clinchem.2019.307140

[25] Pös Z, Pös O, Styk J, Mocova A, Strieskova L, Budis J, et al. Technical and methodological aspects of cell-free nucleic acids analyzes. International Journal of Molecular Sciences. 2020;**21**:8634. DOI: 10.3390/ijms21228634

[26] Bennett CW, Berchem G, Kim YJ, El-Khoury V. Cell-free DNA and next-generation sequencing in the service of personalized medicine for lung cancer. Oncotarget. 2016;**7**:71013-71035. DOI: 10.18632/oncotarget.11717

[27] Lane RE, Korbie D, Hill MM, Trau M. Extracellular vesicles as circulating cancer biomarkers: Opportunities and challenges. Clinical and Translational Medicine. 2018;**7**:14. DOI: 10.1186/s40169-018-0192-7

[28] Ciferri MC, Quarto R, Tasso R. Extracellular vesicles as biomarkers and therapeutic tools: From pre-clinical to clinical applications. Biology. 2021; **10**(5):359. DOI: 10.3390/biology10050359

[29] Liu Q, Yu Z, Yuan S, Xie W, Li C, Hu Z, et al. Circulating exosomal microRNAs as prognostic biomarkers for non-small-cell lung cancer. Oncotarget. 2017;**8**:13048-13058. DOI: 10.18632/oncotarget.14369

[30] Mirzaei H, Fathullahzadeh S, Khanmohammadi R, Darijani M, Momeni F, Masoudifar A, et al. State of the art in microRNA as diagnostic and therapeutic biomarkers in chronic lymphocytic leukemia. Journal of Cellular Physiology. 2018;**233**(2):888-900

[31] Wang H, Hou L, Li A, Duan Y, Gao H, Song X. Expression of serum exosomal microRNA-21 in human hepatocellular carcinoma. BioMed Research International. 2014;**2014**:864894. DOI: 10.1155/2014/864894

[32] Huang L, Rong Y, Tang X, Yi K, Wu J, Wang F. Circular RNAs are promising biomarkers in liquid biopsy for the diagnosis of non-small cell lung cancer. Frontiers in Molecular Biosciences. 2021;**8**:363. DOI: 10.3389/fmolb.2021.625722

[33] Zhou X, Wen W, Shan X, Zhu W, Xu J, Guo R, et al. A six-microRNA panel in plasma was identified as a potential biomarker for lung adenocarcinoma diagnosis. Oncotarget. 2017;**8**:6513-6525. DOI: 10.18632/oncotarget.14311

[34] Matsumoto Y, Kano M, Akutsu Y, Hanari N, Hoshino I, Murakami K, et al. Quantification of plasma exosome is a potential prognostic marker for esophageal squamous cell carcinoma. Oncology Reports. 2016;**36**(5):2535-2543

[35] Işın M, Uysaler E, Özgür E, Köseoğlu H, Şanlı Ö, Yücel ÖB, et al. Exosomal lncRNA-p21 levels may help to distinguish prostate cancer from benign disease. Frontiers in Genetics. 2015;**6**:168. DOI: 10.3389/fgene.2015.00168

[36] Huang Z, Zhang L, Zhu D, Shan X, Zhou X, Qi L-W, et al. A novel serum microRNA signature to screen esophageal squamous cell carcinoma. Cancer Medicine. 2017;**6**:109-119. DOI: 10.1002/cam4.973

[37] Kwon S, Shin S, Do M, Oh BH, Song Y, Bui VD, et al. Engineering approaches for effective therapeutic applications based on extracellular vesicles. Journal of Controlled Release. 2021;**330**:15-30. DOI: 10.1016/j.jconrel.2020.11.062

[38] Chanteloup G, Cordonnier M, Isambert N, Bertaut A, Marcion G, Garrido C, et al. Membrane-bound exosomal HSP70 as a biomarker for detection and monitoring of malignant solid tumours: A pilot study. Pilot and Feasibility Studies. 2020;**6**:35. DOI: 10.1186/s40814-020-00577-2

[39] Gibellini L, de Biasi S, Porta C, lo Tartaro D, Depenni R, Pellacani G, Sabbatini R, Cossarizza A. Single-cell approaches to profile the response to immune checkpoint inhibitors. Frontiers in Immunology. 2020;**11**:490. DOI: 10.3389/fimmu.2020.00490

[40] Buchbinder EI, Desai A. CTLA-4 and PD-1 pathways: Similarities, differences, and implications of their inhibition. American Journal of Clinical Oncology. 2016;**39**:98-106. DOI: 10.1097/COC.0000000000000239

[41] Gibney GT, Weiner LM, Atkins MB. Predictive biomarkers for checkpoint inhibitor-based immunotherapy. The Lancet Oncology. 2016;**17**:e542-e551. DOI: 10.1016/S1470-2045(16)30406-5

[42] Yu H, Boyle TA, Zhou C, Rimm DL, Hirsch FR. PD-L1 expression in lung

cancer. Journal of Thoracic Oncology. 2016;**11**:964-975. DOI: 10.1016/j.jtho.2016.04.014

[43] Parra ER, Villalobos P, Mino B, Rodriguez-Canales J. Comparison of different antibody clones for immunohistochemistry detection of programmed cell death ligand 1 (PD-L1) on non-small cell lung carcinoma. Applied Immunohistochemistry & Molecular Morphology: AIMM. 2018;**26**:83-93. DOI: 10.1097/PAI.0000000000000531

[44] Li Q, Cai S, Li M, Salma KI, Zhou X, Han F, et al. Tumor-derived extracellular vesicles: Their role in immune cells and immunotherapy. International Journal of Nanomedicine. 2021;**16**:5395-5409. DOI: 10.2147/IJN.S313912

[45] Tanaka Y, Kamohara H, Kinoshita K, Kurashige J, Ishimoto T, Iwatsuki M, et al. Clinical impact of serum exosomal microRNA-21 as a clinical biomarker in human esophageal squamous cell carcinoma. Cancer. 2013;**119**(6): 1159-1167

[46] Labani-Motlagh A, Israelsson P, Ottander U, Lundin E, Nagaev I, Nagaeva O, et al. Differential expression of ligands for NKG2D and DNAM-1 receptors by epithelial ovarian cancer-derived exosomes and its influence on NK cell cytotoxicity. Tumour Biology. 2016;**37**(4):5455-5466

[47] Rossin A, Miloro G, Hueber A-O. TRAIL and FasL functions in cancer and autoimmune diseases: Towards an increasing complexity. Cancers. 2019;**11**:639. DOI: 10.3390/cancers11050639

[48] Vetsika E-K, Sharma P, Samaras I, Markou A, Georgoulias V, Whiteside TL, et al. Small extracellular vesicles in pre-therapy plasma predict clinical outcome in non-small-cell lung cancer patients. Cancers. 2021;**13**:2041. DOI: 10.3390/cancers13092041

[49] Fu C, Jiang A. Dendritic cells and CD8 T cell immunity in tumor microenvironment. Frontiers in Immunology. 2018;**9**:3059. DOI: 10.3389/fimmu.2018.03059

[50] Jabalee J, Towle R, Garnis C. The role of extracellular vesicles in cancer: Cargo, function, and therapeutic implications. Cell. 2018;**7**:93. DOI: 10.3390/cells7080093

[51] Whiteside TL. Exosomes and tumor-mediated immune suppression. The Journal of Clinical Investigation. 2016;**126**:1216-1223. DOI: 10.1172/JCI81136

[52] Dong Y, Sun Q, Zhang X. PD-1 and its ligands are important immune checkpoints in cancer. Oncotarget. 2017;**8**:2171-2186. DOI: 10.18632/oncotarget.13895

[53] Paul S, Lal G. The molecular mechanism of natural killer cells function and its importance in cancer immunotherapy. Frontiers in Immunology. 2017;**8**:1124. DOI: 10.3389/fimmu.2017.01124

[54] Zingoni A, Vulpis E, Loconte L, Santoni A. NKG2D ligand shedding in response to stress: Role of ADAM10. Frontiers in Immunology. 2020;**11**:447. DOI: 10.3389/fimmu.2020.00447

[55] Molfetta R, Zingoni A, Santoni A, Paolini R. Post-translational mechanisms regulating NK cell activating receptors and their ligands in cancer: Potential targets for therapeutic intervention. Frontiers in Immunology. 2019;**10**:2557. DOI: 10.3389/fimmu.2019.02557

[56] Yin Z, Fan J, Xu J, Wu F, Li Y, Zhou M, et al. Immunoregulatory roles of extracellular vesicles and associated therapeutic applications in lung cancer. Frontiers in Immunology. 2020;**11**:2024. DOI: 10.3389/fimmu.2020.02024

[57] Markov O, Oshchepkova A, Mironova N. Immunotherapy based on dendritic cell-targeted/−derived extracellular vesicles—A novel strategy for enhancement of the anti-tumor immune response. Frontiers in Pharmacology. 2019;**10**:1152. DOI: 10.3389/fphar.2019.01152

[58] El-Sherbiny YM, Meade JL, Holmes TD, McGonagle D, Mackie SL, Morgan AW, et al. The requirement for DNAM-1, NKG2D, and NKp46 in the natural killer cell-mediated killing of myeloma cells. Cancer Research. 2007;**67**(18):8444-8449

[59] Wahlgren J, De L, Karlson T, Brisslert M, Vaziri Sani F, Telemo E, et al. Plasma exosomes can deliver exogenous short interfering RNA to monocytes and lymphocytes. Nucleic Acids Research. 2012;**40**(17):e130

[60] Su MJ, Aldawsari H, Amiji M. Pancreatic cancer cell exosome-mediated c the role of microRNAs 155 and 125b2 transfection using nanoparticle delivery systems. Scientific Reports. 2016;**6**:30110

[61] Richards KE, Zeleniak AE, Fishel ML, Wu J, Littlepage LE, Hill R. Cancer-associated fibroblast exosomes regulate survival and proliferation of pancreatic cancer cells. Oncogene. 2017;**36**(13):1770-1778

[62] Tkach M, Théry C. Communication by extracellular vesicles: Where we are and where we need to go. Cell. 2016; **164**:1226-1232. DOI: 10.1016/j.cell. 2016.01.043

[63] Dai J, Su Y, Zhong S, Cong L, Liu B, Yang J, et al. Exosomes: Key players in cancer and potential therapeutic strategy. Signal Transduction and Targeted Therapy. 2020;**5**:145. DOI: 10.1038/s41392-020-00261-0

[64] Kugeratski FG, Kalluri R. Exosomes as mediators of immune regulation and immunotherapy in cancer. The FEBS Journal. 2021;**288**:10-35. DOI: 10.1111/febs.15558